T0122375

Advances in Intelligent Systems and Computing

Volume 373

Series editor

Janusz Kacprzyk, Polish Academy of Sciences, Warsaw, Poland
e-mail: kacprzyk@ibspan.waw.pl

About this Series

The series "Advances in Intelligent Systems and Computing" contains publications on theory, applications, and design methods of Intelligent Systems and Intelligent Computing. Virtually all disciplines such as engineering, natural sciences, computer and information science, ICT, economics, business, e-commerce, environment, healthcare, life science are covered. The list of topics spans all the areas of modern intelligent systems and computing.

The publications within "Advances in Intelligent Systems and Computing" are primarily textbooks and proceedings of important conferences, symposia and congresses. They cover significant recent developments in the field, both of a foundational and applicable character. An important characteristic feature of the series is the short publication time and world-wide distribution. This permits a rapid and broad dissemination of research results.

Advisory Board

Chairman

Nikhil R. Pal, Indian Statistical Institute, Kolkata, India
e-mail: nikhil@isical.ac.in

Members

Rafael Bello, Universidad Central "Marta Abreu" de Las Villas, Santa Clara, Cuba
e-mail: rbellop@uclv.edu.cu

Emilio S. Corchado, University of Salamanca, Salamanca, Spain
e-mail: escorchado@usal.es

Hani Hagras, University of Essex, Colchester, UK
e-mail: hani@essex.ac.uk

László T. Kóczy, Széchenyi István University, Győr, Hungary
e-mail: koczy@sze.hu

Vladik Kreinovich, University of Texas at El Paso, El Paso, USA
e-mail: vladik@utep.edu

Chin-Teng Lin, National Chiao Tung University, Hsinchu, Taiwan
e-mail: ctlin@mail.nctu.edu.tw

Jie Lu, University of Technology, Sydney, Australia
e-mail: Jie.Lu@uts.edu.au

Patricia Melin, Tijuana Institute of Technology, Tijuana, Mexico
e-mail: epmelin@hafsamx.org

Nadia Nedjah, State University of Rio de Janeiro, Rio de Janeiro, Brazil
e-mail: nadia@eng.uerj.br

Ngoc Thanh Nguyen, Wroclaw University of Technology, Wroclaw, Poland
e-mail: Ngoc-Thanh.Nguyen@pwr.edu.pl

Jun Wang, The Chinese University of Hong Kong, Shatin, Hong Kong
e-mail: jwang@mae.cuhk.edu.hk

More information about this series at http://www.springer.com/series/11156

Sigeru Omatu · Qutaibah M. Malluhi
Sara Rodríguez González · Grzegorz Bocewicz
Edgardo Bucciarelli · Gianfranco Giulioni
Farkhund Iqba
Editors

Distributed Computing and Artificial Intelligence, 12th International Conference

 Springer

Editors

Sigeru Omatu
Faculty of Engineering
Department of Electronics, Information
    and Communication Engineering
Osaka Institute of Technology
Osaka
Japan

Qutaibah M. Malluhi
College of Engineering
Qatar University
Doha
Qatar

Sara Rodríguez González
Department of Computing Science
Faculty of Science
University of Salamanca
Salamanca
Spain

Grzegorz Bocewicz
Department of Electronics and Computers
Koszalin University of Technology
Koszalin
Poland

Edgardo Bucciarelli
Dep. of Philosophical, Pedagogical
    and Economic-Quantitative Sciences
University of Chieti-Pescara
Pescara
Italy

Gianfranco Giulioni
Dep. of Philosophical, Pedagogical
    and Economic-Quantitative Sciences
University of Chieti-Pescara
Pescara
Italy

Farkhund Iqba
Abu Dhabi Campus
Zayed University
Abu Dhabi
Utd.Arab.Emir.

ISSN 2194-5357   ISSN 2194-5365   (electronic)
Advances in Intelligent Systems and Computing
ISBN 978-3-319-19637-4   ISBN 978-3-319-19638-1   (eBook)
DOI 10.1007/978-3-319-19638-1

Library of Congress Control Number: 2015940025

Springer Cham Heidelberg New York Dordrecht London

© Springer International Publishing Switzerland 2015
This work is subject to copyright. All rights are reserved by the Publisher, whether the whole or part of the material is concerned, specifically the rights of translation, reprinting, reuse of illustrations, recitation, broadcasting, reproduction on microfilms or in any other physical way, and transmission or information storage and retrieval, electronic adaptation, computer software, or by similar or dissimilar methodology now known or hereafter developed.
The use of general descriptive names, registered names, trademarks, service marks, etc. in this publication does not imply, even in the absence of a specific statement, that such names are exempt from the relevant protective laws and regulations and therefore free for general use.
The publisher, the authors and the editors are safe to assume that the advice and information in this book are believed to be true and accurate at the date of publication. Neither the publisher nor the authors or the editors give a warranty, express or implied, with respect to the material contained herein or for any errors or omissions that may have been made.

Printed on acid-free paper

Springer International Publishing AG Switzerland is part of Springer Science+Business Media
(www.springer.com)

Preface

The 12th International Symposium on Distributed Computing and Artificial Intelligence 2015 (DCAI 2015) is a forum to present applications of innovative techniques for solving complex problems in these areas. The exchange of ideas between scientists and technicians from both the academic and industrial sector is essential to facilitate the development of systems that can meet the ever-increasing demands of today's society. The present edition brings together past experience, current work and promising future trends associated with distributed computing, artificial intelligence and their application in order to provide efficient solutions to real problems. This conference is a stimulating and productive forum where the scientific community can work towards future cooperation in Distributed Computing and Artificial Intelligence areas.

Nowadays it is continuing to grow and prosper in its role as one of the premier conferences devoted to the quickly changing landscape of distributed computing, artificial intelligence and the application of AI to distributed systems. This year's technical program will present both high quality and diversity, with contributions in well-established and evolving areas of research. Specifically, 62 papers were submit-ted to the main track from over 18 different, representing a truly "wide area network" of research activity. The DCAI' 15 technical program has selected 46 papers, the best papers will be selected for their publication on the the following special issues in journals such as Neurocomputing, Frontiers of Information Technology & Electronic Engineering, Journal of Artificial Intelligence (IJAI), the International Journal of Imaging and Robotics (IJIR) and the International Journal of Interactive Multimedia and Artificial Intelligence (IJIMAI). These special issues will cover extended versions of the most highly regarded works.

Moreover, DCAI' 15 Special Sessions have been a very useful tool in order to complement the regular program with new or emerging topics of particular interest to the participating community. Special Sessions that emphasize on multi-disciplinary and transversal aspects, such as *AI-driven methods for Multimodal Networks and Processes Modeling* and *Multi-Agents Macroeconomics* have been especially encouraged and welcome.

We thank the sponsors (Indra, INSA - Ingeniería de Software Avanzado S.A., IBM, IEEE Systems Man and Cybernetics Society Spain, AEPIA *Asociación Española para la Inteligencia Artificial,* APPIA *Associação Portuguesa Para a Inteligência Artificial,* CNRS *Centre national de la recherche scientifique*), and finally, the Local Organization members and the Program Committee members for their hard work, which was essential for the success of DCAI'15.

Sigeru Omatu
Sara Rodríguez González
Gianfranco Giulioni
Qutaibah M. Malluhi
Grzegorz Bocewicz
Edgardo Bucciarelli
Farkhund Iqbal
(Eds.)

Organization

Honorary Chairman

Masataka Inoue   President of Osaka Institute of Technology, Japan

Scientific Committee

Sigeru Omatu (Chairman)	Osaka Institute of Technology, Japan
Qutaibah M. Malluhi (Co-Chairman)	Qatar University, Qatar
Adriana Giret	Politechnich University of Valencia, Spain
Alberto Fernández	University Rey Juan Carlos, Spain
Ana Carolina Lorena	Federal University of ABC, Brazil
Ângelo Costa	University of Minho, Portugal
Antonio Moreno	University Rovira y Virgili, Spain
Antonio Manuel de Jesús Pereira	Leiria Polytechnic Institute, Portugal
Araceli Sanchís	University Carlos III of Madrid, Spain
B. Cristina Pelayo García-Bustelo	University of Oviedo, Spain
Bianca Innocenti	University of Girona, Spain
Bogdan Gabrys	Bournemouth University, UK
Bruno Baruque	University of Burgos, Spain
Carina González	University of La Laguna, Spain
Carlos Carrascosa	Politechnich University of Valencia, Spain
Carmen Benavides	University of Leon, Spain
Daniel Glez-Peña	University of Vigo, Spain
David Griol Barres	University Carlos III of Madrid, Spain
Davide Carneiro	University of Minho, Portugal
Dídac Busquets	University of Girona, Spain
Dongshik KANG	Ryukyu University, Japan

Eladio Sanz   Unversity of Salamanca, Spain
Eleni Mangina   University College Dublin, Ireland
Emilio Corchado   University of Burgos, Spain
Eugenio Aguirre   University of Granada, Spain
Eugénio Oliveira   University of Porto, Portugal
Evelio J. González   University of La Laguna, Spain
Faraón Llorens Largo   University of Alicante, Spain
Fernando Díaz   Univesity of Valladolid, Spain
Fidel Aznar Gregori   University of Alicante, Spain
Florentino Fdez-Riverola   University of de Vigo, Spain
Francisco Pujol López   Polytechnic University of Alicante, Spain
Fumiaki Takeda   Kochi Institute of Technology, Japan
Germán Gutiérrez   University Carlos III, Spain
Grzegorz Bocewicz   Koszalin University of technology, Poland
Gustavo Santos   University of Salamanca, Spain
Ivan López Arévalo   Lab. of Information Technology
                                     Cinvestav-Tamaulipas, Mexico
Jamal Dargham   University of Malaysia, Saba, Malaysia
Javier Carbó   University Carlos III of Madrid, Spain
Javier Martínez Elicegui   Telefónica I+D, Spain
Jesús García Herrero   University Carlos III of Madrid, Spain
Joao Gama   University of Porto, Portugal
Johan Lilius   Åbo Akademi University, Finland
José R. Villar   University of Oviedo, Spain
Juan A. Botia   University of Murcia, Spain
Juan Pavón   Complutense University of Madrid, Spain
José M. Molina   University Carlos III of Madrid, Spain
José R. Méndez   University of Vigo, Spain
José V. Álvarez-Bravo   University of Valladolid, Spain
Joseph Giampapa   Carnegie Mellon, USA
Juan Manuel Cueva Lovelle   University of Oviedo, Spain
Juan Gómez Romero   University Carlos III of Madrid, Spain
Kazutoshi Fujikawa   Nara Institute of Science and Technology, Japan
Lourdes Borrajo   University of Vigo, Spain
Luis Alonso   University of Salamanca, Spain
Luis Correia   University of Libon, Portugal
Luis F. Castillo   Autonomous University of Manizales, Colombia
Luís Lima   Polytechnic of Porto, Portugal
Manuel González-Bedia   University of Zaragoza, Spain
Manuel Pegalajar Cuéllar   University of Granada, Spain
Manuel Resinas   University of Sevilla, Spain
Marcilio Souto   Federal University of Rio Grande do Norte, Brazil
Margarida Cardoso   ISCTE, Portugal
Maria del Mar Pujol López   University of Alicante, Spain
Masanori Akiyoshi   Osaka University, Japan
Masaru Teranishi   Hiroshima Institute of Technology, Japan

Michifumi Yoshioka	Osaka Prefecture University, Japan
Miguel Delgado	University of Granada, Spain
Miguel Ángel Patricio	University Carlos III of Madrid, Spain
Miguel Rebollo	University of Vigo, Spain
Miguel A. Vega-Rodriguez	University of Extremadura, Spain
Mohd Saberi Mohamad	University of Technology Malaysia, Malaysia
Michael Zaki	University Rostok, Germany
Noki Mori	Osaka Prefecture University, Japan
Norihisa Komoda	Osaka University, Japan
Paulo Novais	Polytechnic University of Minho, Portugal
Pawel Pawlewski	Poznan University of Technology, Poland
Peter Fobrig	University Rostok, Germany
Ichiro Satoh	National Institute of Informatics, Japan
Rafael Corchuelo	Catholic University of Sevilla, Spain
Ramón Rizo	University of Alicante, Spain
René Leistikow	University Rostok, Germany
Ricardo Campello	University of Sao Paulo, Brazil
Ricardo Costa	Polytechnic of Porto, Portugal
Rubén Fuentes	Complutense University of Madrid, Spain
Rui Camacho	University of Porto, Portugal
Sebastian Ahrndt	Technische Universität Berlin, Germany
Sebastian Bader	University of Rostock, Germany
Seiichi Ozawa	Kobe University, Japan
Seiji Ito	Hiroshima Institute of Technology, Japan
Seyedsaeid Mirkamali	University of Mysore, India
Shanmugasundaram Hariharan	B.S.Abdur Rahman University Vandalur, India
Silvana Aciar	University of Girona, Spain
Xinyu Que	IBM Research, USA
Shiro Tamaki	University of the Ryukyus, Japan
Sofya V. Zhukova	St. Petersburg State University, Russia
Toru Yamamoto	Hiroshima University, Japan
Vicente Botti	Politechnich University of Valencia, Spain
Vicente Julián	Politechnich University of Valencia, Spain
Victor J. Sosa-Sosa	Laboratory of Information Technology (LTI) CINVESTAV, México
Viet-Hoang VU	Hanoi University of Technology, Vietnam
Waldo Fajardo	University of Granada, Spain
Worawan Diaz Carballo	Thammasat University, Thailand
Yi Fang	Purdue University, USA
Yusuke NOJIMA	Osaka Prefecture University, Japan
Yutaka MAEDA	Kansai University, Japan
Zbigniew Pasek	IMSE/University of Windsor, Canada

AI-Driven Methods for Multimodal Networks and Processes Modeling Special Session Committee

Chairs

Grzegorz Bocewicz	Koszalin University of Technology, Poland
Izabela E. Nielsen	Aalborg University, Denmark

Co-chairs

Paweł Pawlewski	Poznan University of Technology, Poland
Zbigniew Banaszak	Warsaw University of Technology, Poland
Peter Nielsen	Aalborg University, Denmark
Sitek Paweł	Kielce University of Technology, Poland
Robert Wójcik	Wrocław University of Technology, Poland
Marcin Relich	University of Zielona Gora, Poland
Justyna Patalas-Maliszewska	University of Zielona Gora, Poland
Krzysztof Bzdyra	Koszalin University of Technology, Poland

Multi-agents Macroeconomics Special Session Committee

Chairs

Edgardo Bucciarelli (Economist)	University of Chieti-Pescara, Italy
Gianfranco Giulioni (Economist)	University of Chieti-Pescara, Italy

Scientific Committee

José Carlos R. Alcantud (Mathematical economist)	University of Salamanca, Spain
Claudia Casadio (Philosopher of science)	University of Chieti-Pescara, Italy

Shu-Heng Chen (Economist)   National Chengchi University, Taipei, Taiwan
S. Barry Cooper   University of Leeds, United Kingdom
   (Mathematical logician)
Felix Freitag (Computer scientist)   Universitat Politécnica de Catalunya, Spain
Jakob Kapeller (Economist)   Johannes Kepler University of Linz, Austria
Amin Khan (Computer scientist)   IST, University of Lisbon, Portugal
Alan Kirman (Economist)   Aix-Marseille Université, France
Nicola Mattoscio (Economist)   University of Chieti-Pescara, Italy
Giulio Occhini   Italian Association for Informatics and Automatic
   (Computer scientist)   Calculation, Milan, Italy
Enrico Rubaltelli   University of Padua, Italy
   (Cognitive psychologist)
Anwar Shaikh (Economist)   The New School for Social Research, New York,
                                        United States
Antonio Teti (Computer scientist)   Italian Association for Informatics and Automatic
                                        Calculation, Milan, Italy
Katsunori Yamada (Economist)   Osaka University, Japan
Stefano Zambelli (Economist)   University of Trento, Italy

Organizing Committee

Sara Rodríguez (Chairman)   University of Salamanca, Spain
Juan F. De Paz (Chairman)   University of Salamanca, Spain
Javier Bajo   Technical University of Madrid, Spain
Fernando De la Prieta   University of Salamanca, Spain
Gabriel Villarrubia González   University of Salamanca, Spain
Javier Prieto Tejedor   University of Salamanca, Spain
Alejandro Hernández Iglesias   University of Salamanca, Spain
Emilio S. Corchado   University of Salamanca, Spain
Belén Pérez Lancho   University of Salamanca, Spain
Angélica González Arrieta   University of Salamanca, Spain
Vivian F. López   University of Salamanca, Spain
Ana de Luís   University of Salamanca, Spain
Ana B. Gil   University of Salamanca, Spain

Review Process

DCAI welcomed the submission of application papers with preference to the topics listed in the call for papers.

All submitted papers followed a thorough review process. For each paper was be refereed by at least two international experts in the field from the scientific committee (and most of them by three experts) based on relevance, originality, significance, quality and clarity.

The review process took place from January 20, 2015 to February 18, 2015 and was carried out using the Easychair conference tool.

The review process granted that the papers consisted of original, relevant and previously unpublished sound research results related to any of the topics of the conference.

All the authors of those papers requiring modifications were required to upload a document stating the changes included in the paper according to the reviewers' recommendations. The documents, together with the final versions of the papers, were revised in detail by the scientific committee chairs.

Contents

Part I: Main Track

Part II: Special Session on AI–Driven Methods for Multimodal Networks and Processes Modeling (AIMPM 2015)

Part III: Special Session on Multi-agents Macroeconomics (MAM 2015)

Part I
Main Track

A Natural Language Interface
to Ontology-Based Knowledge Bases

Mario Andrés Paredes-Valverde[1], José Ángel Noguera-Arnaldos[2],
Cristian Aarón Rodríguez-Enríquez[3], Rafael Valencia-García[1],
and Giner Alor-Hernández[3]

[1] Departamento de Informática y Sistemas,
Universidad de Murcia, Campus de Espinardo, 30100, Murcia, Spain
[2] PROASISTECH, S.L.
Edificio CEEIM, Campus Universitario Espinardo s/n, 30100, Murcia
[3] Division of Research and Postgraduate Studies, Instituto Tecnológico de Orizaba, Mexico
{marioandres.paredes,valencia}@um.es, jnoguera@proasistech.com,
crodriguezen@gmail.com, galor@itorizaba.edu.mx

Abstract. The aim of the Semantic Web is to improve the access, management, and retrieval of information on the Web-based. On this understanding, ontologies are considered a technology that supports all aforementioned tasks. However, current approaches for information retrieval on ontology-based knowledge bases are intended to be used by experienced users. To address this gap, Natural Language Processing (NLP) is deemed a very intuitive approach from a non-experienced user`s perspective, because the formality of a knowledge base is hidden, as well as the executable query language. In this work, we present ONLI, a natural language interface for DBpedia, a community effort to structure Wikipedia's content based on an ontological approach. ONLI combines NLP techniques in order to analyze user's question and populate an ontological model, which is responsible for describing question's context. From this model, ONLI requests the answer through a set of heuristic SPARQL-based query patterns. Finally, we describe the current version of the ONLI system, as well as an evaluation to assess its effectiveness in finding the correct answer.

Keywords: natural language processing, ontology, semantic web.

1 Introduction

The aim of the Semantic Web is to provide well-defined and understandable information not only by humans, but also by computers, allowing these last ones to automate, integrate and reuse high-quality information across different applications. Ontologies are considered a technology which supports all aforementioned tasks. In the context of computer and information sciences, an ontology defines a set of representational primitives allowing to model a domain of knowledge or discourse [1]. Nowadays, data stored in ontology-based knowledge bases has significantly grown, becoming in an important component in enhancing the Web intelligence and in supporting data

© Springer International Publishing Switzerland 2015
S. Omatu et al. (eds.), *Distributed Computing & Artificial Intelligence, 12th Int. Conference,*
Advances in Intelligent Systems and Computing 373, DOI: 10.1007/978-3-319-19638-1_1

integration. Indeed, ontologies are being applied to different domains such as biomedicine [2], finance [3], innovation management [4], cloud computing [5] [6], medicine [7], and human perception [8], among others.

At the same time, Wikipedia, a free-access and free content Internet encyclopedia, has become one of the most popular Web-based knowledge sources, maintained by thousands of collaborators. DBpedia [9] has emerged as a community effort to structure Wikipedia's content and make it available on the Web. Currently, the English version of DBpedia describes 4.22 million of things in a consistent ontology, including persons, places, creative works, organizations, species and diseases.

The most known formal query language for information retrieval from semantic knowledge bases is SPARQL (*Simple Protocol and RDF Query Language*). The use of SPARQL demands a high level of knowledge and expertise about technologies such as RDF (*Resource Description Framework*) and query language expressions. Because of this, several research efforts have been carried out in order to make accessible the Semantic Web information to all kind of users. From this perspective, NLI (*Natural Language Interface*) is deemed to be a very intuitive approach from a user point of view [10] in order to address the gap between knowledge bases systems and end-users. A NLI allows users to access information stored in any repository by formulating requests in natural language [11]. In this work, we present ONLI, a NLI for DBpedia. ONLI processes user's question by means of Natural Language Techniques in order to obtain semantically meaningful words. These elements are queried against knowledge base in order to establish a context. All information obtained is organized in an ontological model, from which the possible ambiguities are managed. Finally, the answer search is carried out, this search is based on a set of heuristic SPARQL-based query patterns. Once possible answers are obtained, these ones are organized and shown back to the user.

This paper is structured as follows: Section 2 presents a set of related works about NLI for knowledge bases. The architecture design, modules and interrelationships of the proposed approach are described in Section 3. Section 4 presents the evaluation of the ONLI system. Finally, conclusions and future work are presented.

2   Related Works

In recent years, several research works concerning to NLI for knowledge bases (NLIKB) have been carried out. These works provide to end-users a mechanism to access knowledge in ontologies, hiding the formality of ontologies and query languages. In this context, there are works focused to query Wikipedia content such as [16] where authors presented a search interface for Wikipedia, which enable users to ask complex queries based on a faceted approach. On the other hand, there are works focused on a specific context such as QACID [12], an ontology-based Question Answering applied to the Cinema Domain. QACID groups user queries into clusters which are associated to a SPARQL-based query. Then, QACID's infer engine deduces between a new query and the clusters in order to associate the query with its corresponding SPARQL-based expression. Other works are focused on portable NLI development. NLP-Reduce [11] identifies triple structures contained in user's question and match them to the synonym-enhanced triple store previously generated from an OWL knowledge base. The identified triples are translated to SPARQL-based statements. Then, NLP-Reduce infers implicit triple statements by using the *Pellet Reasoner*. Finally the results are displayed to the user. NLP-Reduce issue is the limited set of natural language processing techniques used to process question, which limits the identification of triples. ORAKEL [10] is an ontology-based natural language system which works with input queries that contain *wh*-pronouns. The query is parsed to a query logical form, which in turn, is translated to a knowledge representation language. Then, ORAKEL evaluates the query with respect to the knowledge base and presents the results to the user. The main issue of ORAKEL is the need for a lexicon engineer to carry out the adaptation process for a specific domain. Aqualog [13] takes natural language queries and an ontology as input. Then, Aqualog translates queries into ontology-compatible triples. Aqualog takes into account features such as voice and tense in order to facilitate the reasoning about the answer. Aqualog identifies 23 different linguistic categories of intermediate representations of queries. This classification is based on the kind of triple needed. Querix [14] is a domain-independent NLI for the Semantic Web. It processes query and obtains a syntax tree and a query skeleton of word categories such as *Noun*, *Verb*, among others. Then, a synonym search is performed for each element contained in the syntax tree, Querix matches the query skeleton with a small set of heuristic patterns. Next, Querix searches for all matches between synonym-enhanced query, with the resources and their synonyms in the ontology. Finally, Querix composes SPARQL-based queries from the joined triples. SWSNL [15] uses ontologies for storing the domain structure and the description of user queries. SWSNL analyses query by using NLP techniques and then built a semantic and KB-independent representation of query which is translated to a query language.

Most of above presented works were used and tested on different knowledge bases which vary in their size and complexity. Because of this, it is difficult to establish which interface provides the best result. Also, each work employs a different set of natural language processing techniques, in our approach a wider set of these techniques is used in order to obtain a more detailed analysis of the user's question, among these techniques are: POS (Part-of-speech) tagging, lemmatizing, synonym expansion, and NER.

3   ONLI System

In this work, we present a NLI called ONLI, which allows users to obtain answers to questions concerning to DBpedia's content. ONLI system is composed of three main modules: (1) NLP, (2) Question classification, and (3) Answer searching and building. In brief, ONLI system works as follows: User provides a question expressed in natural language. NLP module processes the question with the aim of detecting words potentially belonging to a semantic category. Question classification module queries the obtained elements against DBpedia in order to obtain all semantic information

about them, and then, a question's context is established. All this information is organized in an ontological model called *Question model,* whose main purpose is to describe the user's question so that subsequent modules can use this information to
search the answer. Next, question is classified based on the information contained in
the *Question model.* The Answer Searching and Building module looks for an answer
though a set of heuristic SPARQL-based query patterns. Finally, this module filters
the results obtained and provides an answer to the user. The ONLI's functional architecture, which is based on the generic architecture for NLI proposed by [17], is composed of three different modules that are described in detail in next sections.

3.1   NLP Module

This module carries out the natural language processing of the question in order to
obtain all semantically meaningful of the words for the system. To this end, POS
tagging, Lemmatization, and Named Entity Recognition (NER) techniques are executed. POS-tagging technique allows annotating all question elements with their
lexical category such as verb, proper noun, and adjective, to mention but a few.
Sometimes the word contained in the question is not the same that defines a knowledge entity, e.g. a word could be declared in a different verb time, or in plural.
Lemmatization technique provides the base form for a word, known as *lemma.* The
lemma can be used instead of the word as mentioned in the original question for linking against the knowledge base. The aim of the NER technique is to detect Named
Entities [18], i.e. entities belonging to a predefined semantic category such as person
names, organizations, locations, among others. This task is performed by using gazet-
teers which identifies entity names in the text based on lists, and annotation patterns
i.e., rules which act on annotations assigned in earlier phases, in order to produce
outputs of annotated entities.

Furthermore, NLP module searches for interrogative particles such as *what, where,
when,* and *who,* among others. These particles will help to determine the answer type
expected, thus reducing the search space. Also, NLP module performs a synonym
search for each entity found. This task is supported by the lexical database WordNet
[19]. The synonym search aims to extend the knowledge entities' search through use
of both the entity's lemma and the entity's synonyms.

At the end of NLP phase, there is a set of entities belonging to a specific entity type
(*questionType, individual, objectProperty, datatypeProperty,* and *class* or *concept*),
which in turn have two main properties: *originalContent,* which refers to the word as
it appears in the question, and *lemma.* All these information is mapped to the
Question model.

3.2   Question Classification Module

Once question has been processed, each entity contained in the *Question model* is
queried against DBpedia in order to establish the question's context. According to the
entity type, a SPARQL-based query is executed. In this query, entity's lemma and
entity's synonyms are compared against *rdfs:label* property of each knowledge entity.
The *rdfs:label* property can be used to provide a human-readable version of a resource

name. Hence, each class, property and instance, contained in the knowledge base must have the *rdfs:label* property defined, in order to ensure the correct functioning of the ONLI system. It must be noted that the final result, i.e. the answer, depends of quality of these labels. The knowledge entities' search needs to be flexible, i.e., it does not require to retrieve only entities with a high level of similarity because of this, the present module assigns a score to each knowledge entity using the *Levenshtein* distance metric. This score determines the similarity level between the entity's *lemma* and the knowledge entity's label. The knowledge entity with highest *score* probably is the most crucial to find the correct answer. The question's context establishing does not require all information about a knowledge entity, because these entities are only the base for future queries. Based on this understanding, the properties recovered of each knowledge entity are: *URI*, *rdfs:label*, and *rdf:type*.

On the other hand, most questions contain interrogative particles which provide a guide about the desired response, thus reducing the search space. For instance, when the question contains the interrogative particle *who*, user is generally searching for a person or an organization. In the same context, the interrogative particle *where* makes reference to a location, therefore, the search space can be reduced to knowledge entities belonging to semantic categories such as country, city, and buildings, among others. Taking this into account, this module classifies the question according to the classification of question proposed in [20], furthermore, it determines the answer type expected, which in turn is mapped to the *Question model* as an *answerType* entity. The answer type can be an instance of any class contained in the knowledge base, a set of instances, a date, or a number, among others. All information recovered in this phase is organized in the *Question model*.

3.3 Answer Searching and Building Module

Once question has been processed, and all the entities found have been organized into the *Question model*, the search for an answer is carried out. This module extracts the knowledge entities contained in the *Question model* and it includes each of them in a set of heuristic SPARQL-based query patterns. According to the knowledge entity´s type and question's type, a query pattern is selected. For instance, if the following query is processed: *"How many films has Christian Bale done?"*, a total of two knowledge entities are found, *Film* and *Christian Bale* corresponding to the entity types *Concept* and *Individual* respectively. Also, the *"How many"* question type has been identified, therefore, the user is looking for an amount, i.e. the answer expected corresponds to a *number*. Taking into account the aforementioned information, the Answer Searching and Building module include the entities found in the next heuristic SPARQL-based query pattern:

```
SELECT count (distinct ?uri) as ?count
WHERE {
  ?uri rdf:type <http://dbpedia.org/ontology/Film>.
  ?uri ?prop <http://dbpedia.org/resource/Christian_Bale>.
  ?uri rdfs:label ?label.
  FILTER langMatches( lang(?label), 'EN' )
}
```

Finally, the result obtained from the SPARQL-based query above presented, is shown to the user, i.e. according to the DBpedia's content, Christian Bale has been in 34 movies. It should be noted that each pattern used by this module is focused in different contexts in order to retrieve results consistent with the answer type expected.

4   Evaluation and Results

In order to assess the effectiveness of the present work, we conducted an evaluation focused on the capability of ONLI to provide the correct answer to user queries. For this purpose, we performed the study with a set of five students of Computer Science. The students were introduced to the DBpedia knowledge content, and then they were asked to express 20 queries in natural language. The questions used in this study had to meet the following constraints:

1. The query must be able to be answered through DBpedia's content.
2. The questions must begin with *Wh*-pronouns such as *who*, *what*, *where* and *which*, or contain the expressions *"How many"* and *"How"* followed by an adjective.
3. The question must not have orthographic errors.

Examples of the questions generated are: *"What is the height of Lebron James?"*, *"Where were Rafael Nadal and Roger Federer born?"*, *"What are the movies of Michael Caine?"*, and *"How many astronauts born in Spain?"*

Once natural language queries were defined, these queries were executed by using the ONLI system. The answer provided by ONLI was compared to the correct answer provided by the SPARQL-based query previously built by an experienced user on SPARQL and DBpedia. In this evaluation, we employed the *precision* and *recall* metrics, as well as their harmonic mean, known as *F-measure*. These metrics are commonly applied to information retrieval systems and researches in the context of NLI [10].

The results obtained from the evaluation indicated that ONLI system obtained an F-measure score of 0.82. The global experiments results are reported in Table 1. Despite the fact that the results obtained seem promising, there are yet some issues to be solved. On one hand, the DBpedia has full narrative labels, however we cannot guarantee that all knowledge entities contained in this knowledge base are correctly annotated. On the other hand, once both set of queries (natural language questions used on ONLI, and their corresponding SPARQL-based queries manually-built by an expert user) were analyzed, we ascribe the unsuccessfully answered questions to the complexity of the questions generated i.e., some natural language questions contained more than two knowledge entities of different types, which in turn required more than one triple pattern for their representation on SPARQL-based language. Therefore, the set of heuristic SPARQL-based patterns was a bit limited to deal with this kind of questions.

Table 1. Evaluation results

Precision	Recall	F-measure
0.81	0.84	0.82

5   Conclusions and Future Work

The main contribution of this research is a methodology for the development of a question answering system for ontology-based knowledge bases. Specifically, we present ONLI, an ontology-based NLI developed for answering queries related to the DBpedia's content. Although the results obtained seem promising, ONLI needs to be improved in order to provide a system able to return precise answers to questions, since ONLI system provides a set of possible answers instead of a unique answer.

As subsequent work we plan to develop a method that allows carrying out the lexicalization of the DBpedia ontology, i.e. to obtain a set of words that can be found in the user's question, thus improving the identification of potential ontology concepts, which in turn, can be used to build the SPARQL-based query. Also, we plan to enlarge and enrich the set of heuristic SPARQL-patterns by taking into account a wider set of question and answer types. Other ONLI issue is the lack of robustness, because of all input questions must contain *Wh*-pronouns or expressions *"How many"* and *"How"* followed by and adjective. For solving this, we are considering to extend the question's classification and include a wide set of question patterns that allows establishing the question context and the type of answer expected, avoiding the need to use aforementioned words to express a question.

Acknowledgments. Mario Andrés Paredes-Valverde is supported by the National Council of Science and Technology (CONACYT) and the Mexican government.

References

1. Gruber, T.: Ontology. In: Encyclopedia of Database Systems, vol. 5, p. 3748. Springer
2. Ruiz-Martínez, J.M., Valencia-García, R., Martínez-Béjar, R., Hoffmann, A.: BioOnto-Verb: A top level ontology based framework to populate biomedical ontologies from texts. Knowl.-Based Syst. 36, 68–80 (2012)
3. Lupiani-Ruiz, E., García-Manotas, I., Valencia-García, R., García-Sánchez, F., Castellanos-Nieves, D., Fernández-Breis, J.T., Camón-Herrero, J.B.: Financial news semantic search engine. Expert Syst. Appl. 38(12), 15565–15572 (2011)
4. Hernández-González, Y., García-Moreno, C., Rodríguez-García, M.Á., Valencia-García, R., García-Sánchez, F.: A semantic-based platform for R&D project funding management. Comput. Ind. 65(5), 850–861 (2014)
5. Rodríguez-García, M.Á., Valencia-García, R., García-Sánchez, F., Samper-Zapater, J.J.: Ontology-based annotation and retrieval of services in the cloud. Knowl.-Based Syst. 56, 15–25 (2014)

6. Rodríguez-García, M.Á., Valencia-García, R., García-Sánchez, F., Samper-Zapater, J.J.: Creating a semantically-enhanced cloud services environment through ontology evolution. Future Gener. Comput. Syst. 32, 295–306 (2014)
7. Rodríguez-González, A., Alor-Hernández, G.: An approach for solving multi-level diagnosis in high sensitivity medical diagnosis systems through the application of semantic technologies. Comput. Biol. Med. 43(1), 51–62 (2013)
8. Prieto-González, L., Stantchev, V., Colomo-Palacios, R.: Applications of Ontologies in Knowledge Representation of Human Perception. Int. J. Metadata Semant. Ontol. 9(1), 74–80 (2014)
9. Lehmann, J., Isele, R., Jakob, M., Jentzsch, A., Kontokostas, D., Mendes, P., Hellmann, S., Morsey, M., van Kleef, P., Bizer, C.: DBpedia - A Large-scale, Multilingual Knowledge Base Extracted from Wikipedia. Semantic Web Journal (2014)
10. Cimiano, P., Haase, P., Heizmann, J., Mantel, M., Studer, R.: Towards portable natural language interfaces to knowledge bases – The case of the ORAKEL system. Data Knowl. Eng. 65(2), 325–354 (2008)
11. Kaufmann, E., Bernstein, A.: Evaluating the Usability of Natural Language Query Languages and Interfaces to Semantic Web Knowledge Bases. Web Semant. 8(4), 377–393 (2010)
12. Ferrández, Ó., Izquierdo, R., Ferrández, S., Vicedo, J.L.: Addressing ontology-based question answering with collections of user queries. Inf. Process. Manag. 45(2), 175–188 (2009)
13. Lopez, V., Pasin, M., Motta, E.: AquaLog: An Ontology-Portable Question Answering System for the Semantic Web. In: Gómez-Pérez, A., Euzenat, J. (eds.) ESWC 2005. LNCS, vol. 3532, pp. 546–562. Springer, Heidelberg (2005)
14. Kaufmann, E., Bernstein, A., Zumstein, R.: Querix: A Natural Language Interface to Query Ontologies Based on Clarification Dialogs. In: 5th ISWC, pp. 980–981 (2006)
15. Habernal, I., Konopík, M.: SWSNL: Semantic Web Search Using Natural Language. Expert Syst. Appl. 40(9), 3649–3664 (2013)
16. Hahn, R., Bizer, C., Sahnwaldt, C., Herta, C., Robinson, S., Bürgle, M., Düwiger, H., Scheel, U.: Faceted Wikipedia Search. In: Abramowicz, W., Tolksdorf, R. (eds.) BIS 2010. LNBIP, vol. 47, pp. 1–11. Springer, Heidelberg (2010)
17. Smith, R.W.: Natural Language Interfaces. In: Brown, K. (ed.) Encyclopedia of Language & Linguistics, 2nd edn., pp. 496–503. Elsevier, Oxford (2006)
18. Grishman, R., Sundheim, B.: Message Understanding Conference-6: A Brief History, Stroudsburg, PA, USA, pp. 466–471 (1996)
19. Miller, G.A.: WordNet: A Lexical Database for English. Commun. ACM 38(11), 39–41 (1995)
20. Moldovan, D., Harabagiu, S., Pasca, M., Mihalcea, R., Girju, R., Goodrum, R., Rus, V.: The Structure and Performance of an Open-domain Question Answering System. In: Proceedings of the 38th Annual Meeting on Association for Computational Linguistics, Stroudsburg, PA, USA, pp. 563–570 (2000)

Retweeting Prediction Using Meta-Paths Aggregated with Follow Model in Online Social Networks

Li Weigang and Jianya Zheng

TransLab, Department of Computer Science, University of Brasilia, Brasilia – DF, Brazil
{weigang,zhengdarcy}@cic.unb.br

Abstract. Studying the mechanism of retweeting is useful for understanding the information diffusion in Online Social Networks (OSNs). In this paper, we examine a number of topological features that may affect the retweeting behavior. We apply the *Follow Model* to formulate the user relations and then propose *Relationship Commitment Adjacency Matrix* (*RCAM*) to present the connectivity between users in OSNs. We define three meta-paths to identify the people who may retweet. With these meta-paths, various instance-paths are revealed in the retweeting prediction problem. A framework based on Conditional Random Field model is developed and implemented with the data from Sina Weibo. The case study obtains the results of retweeting prediction with the indices of precision larger than 61% and recall larger than 58%.

Keywords: Follow model, Online Social Network, retweeting prediction, Weibo.

1 Introduction

Online Social Networks (OSNs) have become important social activities for human on the Internet [1-5]. They provide great potential utility for both business marketing and academic research. While business focuses on promoting brand and products, the academic community is interested in user interactions and evolution of network technique. Although the OSN data has great value, how to extract valuable knowledge from the massive data, within a certain time limitation, is still a great challenge.

Many functions have been developed to meet user's demand and facilitate the communication, such as activities *retweet, mention, like* and *comment* in social networks. The information produced by these activities could open up a new space in the study of human behaviors. Retweeting is one of the most popular functions in OSNs, by which the user could forward the message they concerned about to their followers, relatives and friends. In this way, the information carried by the message can spread widely and quickly. The prediction of retweeting action becomes an intense research area because it can help scientists understand better how the information will propagate in the online social networks.

In this paper, based on the concept of meta-path from the study of heterogeneous network [3], we use *mention* information to help predicting the user's retweeting

© Springer International Publishing Switzerland 2015

S. Omatu et al. (eds.), *Distributed Computing & Artificial Intelligence, 12th Int. Conference*,
Advances in Intelligent Systems and Computing 373, DOI: 10.1007/978-3-319-19638-1_2

11

action. Also the composite relationship between users which are represented by the *Follow Model* [4,6] are included to make our approach more effective.

From the conventional graph theory, the relationship between the nodes is represented by *Adjacency Matrix* (AM). In this paper, we apply the *Follow Model* to formulate the user relations and then propose Relationship Commitment Adjacency Matrix (RCAM) to present the connectivity between users in OSNs. We define three meta-paths to identify the people who may retweet. The paper extends the basic idea of [7] and aggregates the followship relations in this matrix presentation to define *Relationship Committed Adjacency Matrix* (RCAM): A_{in} as *Follower Adjacency Matrix* and A_{out} as *Followee Adjacency Matrix*.

In OSNs, the multi-constraint queries such as "who, when and what" are important for mining analysis and retweeting prediction. This paper proposes and implements RCAM associated querying and retweeting prediction framework to a real online social network, Sina Weibo in China (weibo.com). The data involves 58.66 million users with 265.11 million followship relations and 369.80 million messages, of which 51.62% were retweets.

2   Related Work

Follow Model was proposed by Sandes et al. [4, 6] in 2012 to describe the following relationship in OSNs, where the terms of followee, follower and r-friends (the users are their follower and followee with each other) can be represented by three functions: $f_{out}(.)$, $f_{in}(.)$ and $f_r(.)$. Given that objects u and v are associated in the vertices of network, the directed edge set E: $V \times V$ represents the edges:

$f_{out}(u) = \{v | (u,v) \in E\}$, is the followee function to present the subset, V^*, of all followees of user u, $V \to V^*$, $V^* \subset V$;

$f_{in}(u) = \{v | (v,u) \in E\}$, is the follower function to present the subset, V^*, of all followers of user u, $V \to V^*$, $V^* \subset V$;

$f_r(u) = f_{out}(u) \cap f_{in}(u)$, is the r-friend function to present the subset, V^*, of all r-friends of user u, $V \to V^*$, $V^* \subset V$.

These functions collectively form the *Follow Model* [4, 6], which has the following three properties: reverse relationship, compositionality and extensibility.

For retweeting prediction, Peng et al. [8] studied the retweet patterns using Conditional Random Fields (CRFs) method [9] with three types of user-tweet features: content influence, network influence and temporal decay factor. They also investigated the approaches to partition the social graphs and construct the network relations for retweeting prediction. Hong et al. [10] analyzed the problem of predicting the popularity of messages as measured by the number of future retweets and shed some light on what kinds of factors influence the information propagation in Twitter.

Sun et al. [3] proposed meta-path to describe the relationship in a heterogeneous network which consists of multi-typed objects and multi-typed relations. The meta-path can be used to define topological features with different semantic meanings.

For example, the connection path between user u and v in online social network is unique, but the relationship can be various, including following, mentioning, retweeting etc. In order to define the meta-path in this research, we have to identify the target relation in the retweeting prediction. The target relation is the one we wanted in the problem space, and here it is the retweeting relationship. We use $R_{rt} <u,v>$ to represent the action that the user u retweets v.

To establish the target relationship, we also need the similarity relationship R_{sim} as a bridge to connect the originate and the terminate. Sun et al. [3] listed three general forms of meta-path as topological features.

$u\ R_{sim}\ v\ R_{rt}\ w$, means u will $R_{rt}\ w$ if v retweets w and u is similar to v.

$u\ R_{rt}\ v\ R_{sim}\ w$, means u will retweet w if u retweet v and v is similar to w.

$u\ R_1\ v\ R_2\ w$, in this case, R_1 and R_2 are some relations undefined and could be defined by the researchers in the practical needs, for example R_1 may be mention etc.

Using meta-path we extended it into OSN with the combination of *Follow Model* to define the relationship between users [7], the detail is described in Section 3.

3   Who May Retweet?

In this section, we establish three meta-paths considering the target and similarity relations to represent who is likely to retweet based on the methods of [7] with necessary extension.

3.1   The User Re-tweets Others He/She is Concerned with

In the situation of "The user re-tweets others he/she is concerned with", if two users belong to the same category, one may retweet another with a high probability. The formulation of this relation can be represented as follow:

$$v\in f(w)R_{rt}\,w$$

where $f(.)$ is a function to describe the relationship between v and w defined in *Follow Model* [4,6], also see section 2. In this case, user v is likely to retweet w's tweet. We extend the $f(.)$ in more detailed cases:

1) $v \in f_{in}(w)\ R_{rt}\ w$. In w's follower subset $f_{in}(w)$, v is likely to retweet w's message.
2) $v \in f_r(w)\ R_{rt}\ w$. In w's r-friends subset $f_r(w)$, v is more likely to retweet w's message.
3) $v \in m_{out}(w)\ R_{rt}\ w$. Among the people mentioned by w, $m_{out}(w)$, v is likely to retweet w's message after mentioning.

3.2   The User Re-tweets Others He/She is Similar with

The situation of "The user re-tweets others he/she is similar with" can also be found in OSNs. If the similar person of a user retweets a message, then this user may also retweet this message with a high probability. This can be represented as the following:

$$v\,R_{sim}\,f(v)R_{rt}\,w$$

In this scenario, it is represented as R_{sim}, v is likely to retweet w´s tweet if some people similar to v do that. We extend the $f(.)$ in more detail cases:

1) $v\ R_{sim}\ f_{out}(v)\ R_{rt}\ w$. In case of many of v´s followee retweet w´s tweet, v is likely to retweet w´s tweet too.

2) $v\ R_{sim} f_r(v)\ R_{rt} w$. If some of v´s r-friends retweet w´s tweets, then v is likely to do the same thing with w too.

3) $v\ R_{sim}\ f_r^2(v)\ R_T\ w$. Considering of two-step relationships, in this case, in case of v is likely to retweet w´s tweets, many of v´s r-friends of r-friends do that.

4) $v\ R_{sim}\ f_{out}^2(v)\ R_T\ w$, Also considering of two-step relationships, in case of user v is likely to retweet w´s tweets, many of v´s followees of followees retweet the w´s messages.

We can also have more combination situations such as r-friends of followees ($f_r f_{out}(.)$) and followers of r-friends ($f_{out} f_r(.)$) in the cases 3) and 4).

3.3   The User Responses the Same to the Others Who are Similar

This scenario can described as the following equation

$$v\,R_{rt}\ f(w)R_{sim}\ w$$

where, user v may retweet w´s tweets, as v already retweeted the tweets of similarity people to w defined by $f(w)$. We extend $f(w)$ in detail by the following cases:

1) $v\ R_{rt}\ f_r(w)\ R_{sim}\ w$, if v always retweets many of the r-friends of w, v is more likely to retweet w´s messages in the future.

2) $v\ R_T\ f_{in}(w)\ R_{sim}\ w$, if v always retweet many of the followers of w, v is likely to retweet w´s messages in the future.

3) $v\ R_{rt}\ f_r^2(w)\ R_{sim}\ w$, Considering two-step relationships, if v always retweeted messages of the r-friends of r-friends of user w, then it is reasonable for v to retweet w´s tweets.

4) $v\ R_{rt}\ f_{in}^2(w)\ R_{sim}\ w$, Also considering two-step relationships, if v retweets the messages of the followers of the followers of user w, then it is reasonable for v to retweet w´ messages.

Based on the flexibility property of *Follow Model*, we can also have more combination cases such as r-friends of followers ($f_r f_{in}(.)$) and followers of r-friends ($f_{in} f_r(.)$) in the cases 3) and 4).

4   Relationship Committed Adjacency Matrix for OSNs

Follow Model can also be described in the matrix form as OSN is with large dimension [4, 6]. In graph theory, an adjacency matrix is used to represent the adjacency between the members/users. A directed graph $G = (V, E)$ comprising a set V of vertices together with a set E of edges, can be presented by an adjacency matrix A, where $A(u,v) = 1$ if $(u, v) \in E$; otherwise, $A(u, v) = 0$. If there are n vertices in this graph, the matrix A will have $n \times n$ dimensions.

The theory of graph tells us that the distance between vertex u and v in the graph can be calculated by the multiplication of adjacency matrix. K-step neighbors is the $A(u,v)$ entry in the A^k, where $A^k = A\,A \dots A$ is k-th power operation of the matrix A.

Similar to *Follow Model*, a directed and unweighted following relationship network can be presented by a *Follower Adjacency Matrix* (FAM) A_{in}, where $A_{in}(u, v) = 1$, if u follows v and $(u, v) \in E$; otherwise, $A_{in}(u, v) = 0$; where suffix *in* describes the following relationship according to $f_{in}(.)$ function of *Follow Model*. If there are n vertices in this graph, the matrix A_{in} has $n \times n$ dimensions.

In the *Follower Adjacency Matrix* A_{in}, each element represents the following relationship between users. Furthermore, we can get the transpose matrix of A_{in}, A_{in}^T, to reflect the followed relationship in the social network. As indicated by the definition of *Follow Model*, the followed relationship is described by $f_{out}(.)$, here known as the $A_{in}^T = A_{out}$.

In general, we call this type of *Follower/Followee Adjacency Matrix* as *Relationship Committed Adjacency Matrix* (RCAM). With *Follower/Followee Adjacency Matrix*, we can combine many querying operations to get the information from A_{in} and/or A_{out}.

It is observed that the distance between vertex u and v in the graph with following adjacency matrix A_{in} is the (u, v) entry in A_{in}^k, where $A_{in}^k = A_{in} A_{in} \dots A_{in}$. This means that from A_{in}^k we can get the information about the followers of the followers of … of any user within A_{in}. This is shown in the *Follow Model* as $f_{in}^k(.)$.

The distance between vertex u and v in the graph with followee adjacency matrix A_{out} is the (u, v) entry in A_{out}^k, where $A_{out}^k = A_{out} A_{out} \dots A_{out}$ is the multiplication of k times of the matrix A_{out}. This also means that from A_{out}^k we can get the information in k-step of the followees of the followees of … of any user within A_{out}. This is shown in the *Follow Model* as $f_{out}^k(.)$.

In order to know the followees of the followers of any user in A_{in}, we have $A_{out} A_{in}$ of $f_{out}f_{in}(.)$ in the Follow model. Conversely, to get the followers of the followees of any user in A_{in}, we have $A_{in} A_{out}$ shown in $f_{in}f_{out}(.)$ in the Follow model. In these two cases, we can also get more combinations given their relationships in A_{in}.

Considering the analyses of complexity of the multiplications of these matrixes, we have included more ontological commitments to the relations of OSNs. The two-step operation of A_{in}^2 is with the complexity of $O(n^3)$. For a graph with a large number of n connections in OSN, the calculation of A_{in}^2 is time-consuming. Once getting the result, it is useful to conduct more querying operations related in A_{in}^2.

5   Case Study of Sina Weibo

The real data used for our case study is from Sina Weibo which was published in the WISE2012 Challenge (WISE 2012) [4,11]. There were 58.66 million users with 265.11 million following relationships and 369.80 million messages. From this data set, we abstract a subset that consists of 19,142 users and 17,474 retweets from 1,472 unique original tweets using content about Steve Jobs, related to his death.

In the subset of these 17,474 messages, we took 2/3 of them as the training set, and the others were the testing set. All experiments were conducted on a Linux-based machine with four Intel Core 2.80GHz cores and 4G Memory.

Based on the *Follow Model*, RCAM and above meta-paths, the retweeting prediction processes are organized to form a framework with following 5 procedures:

1) **RCAM construction**. From the data set of relationship network, we construct *Relationship Committed Adjacency Matrixes*: A_{in}, A_{out} and A_m (*Mention Adjacency Matrix*). This step provides basic relationship matrices for further computation and can obtain statistical and mining information from OSNs.

2) **RCAM computation**. With the matrices from previous step, two-step relation by multiplication of RCAM can be calculated, such as A_{in}^2, $A_{in}A_{out}$ and $A_{out}A_{in}$ etc.

3) **Information querying**. Based on RCAMs, the most of the queries about the users and tweets can executed, such as the number of followers, the common followers, the friends of friends, and the estimated number of viewing of the tweets etc. These querying results are fundamental information for the prediction of retweeting action.

4) **Feature abstraction**. With the information of the relationship network and the message dataset, we abstract six features which are the basic elements for building the dataset for the experiments of retweeting prediction.

5) **Retweeting prediction**. In this last step, we choose a suitable toolkit, for example CRF++ in this paper, to perform CRFs algorithm [8, 9] for retweeting prediction.

In this study, we designed three experimental cases to evaluate the effectiveness of the proposed method.

1) In the scenarios of *"The user re-tweets others he/she is concerned with; The user re-tweets others he/she is similar with; The user responses the same to the others who are similar"*, where involving A_{in}, we predict the retweeting actions based on the basic relationship, Base RCAM, and other features.

2) The second involves more combinations of RCAMs and two-step multiplication, such as A_{in}^2. We call this case Two-step RCAM procedure.

3) In the last case, we add *Mention Adjacency Matrix* A_m as a feature in the retweeting prediction and call this case the Mention RCAM procedure.

The *precision*, *recall* and *F1 score* are adopted as the indices to measure the prediction performance. The precision is defined as the ratio of the true positive value in our prediction result (the number of correct prediction, of which the users do the retweeting action) to all the positive value in the prediction results. The recall is defined as the ratio of the true positive value in prediction result to the number of all retweet actions in the sample dataset. The performance indices of the prediction are represented in table 1.

In the case of Base RCAM, we take the first 4 features as the basic metrics. From these features we obtained the results of retweeting prediction with the precision of 65.5%, recall 70.3% and F1 score 67.8%.

For Two-step RCAM, we queried the information from two-step multiplication of *Follower Adjacency Matrix*, A_{in}^2, and then studied further the retweeting activity from the followers of followers. From $A_{in}^2 = A_{in} \times A_{in}$, we obtained the information of two-step relationship with comparison between A_{in} and A_{in}^2. The position of matrix with its value is 0 in A_{in} and 1 in A_{in}^2. This means that the users corresponding to the line and the column have two-step relationship.

Table 1. The prediction result using CRF++

	Precision	Recall	F1 score
Base RCAM	65.5%	70.3%	67.8%
Two-step RCAM	62.1%	67.2%	64.5%
Mention RCAM	66.2%	69.7%	67.9%

With A_{in}^2, we added two more features in our dataset which were the friends of the friends and the followers of the followers. With these two features, we got the results with precision of 62.1%, recall of 67.2% and F1 score of 64.5%. This performance was slightly worse than the Base RCAM case, because the relationship network in our dataset was incomplete. The advance of the indirectly relations, such as followers of followers, was not demonstrated as we expected in retweeting prediction.

In the case of Mention RCAM, we appended the mention relationship between users into our dataset. With this feature, we obtained the prediction results with better performance. In this experiment, the indices of precision, recall and F1 score were 66.2%, 69.7% and 67.9% respectively, as shown in table 2. The result was better than Base RCAM with the introduction of the mention relationship. This factor supports our proposal, that is, the mentioner is more likely to retweet the original tweet which mentioned him.

Table 2. The prediction result by CRF++ with modified data

	Precision	Recall	F1 score
Base RCAM	61.1%	58.5%	59.8%
Two-step RCAM	63.2%	59.3%	61.2%
Mention RCAM	63.4%	59.2%	61.2%

From table 2, we note that the two-step RCAM relationship influenced the precision of retweeting prediction. At the same time, with the involvement of the mention relationship, using the Mention RCAM procedure gave us a better performance than the Two-step RCAM procedure. However, more empirical analyses are still needed to verify the proposal. Comparing to the existed tweet prediction models [5, 8, 10], the introduction of Mention RCAM to reflect the mention action in retweeting is a new tentative and obtains better performance.

6   Conclusion

To describe the topological connectivity between users on OSNs, the RCAM aggregated with *Follow Model* was proposed to present the composite relationships for large dimensional social network. We defined the meta-paths in the retweeting prediction with respect to different following functions. Various instance paths were demonstrated to find the person who will retweet. By means of the method of CRFs, a framework was proposed to address this problem.

Based on the real data from Sina Weibo, the case study demonstrated the effectiveness of the proposed methods using RCAM in three human behaviors of retweeting: *a)* The user re-tweets others he/she is concerned with; *b)* The user re-tweets others he/she is similar with; *c)* The user responses the same to the others who are similar. The measurement indices were acceptable with the *precision* larger than 61% and *recall* larger than 58% in the most results of retweeting prediction.

With the complexity of $O(n^3)$ of the operation of A_{in}^2 and other k-step RCAMs, more efficient formulations and algorithms should be developed for large scale networks. The recent development of Singular Value Decomposition (SVD) in matrix completion [12] is more encourage of the application of RCAM. The future work will be in this direction to implement an independent and integrated system instead of using the application of CRF++ software package.

References

1. Cha, M., Haddadi, H., Benevenuto, F., Gummadi, K.: Measuring user influence in twitter: The million follower fallacy. In: Proceedings of 4th International AAAI Conference on Weblogs and Social Media (ICWSM), pp. 10–17 (2010)
2. Kwak, H., Lee, C., Park, H.: What is Twitter, a Social Network or a News Media? In: The 19th International World Wide Web Conference, Raleigh, USA, pp. 591–600. ACM (2010)
3. Sun, Y., Han, J., Aggarwal, C.C., Chawla, N.V.: When Will It Happen?: Relationship Prediction in Heterogeneous Information Networks. In: Proceedings of Int. Conf. on Web Search and Data Mining, WSDM 2012, pp. 663–672 (2012)
4. De Sandes, E.F.O., Weigang, L., de Melo, A.C.M.A.: Logical Model of Relationship for Online Social Networks and Performance Optimizing of Queries. In: Wang, X.S., Cruz, I., Delis, A., Huang, G. (eds.) WISE 2012. LNCS, vol. 7651, pp. 726–736. Springer, Heidelberg (2012)
5. Lu, X., Yu, Z., Guo, B., Zhou, X.: Predicting the content dissemination trends by repost behavior modeling in mobile social networks. Journal of Network and Computer Applications 42, 197–207 (2014)
6. Weigang, L., Sandes, E.F., Zheng, J., de Melo, A.C., Uden, L.: Querying dynamic communities in online social networks. Journal of Zhejiang University SCIENCE C 15(2), 81–90 (2014)
7. Weigang, L., Jianya, Z., Liu, Y.: Retweeting Prediction Using Relationship Commited Adjacency Matrix. In: BraSNAM – II Brazilian Workshop on Social Network Analysis and Mining, Maceio. Proceedings of CSBC 2013, pp. 1561–1572 (2013)
8. Peng, H.K., Zhu, J., Piao, D., Yan, R., Zhang, Y.: Retweet modeling using conditional random fields. In: IEEE 11th International Conference on Data Mining Workshops (ICDMW), pp. 336–343 (2011)

9. Lafferty, J., McCallum, A., Pereira, F.C.N.: Conditional random fields: Probabilistic models for segmenting and labeling sequence data. In: Proceedings of the Eighteenth International Conference on Machine Learning (ICML), pp. 282–289 (2001)

10. Hong, L., Dan, O., Davison, B.: Predicting popular messages in twitter. In: Proceedings of the 20th International Conference Companion on World Wide Web, pp. 57–58 (2011)

11. Ma, H., Wei, J., Qian, W., Yu, C., Xia, F., Zhou, A.: On benchmarking online social media analytical queries. In: First International Workshop on Graph Data Management Experiences and Systems, p. 10 (2013)

12. Hastie, T., Mazumder, R., Lee, J., Zadeh, R.: Matrix completion and low-rank SVD via fast alternating least squares. Journal of Machine Learning Research (to appear, 2015).

Swarm Agent-Based Architecture Suitable for Internet of Things and Smartcities

Pablo Chamoso, Fernando De la Prieta, Francisco De Paz, and Juan M. Corchado

Department of Computer Science and Automation Control, University of Salamanca,
Plaza de la Merced s/n, 37007, Salamanca, Spain
{chamoso,fer,fcofds,corchado}@usal.es

Abstract. Smart cities are proposed as a medium-term option for all cities. This article aims to propose an architecture that allows cities to provide solutions to interconnect all their elements. The study case focuses in locating and optimized regulation of traffic in cities. However, thanks to the proposed structure and the applied algorithms, the architecture is scalable in size of the sensor network, in functionality or even in the use of resources. A simulation environment which is able to show the operation of the architecture in the same way that a real city would, is presented.

Keywords: Agents, Swarm intelligence, Locating Systems, Smart cities

1 Introduction

The social, economic, environmental, and engineering challenges of this transformation will shape the 21st century. The lives of the people living in those cities can be improved –and the impact of this growth on the environment reduced– by the use of "smart" technologies that can improve the efficiency and effectiveness of urban systems. A smart city can be defined as the integration of technology into a strategic approach to achieve sustainability, citizen well-being, and economic development [9]. The smart city offers a coherent vision for bringing together innovative solutions that address the issues facing the modern city, but there are still many challenges to be dealt with.

Through the use of sensor networks, the building automation and control systems, which started with wired technology, have now entered the era of wireless technology, having produced technologies such as ZigBee, Z-Wave, EnOcean, among others. Wireless Sensor Networks (WSN) are used for gathering the information needed by intelligent environments, whether in urban construction and smart cities, home and building automation, industrial applications or smart hospitals [11]. However, the growing heterogeneity of this type of wireless network protocol makes them difficult to use. WSNs make it possible to build a wide range of applications, such as those used to control energy costs, monitor environmental data, implement security and access control in buildings, or automate industrial and home environments, and, Indoor Locating System (ILS).

© Springer International Publishing Switzerland 2015

S. Omatu et al. (eds.), *Distributed Computing & Artificial Intelligence, 12th Int. Conference*,
Advances in Intelligent Systems and Computing 373, DOI: 10.1007/978-3-319-19638-1_3

Although outdoor locating is well covered by systems such as the current GPS or the future Galileo, ILS still needs further development, especially with respect to accuracy and the use of low-cost and efficient infrastructures. Some of the applications of ILSs include among others tracking (people, assets and animals), access control, wander prevention, warning and alert systems, controlling security perimeters and optimization of resources.

Thus, at present there is a growing need for versatile open platforms capable of integrating different sensing technology over wireless technologies, merging data sets from heterogeneous sources, and intelligently managing the generated information. Although there are already some architectures or frameworks that allow the interconnection of sensors, both in academia and in business [5], the reality is that existing platforms are designed for a specific end, using a specific technology stack for each deployment, and offering a very specific set of services whose functionality is very limited. Therefore, current systems are limited by the pre-installation of infrastructure, and integrators have to face the decision of choosing between other technologies or adapting their existing systems and infrastructure. It is also difficult for integrators to combine the information obtained from heterogeneous wireless sensor networks since there are no adequate tools to do so.

This papers is structured as follows, next section introduces the state of the art. Section 3 includes an overview of the proposed architecture. The results of the evaluation of this architecture are presented in Section 4. Finally, a last section contains conclusions and future word.

2   Agents' Theory and Swarm Intelligence

There are many complexities in software agents. In this paper we focus on two completely opposite ends that are described below and take shape in the proposed case study. These ends are lightweight and heavyweight agents.

Lightweight agents have their origins in the study of animal behavior [3] where, compared to the case of ants, the intelligence of a single ant is far from the great skills that the colony shows. Therefore, the interactions between agents are the place where the intelligent part is centralized. The study of behavior has to be done by analyzing the environment of the entities and observing their actions, so that the representation of this behavior is mainly focused on sensory inputs and signals sent to actuators. Computationally, their representation is essentially numerical and it is based on the assignation of values to different parameters. This makes them more efficient than heavyweight agents.

On the other hand, heavyweight agents are based on the cognitive part of artificial intelligence, where each agent tries to reach the level of human intelligence [15]. Therefore, in this case the intelligence resides in each of the agents that take part in the system. They have the advantage over lightweight agents, that the fact of being developed taking into account human knowledge, facilitates understanding and communication with users. For example, the concept of "emergence" of an agent corresponds to the concept of human "emergency" resulting easier to understand.

Other examples of conceptual representations developed by heavyweight agents are beliefs, desires, intentions or goals, plans, etcetera [10].

2.1   Swarm Intelligence

As is known, a swarm is a set of single agents (such as the case of the mentioned lightweight agents) usually identical, cooperating among each other and their environment to complete a common objective or task with the ability of intelligent behavior [2].

In recent years, swarm based algorithms have become more and more popular due to their capabilities to provide solutions to complex problems. Some of these algorithms can be found in [7]. Problems trying to provide solutions swarm intelligence are mainly related to artificial intelligence (AI), data mining [8] and robotics [4]. These problems include optimization tasks, data analysis and pictures, optimal routes, etc. As shown in [2], the main advantages of swarm intelligence are the robustness, adaptability and self-organization, flexibility, scalability and decentralization, getting a good solution to distributed problems.

3   Architecture Overview

To design the platform proposed in this study, special attention is paid to open systems. Open systems exist in dynamic operating environments, in which new components may continuously join the system, or already existing components may leave it, and in which the operation conditions might unpredictability change. Therefore, open systems are characterized by the heterogeneity of their participants, their limited confidence, possible individual goals in conflict and a great probability of non-accordance with specifications.

These large systems can quite naturally be seen in terms of the services offered or demanded, and consequently, in terms of the entities (agents) providing or consuming services. It should be noted that these entities might not have been designed in a joint way or even by the same development team. Entities may enter or leave different organizations in different moments and for different reasons. Moreover, they may form coalitions or organizations among themselves to attain their current goals.

3.1   Platform Layers Description

Under the frame of this work, we proposed a new open platform, especially designed for the integration of heterogeneous sensor, through lightweight agents, and intelligent management of automated environments is means of heavyweight agents. The platform will incorporate advanced algorithms for intelligent information management and novel mechanisms to provide security for the information and communication within the platform. We propose to use a Virtual Organizations (VO) of Multiagent Systems (MAS) [16][12][1] to facilitate the design of the platform, which will be deployed in a cloud environment allowing sustainable growth, and adapted to the needs of deployment.

Since the proposed platform is intended to facilitate major infrastructure management, we propose to develop it within a Cloud Computing (CC) platform that allows resizing the resources and services necessary to manage the platform in an efficient way and according to the computational resources required for each deployment. The proposed development makes it possible to address the deployment of sensor networks in both small and large spaces, regardless of the set of technologies used in each case. This is because the CC environment assumes the computational costs of both the storage and computation of the information collected, treated and subsequently merged. In addition, a CC environment will also assume the subsequent commissioning of services, which the end user will employ. That will allow the information generated in different contexts to be used within the framework of smart city, while maintaining the level of quality specified in the contract for use of the platform with a Utility Computing model.

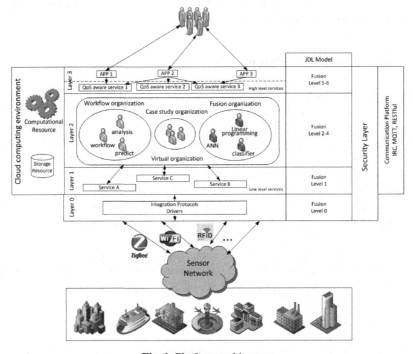

Fig. 1. Platform architecture

The platform is divided into a layered architecture as shown in Figure 1. As we can see, the platform is composed of different layers associated with the different functional blocks. These levels are distributed in the different layers of the architecture shown in Figure 1.

The following section describes the components and main features of the architecture:

- **Layer 0. Integrate sensing/performance technologies**. The platform will facilitate integration sensing technologies currently available, regardless of their nature, and will provide an open environment that allows the dynamic addition of new sensor systems and technologies. In order to accomplish this, the platform will provide data encapsulation mechanisms that standardize the information received from sources such as Wi-Fi, ZigBee, etc. Layer 0 of the platform is a broker that defines communication with sensor networks of different natures (Wi-Fi, ZigBee, Bluetooth, etc.), and obtains the raw data from sensor networks. The main novelty of this layer is the ability to provide the platform and the upper layers with openness regarding the connection to sensor networks of different natures, and thus ensure that upper layers of the architecture have access to information and are able to perform data fusion at different levels.
- **Layer 1. Low-level services**. Given the information exchanged with the environment through layer 0 as described above, the existing functional requirements and a set of low-level services will now be defined; specifically those that depend on the types of networks and technologies integrated into every deployment. After obtaining the raw data, a gateway is provided, defined through adapters that allow the information received to be standardized. In this first stage, the platform provides services such as filtering of signals, normalization services or other treatment services at the basic level signals. These services are provided by the adapters and will be associated with algorithms that perform initial treatment of the data, so that these data can be presented to higher layers in a more homogenized way. Each of these services will expose an API to higher layers which will allow interaction with each low-level service, and thus, with the underlying sensing/performance technologies.
- **Layer 2. Information fusion algorithms**. The platform is structured as a VO of MAS. Each organization includes the roles required to facilitate an intelligent management of the information obtained from the lower levels of the architecture. The MAS incorporates agents specifically designed to interact with low-level services. In addition, we propose the design of intelligent agents specialized in Information Fusion (IF)[6]. For this purpose, roles that allow merging information automatically through supervised learning and previous training have been included. This procedure can be applied in cases such as the mixture of experts. Statistical techniques will be added based on linear programming, neural networks and classifiers. Thus, specialized agents in IF can combine different sensing technologies that provide heterogeneous data and provide more accurate information to the upper layer services. In addition to providing IF techniques, this layer will provide an automatic generation of IF flows between different levels by including the organization of agents. This organization will contain roles such as analysis workflow, prediction workflow and workflow processing. The analysis workflow is responsible for analyzing a given workflow and for creating a new workflow automatically. The workflows will combine different techniques and algorithms integrated into the virtual organization for data processing for a specific case study. To carry out the analysis and prediction of workflow, statistical techniques based on Bayesian networks were applied to estimate the most appropriate

execution flows. These data will be presented to the top layer, as a high-level API specific to each service, so that it is possible to use them in an easy and simple way.

- **Layer 3. High-level services**. The top layer of the platform provides an innovative module that allows management and customization of services to end users, due to the capabilities of the MAS which is deployed over the CC environment. The tasks associated with the man-machine interfaces are performed at these levels in order to adapt according to the characteristics of the user and facilitate decision making by the user. To do this, we will develop a deployment model which will enable seamless access to the lower layers, encapsulating the information merged by the lower layer agents in the form of web services. These web services are developed in a way that is sensitive to context (QoS-aware), making it possible to maintain the quality level regardless of the input data stream and the computational needs of the fusion algorithms. To do this, a business model based on pay-per-use and dynamic contracts could be used to guarantee a reliable service regardless of demand. This intermediate layer of Cloud services allows high-level applications (apps) to use the information from lower layers in a homogenous way, thus facilitating the development of applications in the context of smart cities.

4   Results

Nowadays traffic flow control is a major problem both in large cities as in smaller cities. The main problem in this area is jams formed due to a non-optimal utilization of the pathways' capacity. Within this scope, the proposed architecture is adapted to the application environment, allowing address this problem. It addresses the problem of gathering data from multiple sensors, automate the processing and generate useful information that can be used effectively in managing traffic flow automatically.

To carry out the demonstration, and since test the system in a real environment would be a high cost when necessary to establish the specific sensing or adapt existing ones, an environment that simulates the operation of a real city has been developed. This environment consists of a 3D printed model. With this recreation, real data are received with which we are able to make appropriate tests to validate the system before deploying it in a real city.

Sensor Network

In order to understand the architecture, it is necessary to start with the lower part, the sensor network. The first problem that arises in the case study is the need of tracking the vehicle flow across the stretch of roads that integrate the system.

Although tests will be done with the simulated environment, there are different alternatives sensing adaptable to real environments, able to interact with the system in the same way. To locate vehicles in the simulation environment, a series of magnets are distributed parallel to steel rails on which vehicles circulate. These vehicles are equipped with a hall sensor that detects the presence of those magnets, placed in a specific way and order that defines the streets of the city where they are and the position of each street with an accuracy of 1 to 2 mm.

Each vehicle in the city is part of the swarm and is considered a lightweight agent with his position as the only knowledge.

Layer 0

This layer is responsible for receiving data from different protocols through communication standards used to subsequently serve to a higher layer. It refers to a broker where a specific software for each standard runs. In all cases, software have been developed using different libraries and tools available for each technology. For example, the case study of BLE is based on the Gatttool, which allows us to use BLE services through a series of simple commands. Just a simple configuration that varies depending on the type of communication to use is needed. For example, to BLE one needs to know the MAC address of the device. In the case study BLE is present in the vehicle. So far, in the middleware are integrated communications via Wi-Fi (a, b, g, n and c are treated the same way), RFID, BLE and ZigBee.

Layer 1

This layer translates all data received from the layer 0 to a common protocol. So, at this level, no matter how sensors obtained the data but the data itself, which must be encapsulated to provide it in a well defined manner to the upper layer by low-level services. In this case, the protocol used to serve data in a homogeneous manner is MQTT (Message Queuing Telemetry Transport). This open protocol connectivity M2M (Machine-to-Machine) allows sending telemetry data as a message. It is also a very lightweight protocol. Data are available in a well-defined way in the system hierarchical topics. These topics are conceptually associated with the information they represent, for example, the position of the vehicle is published (served at higher layers) on the topic '/vehicles/location' where 'vehicles' is the root topic of our vehicle system and 'location' is one of the topics that are within 'vehicles' topic. These data represent the information lightweight agents.

Layer 2

The second layer acts directly on the previous layer, obtaining the services corresponding to the needed information. In this layer heavyweight agents are located and they are structured in VO. Thus the system is divided into less complex structures and simplicity.

In this layer, heavyweight agents receive the information. They have more functionality than lightweight agents and they are specialized in IF. Their mission is to communicate this information gained from other agents in this layer, apply the necessary data transformations and serve them to the layer 3. For the present case, moreover, there VO composed of agents that apply statistical techniques and, specifically to this case study, there are agents responsible for implementing algorithm H-ABC (Hierarchical Ant Based Control) 14, based algorithm Ant Based Control (ABC) but it can be applied to large networks as in the case of city streets.

This system is scalable and therefore is applicable both to the demonstration environment, and in a real city. Its implementation is based on the authors presented in 13, but with several differences. The main new proposal is that there is no need for vehicles to carry a GPS sensor, it is the system itself which locates them. Furthermore, this proposal is able to automatically detect events such as accidents or other situations that block access in some way by analyzing the flow, which will be reduced in a sounded way and algorithms based on neural networks will detect the alert, acting accordingly.

The system performs dynamic routing system maintaining a route table for each intersection based on probabilistic results provided by the corresponding agents.

Once the decision is made, the states of the traffic lights will be modified using the services defined high level in the upper layer.

Layer 3
In the top layer of the architecture a series of high-level services are presented to end users to monitor and control the system. System status can be accessed via web services or using the Node.js technology for current study. This allows real-time interaction with the system, either by defining different routes, managing existing intersections or adding new ones or monitoring the state of the whole system. Different apps developed for this purpose.

5   Conclusions and Future Work

As a result, a scalable architecture capable of dealing jointly or independently all kinds of sensor that smart city may need, regardless of the technology used. The size of the swarm formed is not a problem for the architecture and the cost is directly linked to the use of the system, which allows it to be implemented both in small and medium cities.

As future work, we intend to implement the system in a real environment, starting with small neighborhoods or towns of Salamanca (Spain). This will serve to make some studies of the installation, service and maintenance costs when implementing the proposed platform for small and medium cities to be smarter cities and the saving that the system can offer to every kind of town.

Acknowledgements. This work has been supported by the MICINN project TIN2012-36586-C03-03.

References

1. Bajo, J., Fraile, J.A., Pérez-Lancho, B., Corchado, J.M.: The THOMAS architecture in Home Care scenarios: A case study. Expert Systems with Applications 37(5), 3986–3999 (2010)
2. Bin, W., Zhongzhi, S.: A clustering algorithm based on swarm intelligence. In: Proceedings of the 2001 International Conferences on Info-tech and Info-net, ICII 2001, Beijing, vol. 3, pp. 58–66. IEEE (2001)
3. Bonabeau, E., Dorigo, M., Theraulaz, G.: Swarm intelligence: from natural to artificial systems (No. 1). Oxford University Press (1999)

4. Chamoso, P., Perez, A., Rodriguez, S., Corchado, J.M., Sempere, M., Rizo, R., Pujol, M.: Modeling Oil-Spill Detection with multirotor systems based on multi-agent systems. In: 2014 17th International Conference on Information Fusion (FUSION), pp. 1–8. IEEE (July 2014)

5. Daniel, F., Eriksson, J., Finne, N., Fuchs, H., Gaglione, A., Karnouskos, S., Voigt, T.: makeSense: Real-world Business Processes through Wireless Sensor Networks. In: CONET/UBICITEC, pp. 58–72 (April 2013)

6. de la Prieta, F., Pérez-Lancho, B., De Paz, J.F., Bajo, J., Corchado, J.M.: Ovamah: Multiagent-based adaptive virtual organizations. In: 12th International Conference on Information Fusion, FUSION 2009, pp. 990–997. IEEE (July 2009)

7. Marciniak, A., Kowal, M., Filipczuk, P., Korbicz, J.: Swarm Intelligence Algorithms for Multi-level Image Thresholding. In: Korbicz, J., Kowal, M. (eds.) Intelligent Systems in Technical and Medical Diagnostics. AISC, vol. 230, pp. 301–311. Springer, Heidelberg (2013)

8. Martens, D., Baesens, B., Fawcett, T.: Editorial survey: swarm intelligence for data mining. Machine Learning 82(1), 1–42 (2011)

9. Nam, T., Pardo, T.A.: Conceptualizing smart city with dimensions of technology, people, and institutions. In: Proceedings of the 12th Annual International Digital Government Research Conference: Digital Government Innovation in Challenging Times, pp. 282–291. ACM (June 2011)

10. Rao, A.S., Georgeff, M.P.: Modeling rational agents within a BDI-architecture. KR 91, 473–484 (1991)

11. Renuka, N., Nan, N.C., Ismail, W.: Embedded RFID tracking system for hospital application using WSN platform. In: 2013 IEEE International Conference on RFID-Technologies and Applications (RFID-TA), pp. 1–5. IEEE (September 2013)

12. Rodriguez, S., Julián, V., Bajo, J., Carrascosa, C., Botti, V., Corchado, J.M.: Agent-based virtual organization architecture. Engineering Applications of Artificial Intelligence 24(5), 895–910 (2011)

13. Tatomir, B., Rothkrantz, L.: Hierarchical routing in traffic using swarm-intelligence. In: Intelligent Transportation Systems Conference, ITSC 2006, pp. 230–235. IEEE (September 2006)

14. Tatomir, B., Rothkrantz, L.J.M.: H-ABC: A scalable dynamic routing algorithm. Recent Advances in Artificial Life 5, 8 (2005) ISO 690

15. Wooldridge, M.J.: Reasoning about rational agents. MIT Press (2000)

16. Zato, C., Villarrubia, G., Sánchez, A., Barri, I., Rubión, E., Fernández, A., Rebate, C., Cabo, J.A., Álamos, T., Sanz, J., Seco, J., Bajo, J., Corchado, J.M.: PANGEA – Platform for Automatic coNstruction of orGanizations of intElligent Agents. In: Omatu, S., Paz Santana, J.F., González, S.R., Molina, J.M., Bernardos, A.M., Rodríguez, J.M.C. (eds.) Distributed Computing and Artificial Intelligence. AISC, vol. 151, pp. 229–240. Springer, Heidelberg (2012)

Detection of Rice Field
Using the Self-organizing Feature Map

Sigeru Omatu and Mitsuaki Yano

Osaka Institute of Technology,
5-16-1 Omiya, Asahi-ku, Osaka, 535-8585, Japan
omtsgru@gmail.com,yano@elc.oit.ac.jp
http://www.oit.ac.jp

Abstract. We consider a detection method of rice field under rainy conditions by using remote sensing data. The classification method is to use a competitive neural network of self-organizing feature map (SOM) by using remote sensing data observed before and after planting rice in Hiroshima, Japan. Three RADAR Satellites (RADARSAT) and one Satellite Pour l'Observation de la Terre(SPOT)/High Resolution Visible (HRV) data are used to detect rice field. Synthetic Aperture Radar (SAR) reflects back-scattering intensity in rice fields. The intensity decreases from April to May and increases from May to June. It is shown that the competitive neural network of self-organizing feature map is useful for the classification of the SAR data to find the area of rice fields.

Keywords: remote sensing, synthetic aperture radar, area of rice fields.

1   Introduction

The rice is the most important agriculture product in Japan and it is planted in the wide area. Satellite remote sensing images by optical sensors like LAND-SAT TM or SPOT HRV, could be used to know the growth situation of rice field [1][2][3][4]. But it is still difficult to detect rice fields by using an optical remote sensing data since the rice will grow up in a rainy season. Therefore, the monitoring system of the rice crop independent on the weather conditions must be adopted.

SAR penetrates through the cloud and hence, it can observe the land surface unde any weather condition [5][6][7][8]. The back-scattering intensity of C-band SAR images, such as RADARSAT or ERS1/SAR, changes greatly from non-cultivated bare soil condition before rice planting to inundated condition just after rice planting [1]. In addition, RADARSAT images are rather sensitive to the change of rice biomass in a growing period of rice [9][10].

Thus, a rice field detection is expected to be achieved in an early stage. In previous works [2][3][4], we attempted to estimate rice field using RADARSAT fine-mode data in an early stage. The detection accuracy of a rice field by Maximum Lkelihhood Method (MLH) was approximately 40% by comparing with

© Springer International Publishing Switzerland 2015
S. Omatu et al. (eds.), *Distributed Computing & Artificial Intelligence, 12th Int. Conference,*
Advances in Intelligent Systems and Computing 373, DOI: 10.1007/978-3-319-19638-1_4

the estimated area by SPOT multi-spectral data. In this study, we attempt to detect the rice field from RADARSAT data using SOM that is the unsupervised classification [11].

2 Test Area

The test area has a size of about 7.5×5.5 km in Higashi-Hiroshima, Japan. Three merged RADARSAT and one SPOT images in a part of the test site are shown in Figs. 1 and 2. The land surface condition in the rice field of April 8 is a non-cultivated bare soil before rice planting with rather rough soil surface. The surface condition of May 26 is almost smooth water surface just after rice planting, and that of June 19 is a mixed condition of growing rice and water surface. It is found that the rice fields are shown in a dark tone in the RADARSAT image. The RADARSAT raw data were processed using Vexcel SAR Processor (VSARP) and single-look power images with 6.25 meters ground resolution were generated. Then, the images were filtered using median filter with 7×7 moving window. All RADARSAT and SPOT images were overlaid onto the topographic map with 1:25,000 scale. As RADARSAT images are much distorted by foreshortening due to topography, the digital elevation model (DEM) with 50 meters spatial resolution issued by Geographical Survey Institute (GSI) of Japan [2] was used to correct foreshortening of RADARSAT images.

Fig. 1. SPOT-2/HRV image(1999/6/21, R:Band 3, G:Band 2, B:Band 1) where the area enclosed by a white line denotes the test area CNESS 1999

Fig. 2. SPOT-2/HRV image(1999/6/21, R:Band 3, G:Band 2, B:Band 1) in the test site

3   SOM Method

We briefly summarize the algorithm for SOM. The structure of SOM consists of two layers. One is an input layer and the other is a competitive layer. The total input to a neuron j is denoted by net_j and modeled by the following equations:

$$\text{net}_j = \sum_{i}^{n} w_{ji} x_i = (w_{j1}, w_{j2}, \ldots, w_{jn})(x_1, x_2, \ldots, x_n)^t$$

$$= W_j X^t \tag{1}$$

$$W_j = (w_{j1}, w_{j2}, \ldots, w_{jn}), \ \ X = (x_1, x_2, \ldots, x_n) \tag{2}$$

where $(\cdot)^t$ denotes the transpose of $(\cdot)$ and x_i and w_{ji} show an input in the input layer and a weighting function from the input x_i to a neuron j in the competitive layer, respectively. For simplicity, we assume that the norms of X and W_j are equalt to one, that is,

$$\|x\| = 1, \ \ \|W_j\| = 1, j = 1, 2, \ldots, N \tag{3}$$

where $\| \cdot \|$ shows Eucridean norm and N denotes the total number of neurons in the competitive layer.

When an input vector X is applied to the input layer, we find the nearest neighboring weight vector W_c to the input vector X such that

$$\|W_c - X\| = \min_{i} \|W_i - X\|. \tag{4}$$

The neuron c corresponding to the weight vector W_c is called a winner neuron. We select neighborhood neurons within the distance d and a set of indices for neurons located in the neighborhood of c is denoted by N_c. Then, the weighting vectors of the neurons contained in N_c are changed such that those weighting vectors could become similar to the input vector X as close as possible. In other words, the weighting vectors are adjusted as follows:

$$\Delta W_j = \eta(t)(W_j - X) \qquad \forall j \in N_c \tag{5}$$

$$\Delta W_j = 0 \qquad \forall j \notin N_c \tag{6}$$

where

$$\Delta W_j = W_j(\text{new}) - W_j(\text{old}) \tag{7}$$

$$\eta(t) = \eta_0 (1 - \frac{t}{T}), \qquad d(t) = d_0 (1 - \frac{t}{T}). \tag{8}$$

Here, t and T denote an iteration number and the total iteration number for learning, respectively. η_0 and d_0 are positive and denote initial values of $\eta(t)$ and $d(t)$, respectively where $d(t)$ denotes an Euclid distance from the winner neuron c.

4   Evaluation of SOM

We will introduce two criteria, namely, precision and recall to find a suitable SOM. The precision is defined by

$$P_r = \frac{R}{N} \times 100(\%) \tag{9}$$

and the recall is defined by

$$R_r = \frac{R}{C} \times 100(\%) \tag{10}$$

where P_r and R_r denote the precision rate and the recall rate, respectively and N is a trial number, R is a correct classified number, and C is a total correct number.

5   SOM Training

In order to find a suitable size of competitive layer, we assume map size m as 3×3, 4×4, 5×5, 6×6, and 7×7. We take an initila learning rate α_0=0.2 and an initial value of neighborhood d_0=2. Furthermore, we assume initial values of connection weights as random numbers of [0.1, 0.9] and itelation number t=5.

After taining the neural network of SOM, we calculate the P_r, R_r, and f_m for SPOT and RADARSAT remote sensing data. The results are shown in Table 1 and Table 2, respectively. From Table 1, we can see that the classification result of the map size 4×4 is highest value of f_m=84.49. Thus, in case of a small number of category, we can set small number of neurons in the competitive layer. From Table 2, we can see that f_m becomes largest in case of 4×4.

Table 1. P_r, R_r, and f_m for sizes of SPOT remote sensing data

Size of map	P_r	R_r	f_m
3×3	98.95	64.67	78.22
4×4	100	73.15	84.49
5×5	93.62	76.65	84.29
6×6	85.65	68.65	76.21
7×7	96.16	63.00	76.13

6   Iteration for Learning

We consider the iteration t for learning. Here, we count t=1when an input image of 1,800,000 pixels of 1,500pixels per line ×1,200 lines has been trained. In this experiment, we take t=1 (1,800,000 training), t=5 (9,000,000 training), t=10 (18,000,000 training), t=20 (36,000,000 training). Except for t, we take a size of neurons in the competitive layer as 4×4, α_0=0.2, and an initial vector

Table 2. P_r, R_r, and f_m for sizes of RADARSAT remote sensing data

Size of map	P_r	R_r	f_m
3×3	63.08	47.72	54.34
4×4	60.99	51.38	55.77
5×5	62.73	46.41	53.35
6×6	60.48	51.02	55.35
7×7	47.63	64.70	54.87

Table 3. P_r, R_r, and f_m for iterations of SPOT

Iteration	P_r	R_r	f_m
1	100	64.67	78.22
5	100	73.15	84.49
10	100	76.65	84.29
20	100	68.65	76.21

Table 4. The values of P_r, R_r, and f_m for iterations of RADARSAT

Iteration	P_r	R_r	f_m
1	58.40	50.29	54.04
5	60.99	51.38	55.77
10	61.00	51.31	55.74
20	61.01	51.34	55.76

W_0 of weighting functions are random numbers of [0.1, 0.9]. The results are shown in Table 3 and Table 4 for SPOT and RADARSAT remote sensinr data, respectively.

From Table 3, we can see the maximum value of f_m=84.55 is obtained when t=5 for SPOT remote sensing data and from Table 4, the maximum value of f_m=55.77 is obtained when t=5. Therefore, we set t=5 in what follows.

7   Classification Results

A rice field was extracted by using SOM from three temporal RADARSAT images and one SPOT image. SOM is a classification method based on competitive neural networks without teacher. It was applied to the remote sensing data by using the parameters as shown in Table 5. RADARSAT and SPOT images were classified into 16 categories by SOM. Then, we labeled the categories into a rice field, a forest area and an urban area.

8   Experimental Results

In order to make the proposed method effective, we classify the satellite image data by SOM. Fig. 3 shows the classification image by SPOT and Fig. 4 shows the classification image by RADARSAT. By comparing Fig. 3 with Fig. 4, one can see that the rice field by RADARSAT was extracted less than that by SPOT. The speckle noise was still seen in the image of rice field, the majority filter with 7×7 window was applied to the rice extracted images by RADARSAT and SPOT. For the evaluation of the rice field extraction, we defined two indices, True Production Rate (TPR) and False Production Rate (FPR). The TPR and FPR are calculated by

$$TPR = \frac{\alpha}{\alpha + \beta} \times 100 \qquad (11)$$

$$FPR = \frac{\gamma}{\alpha + \gamma} \times 100 \qquad (12)$$

where α means the number of relevant rice field extracted, β means the number of relevant rice field not extracted, and γ means the number of irrelevant rice field extracted. Extraction rice field of SPOT image by supervised MLH was used as reference rice field image.

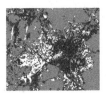

Fig. 3. Classification result of SPOT image by SOM (White: rice, Light gray: forest, Dark gray: urban)

Table 6 shows the results of TPR and FPR by SOM and MLH classification for SPOT data and RADARSAT data. From this results we can see that in case of RADARSAT data, the values of TPR (extraction rate of rice field) are, 50.97% and 60.96% for MLH and SOM, respectively. This means that SOM is better than MLH for extraction of rice field by about 10%. As for FPR (misclassification rate) SOM is also better than MLH by about 3%. The SPOT data is very fine images like 5m per pixel and we can assume the image of SPOT reflects almost all land surface. Using SPOT data TPR and FPR are 70.41% and 21.51%, respectively. Thus, we could extract rice field a certain level in practice fron Table 6.

Table 5. The parameters of SOM

Neurons	Training iterations	η_0	d_0
4 × 4	1080,000	0.03	2

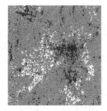

Fig. 4. Classification result of RADARSAT image by SOM (White: rice, Light gray: forest, Dark gray: urban)

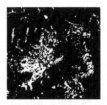

Fig. 5. Extraction result of rice field by SPOT data

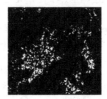

Fig. 6. Extraction result of rice field by multi temporal RADARSAT data

Figs. 5 and 6 show the extraction result of rice field by SPOT and RADARSAT, respectively. The results show that fine rice field could be extracted by using SOM compared with MLH from there figures.

Table 6. The results of TPR and FPR for rice field evaluation

Comparison data	Classification method	TPR(%)	FPR(%)
SPOT	SOM	70.4	21.51
RADARSAT	MLH	50.97	51.44
RADARSAT	SOM	60.97	48.59

9   Conclusions

Rice-planted area detection was attempted using multi-temporal RADARSAT data taken in an early stage of rice growing season by SOM classifications.

The SOM is unsupervised classification whose computational time is shorter than the other supervised classification method like MLH [3] or LVQ [4][11]. We have been engaged with image analysis on remote sensing data analysis. Especially, we are concentrated on SAR data analysis to land cover classification by using neural network classification methods. Since the SAR data is completely different with optical sensor images, it is difficult to obtain usual land cover map although it can observe the earth at any time and under any weather conditions such as rainy or cloudy occasion.

Acknowledgment. This work was supported by JSPS KAKENHI Grant-in-Aid for Challenging Exploratory Research (25630180). The authors would like to thank JSPS to support this research work.

References

1. Suga, Y., Oguro, Y., Takeuchi, S.: Comparison of Various SAR Data for Vegetation Analysis over Hiroshima City. Advanced Space Research 23, 225–230 (1999)
2. Suga, Y., Takeuchi, S., Oguro, Y.: Monitoring of Rice-Planted Areas Using Space-borne SAR Data. In: Proceedings of IAPRS, XXXIII, B7, pp. 741–743 (February 2000)
3. Omatu, S.: Rice-Planted Areas Extraction by RADARSAT Data Using Learning Vector Quantization Algorithm. In: Proceedings of ADVCOMP 2013, pp. 19–233 (September 2013)
4. Konishi, T., Omatu, S., Suga, Y.: Extraction of Rice-Planted Areas Using a Self-Organizing Feature Map. Artificial Life and Robotics 11, 215–218 (2007)
5. Woodhouse, I.H.: Introduction to Microwave Remote Sensing. CRC Press, London (2005) ISBN: 0-415-27123-1
6. Ryan, C.M., Hill, T., Woollen, E., Ghee, C., Mitchard, E., Cassells, G., Grace, J., Woodhouse, I.H., Williams, M.: Quantifying Small-Scale Deforestation and Forest Degradation in African Woodlands Using Radar Imagery. Global Change Biology 18, 1365–2486 (2012) ISSN: 1365-2486
7. Mitchard, E.T.A., Saatchi, S.S., White, L.J.T., Abernethy, K.A., Jeffery, K.J., Lewis, S.L., Collins, M., Lefsky, M.A., Leal, M.E., Woodhouse, I.H., Meir, P.: Mapping Tropical Forest Biomass with Radar and Spaceborne LiDAR in Lope National Park, Gabon: Overcoming Problems of High Biomass and Persistent Cloud. Biogeosciences 9, 179–191 (2012), doi:10.5194/bg-9-179-2012
8. Yoshioka, M., Fujinaka, T., Omatu, S.: SAR Image Classification by Support Vector Machine. In: Chen, C.H. (ed.) Signal and Image Processing, London, ch. 27, pp. 607–619 (January 2007) ISBN: 0-8493-5091-3
9. Bicego, M., Toan, T.L.: Rice Field Mapping and Monitoring with RADARSAT Data. International Journal of Remote Sensing 20, 745–765 (1999)
10. Liew, S.C., Chen, P.: Monitoring Changes in Rice Cropping System Using Space-borne SAR Imagery. In: Proceedings of IGARSA 1999, pp. 741–743 (October 1999)
11. Kohonen, T.: Self-Organizing Maps, pp. 206–217. Springer (1997) ISBN 3-540-62017-6

Ontology-Based Music Recommender System

Miguel Ángel Rodríguez-García[1], Luis Omar Colombo-Mendoza[1],
Rafael Valencia-García[1], Antonio A. Lopez-Lorca[2], and Ghassan Beydoun[3]

[1] Departamento de Informática y Sistemas, Universidad de Murcia, 30100 Murcia, Spain
[2] Faculty of Information and Communication Technologies,
Swinburne University of Technology, Hawthorn, Victoria 3122, Australia
[3] Faculty of Engineering and Information Science, University of Wollongong,
Wollongong, NSW, Australia
{miguelangel.rodriguez,luisomar.colombo,valencia}@um.es,
alopezlorca@swin.edu.au, beydoun@uow.edu.au

Abstract. Recommender systems are modern applications that make sugges-
tions to their users on a variety of items taking into account their preferences in
many domains. These systems use people's opinions to recommend to their end
users items that are likely to be of their interest. They are designed to help users
to decide on appropriate items and facilitate finding them in a very large collection of items. Traditional syntactic-based recommender systems suffer from
several disadvantages, such as polysemy or synonymy, that limit its effectiveness. Semantic technologies provide a consistent and reliable basis for dealing
with data at knowledge level. Adding semantically empowered techniques to
recommender systems can significantly improve the overall quality of recommendations. In this work, a recommender system based on a Music ontology is
presented. A preliminary evaluation of the system shows promising results.

Keywords: Recommender systems, music ontologies, Semantic Web, Knowledge-
based systems.

1 Introduction

In recommender systems, recommendations become more relevant the more the system is used to identify items. Each time new companies adopt these systems to study
their clients' preferences, they adapt their services according to increase their profits.
Interest in recommender systems is gaining momentum. These systems can now be
found in many domains of application such as tourism and leisure (hotels, restaurants,
parks and beaches), advertising, e-commerce and entertainment (film, TV and music),
among others [1; 2; 3].

The increasing amount of data available online makes it more difficult to identify
items, products, services or contents of interest. The Semantic Web has revolutionized
how systems integrate and share data, enabling computational agents to reason about
information and infer new knowledge [4]. The Semantic Web is a very promising
platform for developing knowledge management system. The underlying knowledge

© Springer International Publishing Switzerland 2015

S. Omatu et al. (eds.), *Distributed Computing & Artificial Intelligence, 12th Int. Conference,*
Advances in Intelligent Systems and Computing 373, DOI: 10.1007/978-3-319-19638-1_5

representation mechanism for the Semantic Web is formal ontologies [5]. Ontologies enable the automatic processing of information and allow the use of semantic reasoners to infer new knowledge [6]. In this work, we use the term ontology to refer to "a formal and explicit specification of a shared conceptualization" [5]. Knowledge in ontologies is mainly formalized by defining classes, properties, axioms and instances [7].

Ontologies are being applied to different domains such as biomedicine [8], finance [9], tourism [10; 11], geographic systems, digital contents [12; 13], digital libraries and e-learning [14], to name but a few. More concretely, they have been also used in recommender systems [5; 15; 16; 17]. In the specific field of music, recommender systems are increasingly using ontologies to better address the recommendation problem [18; 19; 20].

This paper similarly uses ontologies to represent music features to enable formalizing a user preference. Moreover, the paper presents an evaluation of closest matching preference derived from the ontology. The remainder of this paper is structured as follows. Firstly, Section 2 describes the system developed in this work. Section 3 presents the evaluation of the proposed framework. Finally, Section 4 closes the paper with its conclusions and future work.

2   Ontology-Based Recommender System

In this section, an overview of the semantic recommender system is presented. The system is composed of two modules, the ontology repository and the knowledge-based recommender system. The system gathers data on demand from LinkedBrainz, which offers a SPARQL end point to query information in RDF for-mat. Users manage their profile by adding and classifying elements (contents or characteristics) according to their personal preferences. The recommender module analyzes user's preferences and suggests matching songs that they might like. In what follows, the ontology repository and recommender engine are described.

2.1   Ontology Repository

Ontology repositories have been considered a key resource for building a global infrastructure to enable the vision of the Semantic Web [21]. Ontology repositories provide an infrastructure where ontologies can be published and shared. Ontology repositories have additional features which distinguish them from other persistent storage systems. For example, they offer an efficient way to store and manage ontologies. They provide several ways of establishing communications using APIs, which provide query languages and high performance (speed of access on search and editions). In this work, Virtuoso[1] has been used to implement the ontology repository.

2.1.1   The Music Ontology
The Music Ontology [22] provides the main concepts and properties for describing the semantic concepts related to the music domain, such as releases, lives events,

---

[1] http://virtuoso.openlinksw.com/

albums, artists and tracks. This ontology is organized on three levels with different granularity. The lower level, of fine granularity, provides a complex vocabulary to describe the characteristics of songs such as, melody line or remarkable events that occurred during the performance. The middle level deals with the information about the performance, such as the recording. Lastly, the higher level has the coarser granularity and provides the vocabulary for editorial information. This ontology is based on MusicBrainz that is an open content music database which contains 300,000 tracks [23]. The Music Ontology provides semantics to the data described by MusicBrainz. For example, it formalizes relations between artists, their works, albums, track titles and length of each track. All entries in MusicBrainz are maintained by volunteer editors who follow specific style guidelines for upload musical information.

The framework presented in this paper mainly focuses on the subset of the Music Ontology corresponding to the class MusicalManifestation and its subclasses. The entity MusicalManifestation is related to the edition/production/publication of a musical expression. Moreover, some of the subclasses of this class relevant for the purposes of the framework are:

- *record*: Manifestation which first aim is to render the product of a recording.
- *release*: A specific release with barcode, box, liner notes, covert art and records.
- *track*: A track on a particular record.

Furthermore, some of the properties that relate individuals belonging to the subclasses of the class MusicalManifestation and relevant for the purpose of the framework are:

- *maker*: Indicates the creator of the individual.
- *genre*: Relates an event to a particular musical genre.
- *producer*: Relates the manifestation of a work to a person or a group of people who produced it.
- *publisher*: Relates a musical manifestation to a person or a group of people who published it.

2.2   Knowledge-Based Recommender System

This system is based on previous works of our research group [2; 15]. The knowledge-based recommender system is the main part of the system and identifies instances that match the users' preferences of the "reference class" of the domain ontology. In case study, the domain ontology is related to the music domain and the "reference class" is the class *MusicalManifestation*. Recommendations produced by the system relate to individuals of this class. Like content, user profiles in this work are represented by means of an ontology, so that user preferences (i.e., ratings) in the form of individuals of a class of this ontology are directly linked to individuals of the reference class or other related classes in the domain ontology. For practical purposes, details about representation of user profiles are not covered in this paper.

In order to achieve its aims, the system maintains a matrix called the Similarity Music-Music Matrix (Fig. 2). This matrix contains the similarity rates between all the instances (n instances) that belong to the *MusicalManifestation* class.

$$music2musicSim = \begin{bmatrix} 1 & SM_{12} & ... & SM_{1n} \\ SM_{21} & 1 & ... & SM_{2n} \\ ... & ... & 1 & ... \\ SM_{n1} & SM_{n2} & ... & 1 \end{bmatrix}$$

Fig. 1. Similarity matrix

Each element of the music2musicSim matrix is calculated according to the similarity function MSimilarity(a,b) equation which is based on the formula Similarity presented in previous works of our research group [2]. Here, MSimilarity(a,b) is the quantitative value of the similarity of the semantic profiles. The algorithm depicted in Fig. 3 is used for determining the ontological similarity between each part of the semantic profile of two musical items.

Algorithm 1: MSimilarity(a, b)- Similarity function

Input: A and B represent set of instances where$\{\{E_1, E_2, E_3, ..., E_n\} | E_x$ is a set of instances of type x$\}\}$; Set Weight =
$\{(p_1, v_1), (p_2, v_2), (p_3, v_3), ..., (p_m, v_m) | p_x$ is a property and v_x its weight in the equation $\}$.
Output: Similarity rate between A and B normalized in [0,1]
subTotal = 0 ; totalWeight = 0
/* for each element of Weight p_i is a property/ relationship, i.e.:
'maker k_i is p_i's weight, i.e.: 0.3*/
Foreach(p_i, k_i) $\in$ Weight **do**
  elementsA = $E_{pi} \in A$ /*Elements related to A through p_i*/
  elementsB = $E_{pi} \in B$ /*Elements related to B through p_i*/
/*If the properties values are numbers, the average formula is used*/
  if isNumber(elementsA and elementsB) **then**
    valueA = getDataPropertyValue(elementsA);
    valueB = getDataPropertyValue(elementsB);
    subTotal = subTotal+ (avg(valueA,valueB)/max(valueA,valueB);
/*If the properties values are strings, the Levenshtein distance
formula is used.*/
  else if(isString(elementsA and elementsB) **then**
    valueA = getDataPropertyValue(elementsA);
    valueB = getDataPropertyValue(elementsB);
    subTotal = subtotal + (calculateLevenshteinDis-
tance(valueA,valueB)/strlength(max(valueA,valueB)));
/*If the properties are relationships, the following formula is
used*/
  else if(isRelationship(elementsA and elementsB) **then**
    subTotal = subTotal + (objectPropertySimilarity(A, B,
p_i)/max(lengthA,lengthB));
  end
  totalWeight = totalWeight + subTotal*weight;
end
return(totalWeight)   /*Similarity normalized in [0,1]*/

Fig. 2. Similarity function algorithm

Fig. 3 depicts the algorithm that we use for calculating the similarity of each approach of the properties of the semantic profiles that represent musical items. Each musical item is defined by a set of properties, some of them are used to calculate the similarity. The algorithm distinguishes between three different types of properties, numerical datatype properties, string datatype properties and object properties. For each type, a different equation is used. Similarity in numerical properties is calculated using the average between the values of this property. If both properties are string-based, similarity is calculated using the Levenshtein distance. Calculation of similarity in the case of object properties is based on the semantic similarity measure proposed in [2].

The recommender system gives the user music item recommendations using the Recon vector shown in Formula 1. Each index in the Recon vector represents a song and its value represents a probability, which reflects the inferred degree of interest that a user might have about such music item.

$$Recon_u = (P_1, P_2, P_3, ..., P_n) \qquad (1)$$

Where, Pa represents the relationship between the musical content ant the user's preferences. The higher the value, the greater the relationship between the content and the user's preferences.

The music2musicSim matrix can be used to adjust and calculate the vector Recon, on the following way: each time that a user wants to add information concerning their preferences (e.g., rate a music item, artist or other property that like or dislike), their Recon vector is adjusted. The adjustment depends on which elements have been rated by the user. Our approach distinguishes between two different kinds of adjustments:

- If a user rates an element that belongs to the reference class, i.e. a song, the system, through the RefSim matrix, searches for all other individuals that are similar to the referred item and adds or subtracts points from these songs in the Recon vector. The value added or subtracted would be directly related to the level of similarity between the contents.
- If a user rates elements from other classes (e.g., MusicArtist, Genre or Publisher) or datatypes properties, the system searches for all the individuals belonging to the reference class (i.e., *MusicalManifestation*) that are directly related to the rated item and adds or subtracts points from these songs in the Recon vector.

3   Evaluation

This section discusses the evaluation of the recommender system. The measures that have been used in this evaluation are the metrics of precision, recall and F1-measure. These evaluation metrics are the one usually applied to information retrieval systems.

As the relevance of a recommendation is subjective to each user, it is not possible to automate the assessment process. Thus, to evaluate the system five students were asked to interact with it. The students set up their personal preferences by introducing a minimum of information in the system. They chose a minimum of 10 songs they liked, 10 that they did not like. Similarly they chose music artists and genders who they liked and dislike. To determine the precision and recall of the system, 30 songs in the database were randomly selected and each participant in the experiment was requested to classify each one of these items between relevant and not relevant to their personal interests. "Relevant" items represent "good" items according to the users' preferences. Conversely "not relevant" represent items that they did not like. The recommendations of the system were then compared with the users' decisions to assess, how many, among these 30 items, were present in the lists of recommended items and how many recommendations were classified as "not relevant" by the user. The result of these measures can be seen in Table 1.

Table 1. Results of the experiment.

User	Relevant			Irrelevant			Total precision
	Total	Correct	Precision	Total	Correct	Precision	
A	15	12	80.00%	15	9	60,00%	70,00%
B	18	15	83.33%	12	8	66,67%	76,67%
C	10	9	90.00%	20	16	80,00%	83,33%
D	8	6	75.00%	22	14	63,64%	66,67%
E	14	12	85.71%	16	12	75,00%	80,00%
Avg	13	10.8	**82.81%**	17	11.8	**69.06%**	**75.33%**

4   Conclusions and Future Work

In this paper we have presented a system for recommending music based on Semantic Web technologies. The system uses knowledge about the music domain to find content similar to the music preferences provided by the user. The recommendation system has shown promising results in our tests in a controlled experiment.

In the music domain, recommender systems help users finding relevant music items such as new artists or musical tracks that they may not know. Most of the available recommender systems use either content or collaborative or social-based approaches or even a hybrid combination of these [24]. In our system, the recommendation algorithm is content-based. The recommender system monitors users' preferences. When the user likes a music track, the system examines its features to elaborate a music profile for each user. Based on this profile, the system is capable of predicting other music tracks that may match the user's preferences.

As future work, we will conduct further studies to gain insight on the causes of the discrepancies in accuracy between like and dislike predictions. From the results of the experiment, which are shown in Table 1, it is possible to conclude that the proposed framework is able to correctly predict 9 out of 10 relevant songs and 8 out of 10 (16 out of 20) non-relevant songs in the best-case scenario (user "C"); these scores are equivalent to precision values of 90% (0,9) and 80% (0,8), respectively. This slight discrepancy may be tied to the quality of the ontology used. By extending the size of the sample, we will discover whether the algorithm is biased towards liking prediction. If this were the case, we will redesign the algorithm to remove this bias and produce better recommendations. We will also apply some ontology evaluation mechanisms as proposed in [25].

In its current version our system focuses on individuals belonging to the *Musical-Manifestation* class. Our next step is to extend its functionality to cover other classes, such as artists or producers. This will require redesigning the similarity algorithm to support various comparison matrixes and to provide several types of similarity depending on the instances which have been compared.

Finally, the prototype has been validated in the music domain. Further validations in other domains, such as tourism/leisure domain, e-commerce or films, would help assessing the adequacy of the recommendation system.

Acknowledgements. Luis Omar Colombo-Mendoza is supported by the National Council of Science and Technology (CONACYT), the Public Education Secretary (SEP) and the Mexican government.

References

1. Rodríguez-González, A., Torres-Niño, J., Jiménez-Domingo, E., Gómez Berbís, J.M., Alor-Hernández, G.: AKNOBAS: A knowledge-based segmentation recommender system based on intelligent data mining techniques. Comput. Sci. Inf. Syst. 9(2), 713–740 (2012)
2. Carrer-Neto, W., Hernández-Alcaraz, M.L., Valencia-García, R., García-Sánchez, F.: Social knowledge-based recommender system. Application to the movies domain. Expert Syst. Appl. 39(12), 10990–11000 (2012)
3. Hyung, Z., Lee, K., Lee, K.: Music recommendation using text analysis on song requests to radio stations. Expert Syst. Appl. 41(5), 2608–2618 (2014)
4. Berners-Lee, T., Hendler, J., Lassila, O.: The Semantic Web. Sci. Am. 284, 34–43 (2001)
5. Studer, R., Benjamins, V.R., Fensel, D.: Knowledge engineering: Principles and methods. Data and Knowledge Engineering 25, 161–197 (1998)
6. Rodríguez-García, M.A., Valencia-García, R., García-Sánchez, F., Samper-Zapater, J.J.: Ontology-based annotation and retrieval of services in the cloud. Knowl.-Based Syst. 56, 15–25 (2014)
7. Gruber, T.R.: A translation approach to portable ontology specifications. Knowl. Acquis. 5, 199–220 (1993)
8. Ruiz-Martínez, J.M., Valencia-García, R., Martínez-Béjar, R., Hoffmann, A.G.: BioOntoVerb: A top level ontology based framework to populate biomedical ontologies from texts. Knowl.-Based Syst. 36, 68–80 (2012)
9. Esteban-Gil, A., García-Sánchez, F., Valencia-García, R., Fernández-Breis, J.T.: Social-BROKER: A collaborative social space for gathering semantically-enhanced financial information. Expert Syst. Appl. 39(10), 9715–9722 (2012)
10. Yılmaz, Ö., Erdur, R.C.: iConAwa – An intelligent context-aware system. Expert Syst. Appl. 39(3), 2907–2918 (2012)
11. Moreno, A., Valls, A., Isern, D., Marin, L., Borràs, J.: SigTur/E-Destination: Ontology-based personalized recommendation of Tourism and Leisure Activities. Eng. Appl. Artif. Intel. 26(1), 633–651 (2013)

12. Alor-Hernández, G., Sánchez-Ramírez, C., Cortes-Robles, G., Rodríguez-González, A., García-Alcaráz, J.L., Cedillo-Campos, M.G.: BROSEMWEB: A brokerage service for e-Procurement using Semantic Web Technologies. Comput. Ind. 65(5), 828–840 (2014)
13. Colomo-Palacios, R., López-Cuadrado, J.L., González-Carrasco, I., García-Peñalvo, F.J.: SABUMO-dTest: Design and Evaluation of an Intelligent collaborative distributed testing framework. Computer Science and Information Systems 11(1), 29–45 (2014)
14. Colomo-Palacios, R., Casado-Lumbreras, C., Soto-Acosta, P., Misra, S.: Providing knowledge recommendations: an approach for informal electronic mentoring. Interact. Learn. Envir. 22(2), 221–240 (2014)
15. Colombo-Mendoza, L.O., Valencia-García, R., Rodríguez-González, A., Alor-Hernández, G., Samper-Zapater, J.J.: RecomMetz: A context-aware knowledge-based mobile recommender system for movie showtimes. Expert Syst. Appl. 42(3), 1202–1222 (2015)
16. Aréchiga, D., Crestani, F., Vegas, J.: Ontology-driven word recommendation for mobile Web search. Knowl. Eng. Rev. 29(02), 186–200 (2014)
17. Luna, V., Quintero, R., Torres, M., Moreno-Ibarra, M., Guzmán, G., Escamilla, I.: An ontology-based approach for representing the interaction process between user profile and its context for collaborative learning environments. Comput. Hum. Behav. (2015)
18. Rho, S., Song, S., Hwang, E., Kim, M.: COMUS: Ontological and Rule-Based Reasoning for Music Recommendation System. In: Theeramunkong, T., Kijsirikul, B., Cercone, N., Ho, T.-B. (eds.) PAKDD 2009. LNCS, vol. 5476, pp. 859–866. Springer, Heidelberg (2009)
19. Celma, Ò., Serra, X.: FOAFing the music: Bridging the semantic gap in music recommendation. Web Semantics 6(4), 250–256 (2008)
20. Mohanraj, V., Chandrasekaran, M., Senthilkumar, J., Arumugam, S., Suresh, Y.: Ontology driven bee's foraging approach based self-adaptive online recommendation system. J. Syst. Software 85(11), 2439–2450 (2012)
21. Hyvönen, E., Viljanen, K., Tuominen, J., Seppälä, K.: Building a national semantic web ontology and ontology service infrastruc-ture—the FinnONTO approach. In: Bechhofer, S., Hauswirth, M., Hoffmann, J., Koubarakis, M. (eds.) ESWC 2008. LNCS, vol. 5021, pp. 95–109. Springer, Heidelberg (2008)
22. Raimond, Y., Abdallah, S.A., Sandler, M.B., Giasson, F.: The Music Ontology. In: ISMIR 2007, pp. 417–422 (2007)
23. Swartz, A.: MusicBrainz: A Semantic Web Service. IEEE Intell. Syst. 17(1), 76–77 (2002)
24. Ricci, F.: Context-aware music recommender systems: workshop keynote abstract. In: WWW 2012 Companion, pp. 865–866. ACM, New York (2012)
25. Beydoun, G., Lopez-Lorca, A., García-Sánchez, F., Martínez-Béjar, R.: How do we measure and improve the quality of a hierarchical ontology? J. Syst. Software 84(12), 2363–2373 (2011)

Data Reordering for Minimizing Threads Divergence in GPU-Based Evaluating Association Rules

Youcef Djenouri[1], Ahcene Bendjoudi[1], Malika Mehdi[2],
Zineb Habbas[3], and Nadia Nouali-Taboudjemat[1]

[1] DTISI, CERIST Research center, rue des freres Aissou, Benaknoun, Algiers, Algeria
[2] MoVeP Lab, USTHB University, Algiers, Algeria
[3] LCOMS, University of Lorraine Ile du Saulcy, 57045 Metz Cedex, France
{y.djenouri,malika.mehdi}@gmail.com, {abendjoudi,nnouali}@cerist.dz,
zineb@univ-lorraine.fr

Abstract. This last decade, the success of Graphics Processor Units (GPUs) has led researchers to launch a lot of works on solving large complex problems by using these cheap and powerful architecture. Association Rules Mining (ARM) is one of these hard problems requiring a lot of computational resources. Due to the exponential increase of data bases size, existing algorithms for ARM problem become more and more inefficient.Thus, research has been focusing on parallelizing these algorithms. Recently, GPUs are starting to be used to this task. However, their major drawback is the threads divergence problem. To deal with this issue, we propose in this paper an intelligent strategy called Transactions-based Reordering "TR" allowing an efficient evaluation of association rules on GPU by minimizing threads divergence. This strategy is based on data base re-organization. To validate our proposition, theoretical and experimental studies have been carried out using well-known synthetic data sets. The results are very promising in terms of minimizing the number of threads divergence.

Keywords: Association Rules Mining, GPU Computing, Threads Divergence.

1 Introduction

Finding hidden patterns from a large data base is a challenging problem in articficial inteligence community (AI). Association Rules Mining (ARM) is one of the most used techniques allowing the extraction of useful rules among a transactional data base. Because of the exponential increasing of data bases size, the existing ARM algorithms are highly time consuming. Parallel computing is then used to reduce the computation time of highly complex ARM algorithms. Recently, Graphics Processor Units (GPUs) are considered as one of the most used parallel hardware to solve large scientific complex problems. Indeed, nowaday computer architectures are hybrid. Hybrid architectures are mainly composed of

S. Omatu et al. (eds.), *Distributed Computing & Artificial Intelligence, 12th Int. Conference,*
Advances in Intelligent Systems and Computing 373, DOI: 10.1007/978-3-319-19638-1_6

two components, a CPU containing a multi-core processor and a main memory and the GPU host composed of hundreds of computing cores (threads) and a hierarchy of memories with different carachteristics. The GPU threads are grouped on blocs and a bloc is scheduled on several warps containing generally 32 threads executed in parallel. Threads within the same warp follow the SIMD model (Single Instruction Multiple Data) and should execute the same instruction at any time. This property can be the source of threads divergence when distinct threads of the same warp are constrained to execute different instructions.

Several GPU-based ARM algorithms have been proposed in the literature ([1-3, 6 and 8]). These algorithms reduce perfectly the execution time of the ARM process. However their major drawback is the threads divergence problem. In association rules mining, the rule evaluation is time consuming. The evaluation of the rules consists on calculating two statistical measures (the support and the confidence of each rule). To determine these measures, transactional data bases must be entirely scanned for each evaluation. In [8], the authors proposed an efficient threads mapping strategy to evaluate the rules. Each thread is mapped with one transaction and a single rule. After that, sum reduction is used to compute the global evaluation of such rule. This technique induces threads divergence between threads because the transactions have in general different data values with different sizes (see Section 3 for more details).

The main goal of this paper is to propose an efficient strategy minimizing GPU-threads divergence in the rules evaluation process. The idea is to perform a pre-processing step before the mining process. The transactional data base is re-organized according to the length of each transaction. This operation allows grouping the set of transactions having the same size and then assign them to the same warp. However, it causes unload balancing between GPU blocs. To deal with this issue, we propose a new strategy to be added to the pre-processing step. The transactions are first sorted according to the number of items, then assigned successively to the threads. To validate the proposed approaches, several experiments have been carried out on well-known synthetic data sets. The results are very promising in terms of the number of threads divergence and then improving the overall execution time.

The remainder of the paper is as follows, the next section presents a state-of-the-art on GPU-based ARM algorithms. Section 3 explains the GPU-threads divergence in the evaluation rules process. Section 4 presents the proposed strategy to avoid threads divergence. Analytical and experimental study of the suggested strategy are detailed in Section 5. Finally, we conclude this work in Section 6.

2   Related Work

Association rule mining attempts to discover a set of rules covering a large percentage of data and tends to produce an important number of rules. However, since the data bases are increasingly large, ARM process has became a high intensive computing.

ARM community has investigated on GPU architectures. In the last decade, several algorithms based on GPU have been developed. The parallel ARM on GPU was first introduced by Wenbin *et al.* in [1]. They proposed two parallel versions of Apriori algorithm called PBI (Pure Bitmap Implementation) and TBI (Trie Bitmap Implementation) respectively. In [2], a new algorithm called GPU-FPM is developed. It is based on Apriori-Like algorithm and uses a vertical representation of the data set because of the limitation of the memory size in the GPU. It builds Mempack structure to store different types of data. Syed *et al.* in [3] proposed a new Apriori algorithm for GPU architecture which performs two steps. In the first step, the generation of itemsets is done on GPU host where each thread bloc computes the support of a set of itemsets. In the second step, the generated itemsets are sent back to the CPU in order to generate the rules corresponding to each itemset and to determine the confidence of each rule. The main drawback of this algorithm is the cost of the CPU/GPU communications. In [4], the authors proposed a parallel version of DCI algorithm [5] in which the two most frequent operations of DCI (intersection and computation operators) are parallelized. In [7], a GPApriori algorithm is proposed using two data structures in order to accelerate the itemsets counting.

All the above algorithms used the vertical representation in the GPU support counting. Threads divergence is then detected because each item belongs to different number of transactions. This issue is solved by transforming the vertical representation in bitmap ones. This leads to two limitations, the first one is the multi-transformation process (Horizontal to Vertical and then Vertical to Bitmap) because the existing transactional data bases are designed using the horizontal representation. Furthermore, when using bitmap representation, all bits are needed to be scanned, which reduce the performance of such algorithms.

Two approaches with horizontal representation have been recently developed for GPU support computing [6,8]. Cuda-Apriori algorithm is proposed in [6]. First, the transactional dataset is divided among different threads. Then, k-candidate itemsets are generated and allocated to a global memory. Each thread handles one candidate using only the portion of the dataset assigned to its block. At each iteration, a synchronization between blocks is done in order to compute the global support of each candidate. The second one is PEMS [8], it is the parallel version of BSO-ARM algorithm [9]. The evaluation of multiple itemsets is done on GPU where each bloc is used by a single itemset. Then, each thread checks the current itemset with respect to the set of transactions assigned to it. The two algorithms outperform the vertical GPU-based algorithms in terms of execution time. However, they are still time consuming when dealing with real transactional data bases. Indeed, the most real data bases require the irregular distribution of items among the set of transactions. The evaluation rules process may generate threads divergence (see Section 3), that degrades the performance of such algorithms. In this paper, we deal with this issue by proposing a new strategy for minimizing threads divergence.

3 Threads Divergence Analysis in GPU-Based Rules Evaluation

Let us recall that in ARM problem, a data base is a set of transactions with different number of items. As shown in Figure 1, to evaluate a single rule on GPU, the different threads have to scan all the items of such rule, each of which compares them to the transaction it is mapped with. Thus, two situations may lead to threads divergence. First, each thread handles different number of items. In this case, there are threads that finish before others. Second, the comparison process of a given thread is stopped when it does not find a given item of the considered rule in the transaction.

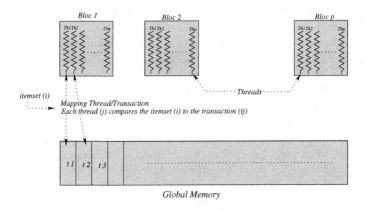

Fig. 1. Rules Evaluation on GPU

The number of threads divergence (TD) caused by these two situations is computed according to the number of comparisons done by the different threads as follows:

$$TD = max\{max\{(|t_{(r*w)+i}| - |t_{(r*w)+j}|\}/(i,j,r) \in [1...w]^3\}. \qquad (1)$$

Where
$|t_{(r*w)+i}|$ is the size of the $(r*w) + i^{th}$ transaction assigned to the ith thread allocated to the rth warp. w is the number of warps. Furthermore, threads divergence can be computed according to the distribution of items in transactional data bases. We distinguish the following two cases

1. Irregular Distribution of Items: The transactions are highly different in size. In the worst case, we find on the same warp one transaction containing all items and another containing only a single item. For this, we can approximate threads divergence to the maximal number of items minus one and we can write:

$$\lim_{M \to +\infty} TD(M) = N - 1. \qquad (2)$$

2. Regular Distribution: Here we notice a slight difference between the size of transactions. We consider r_1 is the variation between transactions so:

$$\lim_{M \to +\infty} TD(M) = r_1. \tag{3}$$

4   The Proposed Strategy

In the following, we propose a new strategy called Transactions-based Reordering for minimizing threads divergence.

4.1   Algorithm

The transactions are sorted according to the number of items. Then, each one is assigned to one thread where the ith transaction is handled by the ith thread. The pseudo algorithm of this strategy is given in Algorithm 1. We first consider T as a vector of transactions where each element represents the number of items in a given transaction. Then, a quick sort algorithm is applied to sort the transactions according to the number of items of each transaction. Finally, we assign the transactions to the threads taking into account the number of items in each transaction. Using TR strategy, all blocs have the same number of transactions, so the load balancing problem does not arise.

Algorithm 1. Transactions-based Reordering (TR) algorithm

1: **for** each transaction $t \in$ T **do**
2:     Calculate_Number_Of_Items(t)
3: **end for**
4: Quick_Sort(T)
5: i $\leftarrow$ 0
6: **for** each transaction $t \in$ T **do**
7:     assigned t to the ith thread
8:     i $\leftarrow$ i + 1
9: **end for**

4.2   Example

Consider the transactional data base containing five different items {A, B, C, D, E} and ten transactions:

$t_1 : A, B, C$ $t_2 : A, C$
$t_3 : B, C$ $t_4 : A, B, C, D, E$
$t_5 : C, D, E$ $t_6 : D, E$
$t_7 : B, D$ $t_8 : A, E$
$t_9 : A, B, C, D, E$ $t_{10} : A, D, E.$

We first calculate the number of items of each transaction, then, we sort the transactions according to the number of items as

t_2	t_3	t_7	t_8	t_1	t_5	t_6	t_{10}	t_4	t_9
2	2	2	2	3	3	3	3	5	5

After that, the transactions are assigned one by one. By considering two blocs for this scenario, the transactions (t_2, t_3, t_7, t_8, t_1) are assigned to the first bloc, the transactions (t_5, t_6, t_{10}, t_4, t_9) are assigned to the second bloc. The final status of the blocs is given in Figure 2.

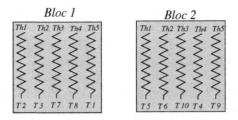

Fig. 2. Assignment of transactions by TR strategy

The load balancing is respected, each bloc deals with five transactions. However, if t_1 and t_8 or t_4 and t_6 are allocated to two different warps, thread divergence between threads will occur.

4.3   Complexity

Proposition
The complexity of "TR" strategy is $O(M^2 + 2 \times M \times N)$.
Proof
The complexity of "TR" is the sum of three functions:

- Calculate_Number_Of_Items(t): In the worst case, this procedure must be repeated N times for each transaction so its theoretical cost is $O(N \times M)$.
- Quick_Sort(T): At each step of the Quick_Sort algorithm, an element called *Pivot* is picked, the vector is re-arranged in such a way that all elements less than the *Pivot* appear to the left, and all elements that greater than the *Pivot* appear on the right side. After that, the same process is repeated with the two sub-vectors. In the worst case, the complexity of sorting a vector of M elements is $O(M^2)$
- Assignment(t): The last function of this strategy is to assign each transaction to the corresponding thread, this operation needs the full scan of T which requires $O(N \times M)$.

The global complexity of TR strategy is then $O(M^2 + 2 \times M \times N)$.

5   Performance Evaluation

In this section, the evaluation of the proposed strategy is performed using well-known synthetic data sets used in [8]. The suggested strategy is based on the re-organization of the transactions. In this experimentation, we focus on the effect of the re-organization in the threads divergence. First, we prove that the re-organization's runtime is ignored compared to the mining process. Then, we calculate threads divergence before and after the sorting process. Table 1 presents the runtime of the re-organization step and the association rules mining process with different data instances. This table reveals that the re-organization CPU time is no significant compared to the whole mining time on GPU. Furthermore, the ratio is less than 0.0001 for all tested instances. The re-organizing step is performed only one time before the mining process as pre-processing stage. Again, with connect data set (100000) transactions, the re-organization process consumes only 27 seconds. Fig 3 observes threads divergence results before and after sorting process and shows clearly the benefit of sorting process reliability to the threads divergence.

Table 1. Runtime (Sec) of the re-organization step (CPU) compared to the mining step (GPU)

Data Sets Name	Pre-Processing	Mining	Ratio(%)
IBM Q.std	0.005	545	<<< 1
Quake	0.03	495	<<< 1
Chess	0.07	845	<<< 1
Mushroom	0.46	1046	<<< 1
Pumbs_Star	10.2	2475	<<< 1
BMS-WebView2	17.66	4177	<<< 1
Connect	27.34	5815	<<< 1

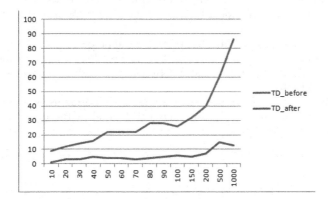

Fig. 3. The number of threads divergence before and after the re-organization of transactions with different number of items

6   Conclusion

In this paper, we proposed an intelligent strategy minimizing the threads divergence caused by evaluating single association rules on GPU architecture. This strategy is based on the re-organization of the input transactions. The proposed strategy sorts the transactions according to the number of items and assigns the ith transaction to the ith thread. To validate our strategy, several experimentations have been performed after and before the re-organization of the transactions. The results show that the re-organization step reduces significantly the threads divergence.

References

1. Fang, W., et al.: Frequent itemset mining on graphics processors. In: Proceedings of the Fifth International Workshop on Data Management on New Hardware. ACM (2009)
2. Zhou, J., Yu, K.-M., Wu, B.-C.: Parallel frequent patterns mining algorithm on GPU. In: 2010 IEEE International Conference on Systems Man and Cybernetics (SMC). IEEE (2010)
3. Adil, S.H., Qamar, S.: Implementation of association rule mining using CUDA. In: International Conference on Emerging Technologies, ICET 2009. IEEE (2009)
4. Silvestri, C., Orlando, S.: gpudci: Exploiting gpus in frequent itemset mining. In: 2012 20th Euromicro International Conference on Parallel, Distributed and Network-Based Processing (PDP). IEEE (2012)
5. Orlando, S., et al.: Adaptive and resource-aware mining of frequent sets. In: Proceedings of the 2002 IEEE International Conference on Data Mining, ICDM 2003. IEEE (2002)
6. Cui, Q., Guo, X.: Research on Parallel Association Rules Mining on GPU. In: Yang, Y., Ma, M. (eds.) Proceedings of the 2nd International Conference on Green Communications and Networks 2012 (GCN 2012): Volume 2. LNEE, vol. 224, pp. 215–222. Springer, Heidelberg (2013)
7. Zhang, F., Zhang, Y., Bakos, J.: Gpapriori: Gpu-accelerated frequent itemset mining. In: 2011 IEEE International Conference on Cluster Computing (CLUSTER). IEEE (2011)
8. Djenouri, Y., Bendjoudi, A., Mehdi, M., Nouali-Taboudjemat, N., Habbas, Z.: Parallel Association Rules Mining Using GPUs and Bees Behaviors. In: Proceeding of 6th International Conference on Pattern Recognition and Soft Computing, Tunis, Tunisia. IEEE (2014)
9. Djenouri, Y., Drias, H., Habbas, Z.: Bees swarm optimisation using multiple strategies for association rule mining. International Journal of Bio-Inspired Computation 6(4), 239–249 (2014)

Checking RTECTL Properties of STSs
via SMT-Based Bounded Model Checking*

Agnieszka M. Zbrzezny and Andrzej Zbrzezny

IMCS, Jan Długosz University,
Al. Armii Krajowej 13/15, 42-200 Częstochowa, Poland
{agnieszka.zbrzezny,a.zbrzezny}@ajd.czest.pl

Abstract. We present an SMT-based bounded model checking (BMC) method for Simply-Timed Systems (STSs) and for the existential fragment of the Real-time Computation Tree Logic. We implemented the SMT-based BMC algorithm and compared it with the SAT-based BMC method for the same systems and the same property language on several benchmarks for STSs. For the SAT-based BMC we used the PicoSAT solver and for the SMT-based BMC we used the Z3 solver. The experimental results show that the SMT-based BMC performs quite well and is, in fact, sometimes significantly faster than the tested SAT-based BMC.

1 Introduction

Verification of soft real-time systems is an actively developing field of research [2,9,10]. Popular models of such systems include, among others, timed automata [1], and simply-timed systems (STSs) [5], i.e., Kripke models where each transition holds a duration, which can be any integer value (including zero).

The fundamental thought behind bounded model checking (BMC) is, given a system, a property, and an integer bound $k \geq 0$, to define a formula such that the formula is satisfiable if and only if the system has a counterexample (of the length at most k) violating the property. The bound is incremented until a satisfiable formula is discovered or a completeness threshold is reached without discovering any satisfiable formulae. The SMT problem [3] is a generalisation of the SAT problem, where Boolean variables are replaced by predicates from various background theories, such as linear, real, and integer arithmetic. SMT generalises SAT by adding equality reasoning, arithmetic, fixed-size bit-vectors, arrays, quantifiers, and other useful first-order theories.

The original contribution of this paper consists in defining a SMT-based BMC method for the existential fragment of RTCTL (RTECTL) interpreted over simply-timed systems (STSs) generated by simply-timed automata with discrete data (STADDs). We implemented our SMT-based BMC algorithm and we compared it with the SAT-based BMC method for RTCTL and STSs. For a

---

* The study is co-funded by the European Union, European Social Fund. Project PO KL "Information technologies: Research and their interdisciplinary applications", Agreement UDA-POKL.04.01.01-00-051/10-00.

constructive evaluation of our SMT-based BMC method we have used two scalable benchmarks: a modified *bridge-crossing problem* [8] and a modified *generic pipeline paradigm* [7].

2   Preliminaries

In this section we first define simply-timed automata with discrete data and simply-timed systems, and next we introduce syntax and semantics of RTECTL. The formalism of STADD was introduced in [12] and formalism of STS in [10].

Simply-Timed Automata with Discrete Data and Simply-Timed Systems. Let $\mathbb{Z}$ be the set of integer numbers, $\mathcal{Z}$ a finite set of integer variables $\{z_1, \ldots, z_n\}$, $c \in \mathbb{Z}$, $z \in \mathcal{Z}$, and $\oplus \in \{+, -, *, mod, div\}$. Then, the set $Expr(\mathcal{Z})$ of all the *arithmetic expressions* over $\mathcal{Z}$ is defined by the following grammar: $ae ::= c \mid z \mid ae \oplus ae \mid -ae \mid (ae)$. Next, for $ae \in Expr(\mathcal{Z})$ and $\sim \in \{=, \neq, <, \leq, \geq, >\}$, the set $BoE(\mathcal{Z})$ of all the *Boolean expressions* over $\mathcal{Z}$ is defined by the following grammar: $\beta ::= true \mid ae \sim ae \mid \beta \wedge \beta \mid \beta \vee \beta \mid \neg \beta \mid (\beta)$. For $z \in \mathcal{Z}$, $ae \in Expr(\mathcal{Z})$, ϵ denoting the empty sequence, the set $SimAss(\mathcal{Z})$ of all the *simultanoues assignments* over $\mathcal{Z}$ is defined as: $\alpha ::= \epsilon \mid z_{i_1}, \ldots, z_{i_m} := ae_{i_1}, \ldots, ae_{i_m}$, where $i_j \in \{1, \ldots, n\}$ and any $z \in \mathcal{Z}$ appears on the left-hand side of „:=" at most once.

A *variables valuation* is a total mapping $\mathbf{v} : \mathcal{Z} \to \mathbb{Z}$. We extend this mapping to expressions of $Expr(\mathcal{Z})$ in the usual way. Furthermore, we assume that a domain of values for each variable is finite. Satisfiability of a Boolean expression $\beta \in BoE(\mathcal{Z})$ by a variable valuation $\mathbf{v}$, denoted $\mathbf{v} \models \beta$, is defined inductively in the standard way. Given a variables valuation $\mathbf{v}$ and an instruction $\alpha \in SimAss(\mathcal{Z})$, we denote by $\mathbf{v}(\alpha)$ a valuation $\mathbf{v}'$ such that: if $\alpha = \epsilon$, then $\mathbf{v}' = \mathbf{v}$; if $\alpha = (z_{i_1}, \ldots, z_{i_m} := ae_{i_1}, \ldots, ae_{i_m})$, then for all $j \in \{1, \ldots, n\}$ it holds $\mathbf{v}'(z_j') = \mathbf{v}(ae_j)$ if $j \in \{i_1, \ldots, i_m\}$, and $\mathbf{v}'(z_j') = \mathbf{v}(z_j')$ otherwise.

Definition 1. *Let $\mathcal{PV}$ be a set of atomic propositions. A simply-timed automaton with discrete data (STADD) is a tuple $\mathcal{A} = (\Sigma, L, l^0, \mathcal{Z}, E, d, \mathcal{V_A})$, where Σ is a finite set of actions, L is a finite set of locations, l^0 is an initial location, $\mathcal{Z}$ is a finite set of integer variables, $E \subseteq L \times \Sigma \times BoE(\mathcal{Z}) \times SimAss(\mathcal{Z}) \times L$ is a transition relation, $d : \Sigma \to \mathbb{N}$ is a duration function, and $\mathcal{V_A} : L \to 2^{\mathcal{PV}}$ is a valuation function that assigns to each location a set of propositional variables that are assumed to be true at that location.*

The semantics of the STADD is defined by associating to it a *simply-timed system* as defined below.

Definition 2. *Let $\mathcal{PV}$ be a set of atomic propositions, $\mathbf{v}^0 : \mathcal{Z} \to \mathbb{Z}$ an initial variables valuation, and $\mathcal{A} = (\Sigma, L, \ell^0, \mathcal{Z}, E, d, \mathcal{V_A})$ a simply-timed automaton with discrete data. A simply-timed system (or a model) for $\mathcal{A}$ is a tuple $M = (\Sigma, S, \iota, T, d, \mathcal{V})$, where Σ is a finite set of actions of $\mathcal{A}$, $S = L \times \mathbb{Z}^{|\mathcal{Z}|}$ is a set of states, $\iota = (\ell^0, \mathbf{v}^0) \in S$ is the initial state, $T \subseteq S \times \Sigma \times S$ is the smallest simply-timed transition relation defined in the following way: for $\sigma \in \Sigma$, $(\ell, \mathbf{v}) \xrightarrow{\sigma} (\ell', \mathbf{v}')$*

iff there exists a transition $(\ell, \sigma, \beta, \alpha, \ell') \in E$ *such that* $\mathbf{v} \models \beta$, $\mathbf{v}' = \mathbf{v}(\alpha)$. *We assume that the relation* T *is total, i.e., for any* $s \in S$ *there exists* $s' \in S$ *and* $\sigma \in \Sigma$ *s.t.* $(s, \sigma, s') \in T$ *(or* $s \xrightarrow{\sigma} s'$*).* $d : \Sigma \to \mathbb{N}$ *is the duration function of* $\mathcal{A}$ *and* $\mathcal{V} : S \to 2^{\mathcal{PV}}$ *is a valuation function defined as* $\mathcal{V}((\ell, \mathbf{v})) = \mathcal{V}_{\mathcal{A}}(\ell)$.

A *path* in M is an infinite sequence $\pi = s_0 \xrightarrow{\sigma_1} s_1 \xrightarrow{\sigma_2} s_2 \xrightarrow{\sigma_3} \dots$ of transitions. For such a path, and for $m \in \mathbb{N}$, by $\pi(m)$ we denote the m-th state s_m. For $j \leq m \in \mathbb{N}$, $\pi[j..m]$ denotes the finite sequence $s_j \xrightarrow{\sigma_{j+1}} s_{j+1} \xrightarrow{\sigma_{j+2}} \dots s_m$ with $m - j$ transitions and $m - j + 1$ states. The (cumulative) duration $D\pi[j..m]$ of such a finite sequence is $d(\sigma_{j+1}) + \dots + d(\sigma_m)$ (hence 0 when $j = m$). By $\Pi(s)$ we denote the set of all paths starting at $s \in S$.

RTECTL: An Existential Fragment of a Soft Real-Time Temporal Logic. In the syntax of RTECTL we assume the following: $p \in \mathcal{PV}$ is an atomic proposition, and I is an interval in $\mathbb{N} = \{0, 1, 2, \dots\}$ of the form: $[a, b)$ or $[a, \infty)$, for $a, b \in \mathbb{N}$ and $a \neq b$. The RTECTL formulae are defined by the following grammar:

$$\varphi ::= \mathbf{true} \mid \mathbf{false} \mid p \mid \neg p \mid \varphi \wedge \varphi \mid \varphi \vee \varphi \mid \mathrm{EX}\varphi \mid \mathrm{E}(\varphi \mathrm{U}_I \varphi) \mid \mathrm{E}(\varphi \mathrm{R}_I \varphi).$$

An RTECTL formula φ is *true* in the model M (in symbols $M \models \varphi$) iff $M, \iota \models \varphi$ (i.e., φ is true at the initial state of the model M). For every $s \in S$ the relation $\models$ is defined inductively as follows:

- $M, s \models \mathrm{EX}\alpha$ iff $(\exists \pi \in \Pi(s))(M, \pi(1) \models \alpha)$,
- $M, s \models \mathrm{E}(\alpha \mathrm{U}_I \beta)$ iff $(\exists \pi \in \Pi(s))(\exists m \geq 0)(D\pi[0..m] \in I$ and $M, \pi(m) \models \beta$ and $(\forall j < m)M, \pi(j) \models \alpha)$,
- $M, s \models \mathrm{E}(\alpha \mathrm{R}_I \beta)$ iff $(\exists \pi \in \Pi(s))((\exists m \geq 0)(D\pi[0..m] \in I$ and $M, \pi(m) \models \alpha$ and $(\forall j \leq m)M, \pi(j) \models \beta)$ or $(\forall m \geq 0)(D\pi[0..m] \in I$ implies $M, \pi(m) \models \beta))$.

3   SMT-Based Bounded Model Checking

In this section we define the SMT-based BMC method for the existential fragment of RTCTL (RTECTL) [9]. Similarly to SAT-based BMC, the SMT-based BMC is based on the notion of the *bounded semantics* for RTECTL ([10]) in which one inductively defines for every $s \in S$ the relation $\models_k$. Let M be a model, $k \geq 0$ a bound, φ an RTECTL formula, and let $M, s \models_k \varphi$ denotes that φ is k-true at the state s of M. The formula φ is k-true in M (in symbols $M \models_k \varphi$) iff $M, \iota \models_k \varphi$ (i.e., φ is k-true at the initial state of the model M).

The *bounded model checking problem* asks whether there exists $k \in \mathbb{N}$ such that $M \models_k \varphi$. The following theorem states that for a given model and an RTECTL formula there exists a bound k such that the model checking problem ($M \models \varphi$) can be reduced to the BMC problem ($M \models_k \varphi$). The theorem can be proven by induction on the length of the formula φ.

Theorem 1 ([10]). *Let* M *be a model and* φ *an RTECTL formula. Then, the following equivalence holds:* $M \models \varphi$ *iff there exists* $k \geq 0$ *such that* $M \models_k \varphi$.

Translation to SMT. The translation to SMT is based on the bounded semantics. Let M be a simply-timed model, φ an RTECTL formula, and $k \geq 0$ a bound. The presented SMT encoding of the BMC problem for RTECTL and for STS is based on the SAT encoding of the same problem [10,11], and it relies on defining the quantifier-free first-order formula $[M, \varphi]_k := [M^{\varphi, \iota}]_k \wedge [\varphi]_{M,k}$ that is satisfiable if and only if $M \models_k \varphi$ holds.

The definition of the formula $[M^{\varphi, \iota}]_k$ assumes that states of the model M are encoded in a symbolic way. Such a symbolic encoding is possible, since the set of states of M is finite. In particular, each state s can be represented by a vector w (called a *symbolic state*) of different individual variables ranging over the natural numbers (called *individual state variables*).

The formula $[M^{\varphi, \iota}]_k$ encodes a rooted tree of $k-$paths of the model M. The number of branches of the tree depends on the value of the auxiliary function $f_k : \text{RTECTL} \to \mathbb{N}$ defined in [10].

Given the above, the j-th symbolic k-path π_j is defined as the following sequence $((d_{0,j}, w_{0,j}), \ldots, (d_{k,j}, w_{k,j}))$, where $w_{i,j}$ are symbolic states and $d_{i,j}$ are *symbolic durations*, for $0 \leq i \leq k$ and $0 \leq j < f_k(\varphi)$. The *symbolic duration* $d_{i,j}$ is a individual variable.

Let w and w' (resp., d and d') be two different symbolic states (resp., durations). We assume definitions of the following auxiliary quantifier-free first-order formulae: $I_\iota(w)$ - encodes the initial state of the model M, $\mathcal{T}((d, w), (d'w'))$ - encodes the transition relation of M, $p(w)$ - encodes the set of states of M in which $p \in \mathcal{PV}$ holds, $H(w, w')$ - encodes equality of two global states, $\mathcal{B}_k^I(\pi_n)$ - encodes that the duration time represented by the sequence $d_{1,n}, \ldots, d_{k,n}$ of symbolic durations is less than $right(I)$, $\mathcal{D}_j^I(\pi_n)$ - encodes that the duration time represented by the sequence $d_{1,n}, \ldots, d_{j,n}$ of symbolic durations belongs to the interval I, $\mathcal{D}_{k;l,m}^I(\pi_n)$ for $l \leq m$ - encodes that the duration time represented by the sequences $d_{1,n}, \ldots, d_{k,n}$ and $d_{l+1,n}, \ldots, d_{m,n}$ of symbolic durations belongs to the interval I.

The formula $[M^{\varphi, \iota}]_k$ encoding the unfolding of the transition relation of the model M $f_k(\varphi)$-times to the depth k is defined as follows:

$$[M^{\varphi, \iota}]_k := I_\iota(w_{0,0}) \wedge \bigwedge_{j=0}^{f_k(\varphi)-1} \bigwedge_{i=0}^{k-1} \mathcal{T}((d_{i,j}, w_{i,j}), (d_{i+1,j}, w_{i+1,j})) \qquad (1)$$

For every RTECTL formula φ the function f_k determines how many symbolic k-paths are needed for translating the formula φ. Given a formula φ and a set A of k-paths such that $|A| = f_k(\varphi)$, we divide the set A into subsets needed for translating the subformulae of φ. To accomplish this goal we need the auxiliary functions $h_n^U(A, e)$ and $h_n^R(A, e)$ that were defined in [11].

Below we show the translation for the temporal operators EU_I and ER_I only. Let φ be an RTECTL formula, M a model, and $k \in \mathbb{N}$ a bound. The quantifier-free first-order formula $[\varphi]_{M,k} := [\varphi]_k^{[0,0,F_k(\varphi)]}$, where $F_k(\varphi) = \{j \in \mathbb{N} \mid 0 \leqslant j < f_k(\varphi)\}$, encodes the bounded semantics for RTECTL, and it is defined inductively as shown below. Namely, let $0 \leqslant n < f_k(\varphi)$, $m \leqslant k$, $n' = min(A)$, $h_k^U = h_k^U(A, f_k(\beta))$, and $h_k^R = h_k^R(A, f_k(\alpha))$, then:

$$[E(\alpha U_I \beta)]_k^{[m,n,A]} := H(w_{m,n}, w_{0,n'}) \wedge \bigvee_{i=0}^{k}([\beta]_k^{[i,n',h_k^U(k)]} \wedge D_i^I(\pi_{n'}) \wedge$$
$$\bigwedge_{j=0}^{i-1}[\alpha]_k^{[j,n',h_k^U(j)]}),$$

$$[E(\alpha R_I \beta)]_k^{[m,n,A]} := H(w_{m,n}, w_{0,n'}) \wedge (\bigvee_{i=0}^{k}([\alpha]_k^{[i,n',h_k^R(k+1)]} \wedge D_i^I(\pi_{n'}) \wedge \bigwedge_{j=0}^{i}$$
$$[\beta]_k^{[j,n',h_k^R(j)]}) \vee (\neg \mathcal{B}_k^I(\pi_{n'}) \wedge \bigwedge_{j=0}^{k}(D_j^I(\pi_{n'}) \rightarrow [\beta]_k^{[j,n',h_k^R(j)]}))$$
$$\vee (\mathcal{B}_k^I(\pi_{n'}) \wedge \bigwedge_{j=0}^{k}(D_j^I(\pi_{n'}) \rightarrow [\beta]_k^{[j,n',h_k^R(j)]}) \wedge \bigvee_{l=0}^{k-1}$$
$$[H(w_{k,n'}, w_{l,n'}) \wedge \bigwedge_{j=l}^{k-1}(D_{k;l,j+1}^I(\pi_{n'}) \rightarrow [\beta]_k^{[j,n',h_k^R(j)]})])).$$

The theorem below states the correctness and the completeness of the presented translation. It can be proven by induction on the complexity of the given RTECTL formula.

Theorem 2. *Let M be a model, and φ an RTECTL formula. Then for every $k \in \mathbb{N}$, $M \models_k \varphi$ if, and only if, the formula $[M, \varphi]_k$ is satisfiable.*

4   Experimental Results

Our SAT-based and SMT-based BMC algorithms were implemented as standalone programs written in the programming language **C++**. For the SAT-based BMC module we used the state of the art SAT-solver PicoSAT [4], and for our SMT-based BMC module we used the state of the art SMT-solver Z3 [6].

The Bridge-Crossing Problem (BCP) [8] is a famous mathematical puzzle. To generate experimental results we have tested BCP system defined in [10]. We have five automata that run in parallel and synchronised on actions LR_i, RL_i, and F_{ij}, for $i \neq j$ and $i, j \in \{1, \ldots, 4\}$. The action LR_i (respectively, RL_i) means that the i-th person goes from the left side of the bridge to its right side (respectively, from the right side of the bridge to its left side) bringing back the lamp. The action F_{ij} with $i < j$ (respectively, F_{ij} with $i > j$) means that the persons i and j cross the bridge together from its left side to its right side (respectively, from its right side to its left side). Four automata (those with states named as Li and Ri, for $1 \leq i \leq 4$) represent persons, and one represents a lamp that keeps track of the position of the lamp, and ensures that at most two persons cross in one move. Let Min denote the minimum time required to cross the bridge, $N \geq 2$ be the number of persons, and $right = (2 \cdot N - 3) \cdot (t_1 + (N - 1) \cdot 3)$. We have tested BCP for $N \geq 4$ persons, $t_i = t_1 + (i - 1) \cdot 3$ with $1 \leq i \leq N$ and $t_1 \geq 10$, on the following RTECTL formulae: $\varphi_{1BCP} = EF_{[Min,Min+1]}(\bigwedge_{i=1}^{N} Ri)$ and $\varphi_{2BCP} = EG_{[0,right]}(\bigvee_{i=1}^{N} \neg Ri)$; the formulae are true in the model for BCP.

Generic Simply-Timed Pipeline Paradigm. We adapted the benchmark scenario of *a generic pipeline paradigm* [7], and we called it the *generic simply-timed pipeline paradigm* (GSPP). The model of GSPP involves Producer producing data, Consumer receiving data, and a chain of n intermediate Nodes that transmit data produced by Producer to Consumer. Producer, Nodes, and Consumer have different producing, sending, processing, and consuming times.

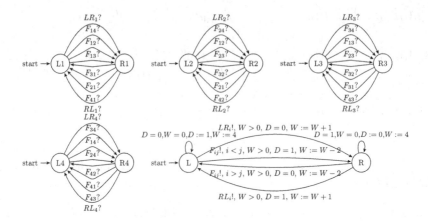

Fig. 1. A network of STADD automata that models BCP

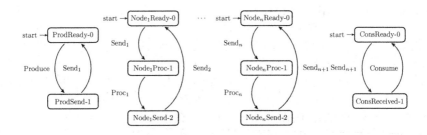

Fig. 2. A network of STADD automata that models GSPP

A STADD automata model of GSPP is shown in Fig. 2. We have $n + 2$ automata (n automata representing Nodes, one automaton for Producer, and one automaton for Consumer) that run in parallel and synchronise on actions $Send_i$ ($1 \leqslant i \leqslant n + 1$). Action $Send_i$ ($1 \leqslant i \leqslant n$) means that i-th Node has received data produced by Producer. Action $Send_{n+1}$ means that Consumer has received data produced by Producer. Action $Proc_i$ ($1 \leqslant i \leqslant n$) means that i-th Node processes data. Action $Produce$ means that Producer generates data. Action $Consume$ means that Consumer consumes data produced by Producer. Let $1 \leqslant i \leqslant n$. We have tested the GSPP problem with the following basic durations: $d(Produce) = 2$, $d(send_i) = 2$, $d(Proc_i) = 4$, $d(Consume) = 2$, and their multiplications by 50, 100, 150, etc., on the following RTECTL formulae: $\varphi_{1GSPP} = EF_{[Min,Min+1)}ConsReceived$, $\varphi_{2GSPP} = EG_{[0,\infty)}(\neg ProdSend \vee EF_{[0,Min-d(Produce)+1)}ConsReceived)$, $\varphi_{3GSPP} = EG_{[0,\infty)}(\neg ProdSend \vee EG_{[0,Min-d(Produce))}ConsReady)$, where Min denotes the minimum time required to receive by Consumer the data produced by Producer. Note that the φ_{2GSPP} and φ_{3GSPP} are properties, respectively, of the type the existential *bounded-response* and existential *bounded-invariance*. All the above formulae are true in the model for GSPP.

Performance Evaluation. The evaluation of both the BMC algorithms is given by means of the running time and the memory used. In most cases, the experimental results show that the SMT-based BMC method is significantly faster than the SAT-based BMC method. From Fig. 3 we can notice that for the GSPP system and all considered formulae the SMT-based BMC is faster than the SAT-based BMC, however, the SAT-based BMC consumes less memory. Moreover, the SMT-based method is able to verify more nodes for all the tested formulae. In particular, in the time limit set for the benchmarks, the SMT-based BMC is able to verify the formula φ_{1GSPP} for 54 nodes while the SAT-based BMC can handle 40 nodes, for the formula φ_{2GSPP} respectively 25 nodes and 21 nodes. For φ_{3GSPP} the SMT-based BMC is still more efficient - it is able to verify 20 nodes, whereas the SAT-based BMC verifies only 17 nodes for $t = 1$ and 19 nodes for $t = 1000000$.

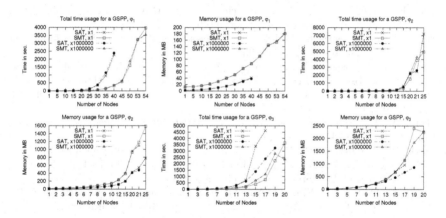

Fig. 3. SAT/SMT-BMC: GSPP scaling up both the number of nodes and durations

As one can see from the line charts for the BCP system (Fig. 4), in the case of this benchmark the SMT-based BMC and SAT-based BMC are complementary. In the case of the formula φ_{1BCP} SMT-based BMC is able to verify system with 10 persons while the SAT-based BMC can handle 11 persons. For φ_{2BCP} the SMT-based BMC is more efficient - it is able to verify 31 persons, whereas the SAT-based BMC verifies only 27 nodes for $t = 1$ and 29 nodes for $t = 1000000$, but the SAT-based BMC consumes less memory.

We have proposed an SMT-based BMC verification method for model checking RTECTL properties interpreted over the simply-time systems that are generated for simply-timed automata with discrete data. We have provided a preliminary experimental results showing that our method is worth interest. For the analysis of (soft) real-time systems, this extends the verification facilities that are offered in VerICS, since our algorithm is implemented as a new module of the tool.

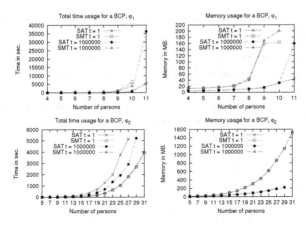

Fig. 4. SAT/SMT BMC: BCP scaling up both the number of persons and durations

References

1. Alur, R.: Timed Automata. In: Halbwachs, N., Peled, D.A. (eds.) CAV 1999. LNCS, vol. 1633, pp. 8–22. Springer, Heidelberg (1999)
2. Alur, R., Courcoubetis, C., Dill, D.: Model checking in dense real-time. Information and Computation 104(1), 2–34 (1993)
3. Barrett, C., Sebastiani, R., Seshia, S., Tinelli, C.: Satisfiability modulo theories. In: Biere, A., Heule, M.J.H., van Maaren, H., Walsh, T. (eds.) Handbook of Satisfiability. Frontiers in Artificial Intelligence and Applications, vol. 185, ch. 26, pp. 825–885. IOS Press (2009)
4. Biere, A.: PicoSAT essentials. Journal on Satisfiability, Boolean Modeling and Computation (JSAT) 4, 75–97 (2008)
5. Markey, N., Schnoebelen, P.: Symbolic model checking for simply-timed systems. In: Lakhnech, Y., Yovine, S. (eds.) FORMATS 2004 and FTRTFT 2004. LNCS, vol. 3253, pp. 102–117. Springer, Heidelberg (2004)
6. de Moura, L., Bjørner, N.S.: Z3: An efficient SMT solver. In: Ramakrishnan, C.R., Rehof, J. (eds.) TACAS 2008. LNCS, vol. 4963, pp. 337–340. Springer, Heidelberg (2008)
7. Peled, D.: All from one, one for all: On model checking using representatives. In: Courcoubetis, C. (ed.) CAV 1993. LNCS, vol. 697, pp. 409–423. Springer, Heidelberg (1993)
8. Cook, E.E., Levmore, S.X.: Super Strategies for Puzzles and Games. Doubleday, Garden City (1981)
9. Woźna-Szcześniak, B., Zbrzezny, A., Zbrzezny, A.: The BMC method for the existential part of RTCTLK and interleaved interpreted systems. In: Antunes, L., Pinto, H.S. (eds.) EPIA 2011. LNCS, vol. 7026, pp. 551–565. Springer, Heidelberg (2011)
10. Woźna-Szcześniak, B., Zbrzezny, A.M., Zbrzezny, A.: SAT-based bounded model checking for RTECTL and simply-timed systems. In: Balsamo, M.S., Knottenbelt, W.J., Marin, A. (eds.) EPEW 2013. LNCS, vol. 8168, pp. 337–349. Springer, Heidelberg (2013)
11. Zbrzezny, A.: Improving the translation from ECTL to SAT. Fundamenta Informaticae 85(1-4), 513–531 (2008)
12. Zbrzezny, A., Półrola, A.: SAT-based reachability checking for timed automata with discrete data. Fundamenta Informaticae 79(3-4), 579–593 (2007)

Multi-agent System for Tracking and Classification of Moving Objects

Sergio Sánchez[1], Sara Rodríguez[1], Fernando De la Prieta[1],
Juan F. De Paz[1], and Javier Bajo[2]

[1] Computer and Automation Department, University of Salamanca,
Plaza de la Merced s/n, 37008, Salamanca, Spain
{sergio_sg,srg,fer,fcofds}@usal.es
[2] Artificial Intelligence Department, Polytechnic University of Madrid,
Campus Montegancedo, Boadilla del Monte, Madrid 28660, Spain
jbajo@fi.upm.es

Abstract. In the past, computational barriers have limited the complexity of video and image processing applications but recently, faster computers have enabled researchers to consider more complex algorithms which can deal successfully with vehicle and pedestrian detection technologies. However, much of the work only pays attention to the accuracy of the final results provided by the systems, leaving aside the computational efficiency. Therefore, this paper describes a system using a paradigm of multi-agent system capable of regulating itself dynamically taking into account certain parameters pertaining to detection, tracking and classification, to reduce the computational burden as low as possible at all times without this in any way compromise the reliability of the result.

Keywords: Computer Vision, Agents, Multi-Agent System, Vehicle Detection, Vehicle Counting, Pedestrian Detection, Classifiers, Video-Surveillance.

1 Introduction

Computer Vision is a field of Artificial Intelligence that has captured the attention of many researchers from diverse communities in recent years. This field is intended to achieve a computational model of the human sense of vision in order to produce numerical or symbolic information to enable decision making . For this, the field includes methods for the acquisition, processing, analysis and understanding of real-world images, whether maps dimensional pixels, combining images from stereo to obtain 3D models cameras and even multidimensional data to work with any type dynamic scene. In this sense, intelligent approaches based on multi-agent systems (MAS) combined with information fusion process have been recently emerging [14]. MAS [20] are increasing in importance in the research line of distributed and dynamic intelligent environments. These intelligent systems offer a high-level tool to support a framework for intelligent information fusion and management. There are now many

© Springer International Publishing Switzerland 2015
S. Omatu et al. (eds.), *Distributed Computing & Artificial Intelligence, 12th Int. Conference*,
Advances in Intelligent Systems and Computing 373, DOI: 10.1007/978-3-319-19638-1_8

lines within the field of artificial vision in which it is necessary to manage different information of different kind of sensors such as systems video surveillance and monitoring, robotics and autonomous navigation, instrumentation for medicine, security systems, satellites and military applications, image processing, data mining large volumes of images or quality control in manufacturing chain[7][17][14].

For all these reasons, it is possible to say that MAS are an ideal option to create and develop the open and heterogeneous systems such as those normally found in the computer vision and information fusion process. This paper presents an intelligent multi-agent system that incorporates new information fusion techniques to automatically locate objects through devices such as surveillance cameras to facilitate tracking or location in a dynamic environment. This article is structured as follow: the next section describes the general background in vehicle detection and tracking focus on optimizing computational efficiency, information fusion and multi-agent systems. The third section is a description of the components of the proposed system. Finally, the fourth section presents the results and conclusions obtained, measured mainly from the point of view of the computational savings offered by the use of agents.

2   Background

The computational techniques used in this work are described in the following paragraphs, focusing on Pedestrian and Vehicle tracking, Multiagent systems and Computational Efficiency Applied to Computer Vision.

Regarding **Pedestrian and Vehicle Tracking**, in this work, we use Optical flow algorithm [8] to detect coming traffics. Given the two consecutive frames, we find corner points with corner detector [15] and these features are matched by the Optical flow algorithm, which assumes that important features are detected in both frames. We fuse some features into a rectangle if the Euclidian distance of two features (the location of feature and the direction of optical flow) are small. The optical flow algorithms find the correspondence within a reasonable time so that we can use the algorithms for real-time application. However, some of optical flows are generated by some cracks of the road and road signs. Haar-like feature detector [18] is used to detect traffics in the same direction, but the optical flow algorithm is not so appropriate for this goal because they do not have any salient movement in most case. The shapes mostly show the rear side of the car. To detect the rear shape we choose to use Haar-like feature detector because it is fast and efficient.

Much of systems using Artificial Vision only pays attention to the accuracy of the final results provided by the system, leaving aside the **computational efficiency**. Research suggests that the computational efficiency and the accuracy of the system are directly linked, stating that it is impossible to have reliable results in dynamic scenarios if the computational effort of the system is reduced; but not impossible in controlled settings [10]. Finally, these studies end up veering towards parallelization hardware for increased performance. Focusing on our work, there are a large number of studies that make comparisons on specific parts of similar systems, such as comparisons between different methods of "Background subtraction" [11], but never come to relate it to the scene, no even with the other phases of a typical detection and tracking system. The work presented in this paper will demonstrate that it is possible to have a system of collaborative **agents** that fuse certain information to suit the status of each scene and may be reduced at all times the computational effort to a minimum where accuracy does not decrease, by regulating a number of parameters described in the following sections.

The information obtained from multiple vision sensors needs to be fused because no single sensor can get all the information, and the information from different sensors may be uncertain, inaccurate, or even conflicting [9]. This is the reason why information fusion is a fundamental part of a computer vision management. There is a considerable variety of sensors that can observe user contexts and behaviours and multi-agent architectures that utilize data merging to improve their output and efficiency [7]. The adequacy of **MAS** applied to information fusion and dynamic environments has been deeper discussed in our previous work [13]. Moreover, various works inspire the proposed architecture, among which we mention [4].

On this background, in this work, the above techniques are combined, using a cascade of stages on the geometric characteristics of the objects to detect, track, count and classify the detected features, and the influence of various parameters on the behavior of the system is studied.

3   System Overview

Our goal is to develop a core technology on which to build custom applications that require the use of computer vision for detecting, tracking and counting people and vehicles. To do this, we have set the following technical objectives: (i) Develop an optimal algorithm for detecting, tracking, counting and classification of moving objects, which combine both static and dynamic techniques in image processing; (ii) Develop a graphical user interface for visualizing and modifying some parameters of the image processing phases; (iii) Develop a module for capturing images from IP camera to ensure interoperability with existing video-surveillance systems.

The use of MAS provides a mechanism that allows individual units called agents to perform tasks concurrently. These agents can be software units that undertake simple tasks to reach a common goal of the overall system. This provides a number of important advantages. First, it implies a decentralization of the complete system, so detection calculations tasks, based on a complex Gaussian Filter, could be performed in another computer. It also provides great modularity and scalability, which makes it possible to add new functionality by adding more phases, or combine the input of a few cameras. Finally, the MAS provide a communication system that allows each of the agents to exchange information following the protocol defined by the platform.

There are multiple platforms when making a practical application related to these MAS, such as JADE (Java Agent DEvelopment framework)[2] or PANGEA [20] (*Platform for Automatic Construction of Organizations of Intelligent Agents*). The latter has been used for the deployment and communication of agents that will perform the full functionality needed for the case study.

3.1   Phases of the System

To accomplish the above objectives, it was decided to design a system comprising a number of cascade-connected stages. Each of these phases is in charge of a task, so the output of one stage is the input of the next stage. In addition, each phase is associated with a type of agent that communicates with the agents associated with other phases of the system. The purpose of this communication between agents is dynamically set the appropriate values for a number of parameters, which depend of the state of the scene. This will reduce the computational effort with respect to set all this parameters statically.

The agents involved and therefore the computational details of these phases are described below. As will be shown throughout this article, the average improvement of the computational efficiency at each stage is between 24.90% and 78.34% depending on the three-dimensional structure and congestion level of each particular scene.

Processing Cascade consists of the following phases: stabilization, detection, tracking, classification and group. Below it is explained in details how each of the last four phases work, making an especial emphasis to the improvements made at each stage.

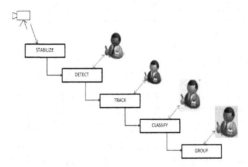

Fig. 1. Block diagram of the Processing Cascade System

Detection Agent

Once the camera has been calibrated and the image stabilized, the next step is the detection of changes in the sequence of frames that would indicate some kind of movement. Because the cameras that provide images are not moving, a background subtraction method for detection motion is used. The operation of this type of methods is based on the error between an estimate of the image without moving objects and the current image. The numerous approaches to this problem differ in the type of background model used and the procedure used to update the model.

There is much research concerning the different background subtraction methods. This researches qualify numerically the behavior and robustness of each of this background subtraction methods according with the type of video input, memory needed or computational effort required.

It has been demonstrated that the static and simple background subtraction methods, as MinMax, whose operation is based on a threshold on a grayscale, offer worse results than any of the dynamic methods [3]. However, more sophisticated methods such as

KDE or dynamic Eigen not always produce more accurate results, especially in noisy images, and due to their consumption of CPU and memory are not suitable for image processing in real time [3], so we have chosen the Gaussian Mixture Model (GMM) method as the basis of this detection phase [16]. This filtering process consists of enclosing the area of mobile contours considered valid between a maximum and a minimum values to ensure that the areas in which motion has been detected by size may correspond with pedestrians or vehicles. If not, the frame will not go into other phases of the Processing Cascade and will not be unnecessarily processed if there were no eligible contours into it. The Detector Agent has the self-learning ability needed to establish the maximum and minimum values using information that other agents have processed and reported from other stages in the cascade as follows:

- When the system starts operating, because of the absence of historical information, all the frames in which motion has been detected will advance to the next stage of the Processing Cascade.
- When a target gets through all stages of the Processing Cascade, the Group Agent communicates with the Detector Agent, sending him information on the average area of this category.
- The higher the number of targets that went through all stages, then we will have more information to affirm with greater certainty that the area of potential targets in that category will be closest to the average area of all the targets that are already part of the historical of that category on this particular scene.

This allows the system to enclose the valid area more precisely when the Group Agent has stored more data. It means that we are providing the system with self-learning, since the longer the system is running, the greater efficiency we are having for this stage. According to the values shown in the training video datasets, the system has been modeled so that a single element in the historical of this category, the maximum possible area is twice that middle area and as you increase the number of historical data, the maximum allowable area will get closer to average historical area of the category.

N_C being the number of objectives that have been classified in a category C, we have the following máximum and mínimum área values for category C:

$$MaxArea_C = AverageArea_C * 1.1 + (AverageArea_C)/(N_C) \qquad (1)$$

$$MinArea_C = AverageArea_C * 0.9 - (AverageArea_C)/(N_C) \qquad (2)$$

The 10 percent additional margin of these two values is due to the Area of a track depends on its distance from the camera, which will vary over time. It has been demonstrated with the training datasets that this variation of the track relative size will never exceed 10 percent of its original value, so that adds and subtracts this value to the maximum area and minimum area respectively. It is considered that an objective O is eligible to be part of a category C if and only if the following condition is met, discarding the frame if it contains no eligible contours:

$$MinArea_C \leq Area_O \leq MaxArea_C \qquad (3)$$

The following figure shows graphically the learning capacity of the system. As the size of the historical increases, the eligible area is bounded more severely and the efficiency of the system increases:

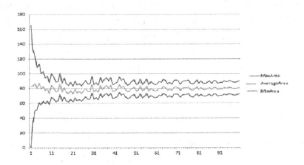

Fig. 2. Eligible area of Category C is bounded more severely while the amount of historical data increases

To demonstrate the computational savings offered by this improvement to the overall process, which will continue to provide the same result, in the following table it is shown the number of discarded frames, which ultimately depends on the degree of influx of pedestrian and vehicles passing in front of the camera:

Table 1. Improving the computational efficiency due to the filtering by area

Training Data: 12,000 Frames for each situation		
Congestion charge	Frames that pass the filter	Frames that do not pass the filter
High	10903	90.85%
Medium	4772	39.76%
Low	692	5.76%

Tracker Agent

Once detected every eligible target, at this stage the system will obtain their position over time through small changes in the image, which are tracked by the technique known as Optical Flow, described in the paper [1]. For this technique to work, it must be defined a zone of influence around the target with a suitable dimension so that there must be a real correspondence between small changes in the frames and the movement of the real targets. Thus if the new motion is detected at a distance greater than the limit of the former movement, the system will track it as another target, rather than treating it as a movement of the previous target. In short, this technique can only be applied reliably if the input video has a number of Frames per Second (FPS) enough to make this association in the area of influence.

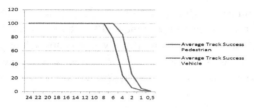

Fig. 3. Average Track Success for pedestrian and vehicles in relation with the number of FPS

The previous graph shows the success rate in relation with the number of FPS for pedestrians and vehicles with a training dataset of 12000 frames. From the above figure can be drawn that the result of the system will be equally reliable if is reduced from 16 FPS, typically offered by the video surveillance cameras [19], up to 8 FPS in the case of vehicles and 6 FPS in the case of pedestrians, because the movement speed is lower. While it is true that setting a static limit of 8 FPS would have a reliable result for both cases, the use of a Tracker Agent to set the FPS rate for each scenario dynamically bring further improvement in performance. The operation of the Tracker Agent is based on reducing the FPS rate to a minimum, so that in each frame the new position of the track approach as much as possible the edge of the area of influence of that track in the previous frame, without never exceed that limit, because it would led to an erroneous result.

The way in which the tracking algorithm is programmed to provide valid results, if the area of a track of a pedestrian is W * H, then its influence area would be a circle whose area would be 0.6*W*H:

$$\text{Influence radius} = \text{Sqrt}((W*H)/(pi)) * 0.6 \qquad (4)$$

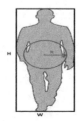

Fig. 4. Influence Radius of a track while using the Optical Flux Method

For a given track, we can immediately calculate the Influence Radius, and compare it with to the amount of movement from the previous frame, which could be defined as:

$$\text{Movement} = \text{sqrt}\ (\Delta x^2 + \Delta y^2) \qquad (5)$$

By linking the two measures, the Tracker Agent can modify the FPS Rate that each track needs, in order to minimize the computational effort as follows:

$$FPS_T = \text{Desplazamiento/RadioMovimiento} * FPS_{T-1} \qquad (6)$$

This measure is directly proportional to the speed of the different tracks, so the Tracker Agent set for this moment the highest FPS rate of all the active tracks belonging to the scene. Ultimately, as the aim of the system is not seeing a fluid video to the human eye, but it can be processed reliably to extract the desired information. The Tracker Agent allows us to improve the efficiency of the system as shown in the following table:

Table 2. Improved system efficiency due to the dynamic regulation of the number of FPS

Training Data 12,000 Frames in the initial video			
Congestion charge	Average FPS	Standard FPS	Percentage of FPS needed.
High	7.58	16	47.37%
Medium	6.31	16	39.43%
Low	4.42	16	27.62%

Classifier Agent

At this point of the cascade stages, you must perform a classification of the target to determine whether it is a pedestrian, a vehicle or unclassifiable goal. This phase is reached only if there is a possible moving target that has not yet been definitively classified in the category of pedestrian or vehicle at the scene. Because the technique classifiers cascade introduces noise in the form of false positives [12], it is not enough that the target is classified in one of two possible categories in a particular frame, because it could be a false positive.

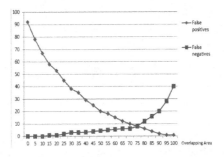

Fig. 5. Point balance between false negatives and false positives

To decide whether a goal has not yet been rated definitely belongs to one of the two categories in a given frame, once the image has been processed by the classifier, the rectangle associated with track and the rectangle in which was detected a person or vehicle must have a certain percentage of their areas overlapped. The ideal percentage overlap was determined by testing with a total of 12000 frames of real video sequences. If the percentage of overlapping is too small, the system will get many false positives, but if that percentage is too large some tracks will not be taken as positive, but they are actually positives.

As seen in the above graph, facing the percentage of overlap in relation to false positives and negatives, the balance point is found in an overlapping area of 70%. Regarding the treatment of false negatives, after a frequency analysis with videos in which appear both pedestrians and vehicles, the following results were obtained:

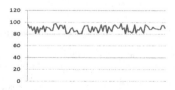

Fig. 6. Percentage of success in the classification of each detected track

According to this analysis, we can ensure that the number of times the track is classified incorrectly will never exceed 20%. Based on this data, the Classifier Agent dynamically set the total number of times each track must be classified into one of two categories to definitely belong to that category. Using this Classifier Agent will stop executing this phase if all the targets of the scene are definitely classified, again causing computational savings. The number of positive enough will be calculated by the Classifier Agent, that communicate with the Group Agent to know at all times the average number of frames that the targets of this category remain on the scene.

As the noise in the classification will never be higher than 20%, a frame definitely belongs to one of the two categories if the number of times it has been classified in that category exceeds the following threshold:

"Classification threshold" = 0.2 * Average frames on stage for that category.   (7)

This means that in the best case we would be saving the classification process in 80% of cases, but this would only be true if there were never more than two targets on stage simultaneously. For actual results, in the following table you can see how the computational savings in the classification process is greater with decreasing the degree of congestion.

Table 3. Improved system efficiency due to dynamic threshold for classification

Training Data: 12,000 Frames in the initial video			
Congestion charge	Frames in which the classifier is used	Frames with detected activity	Percentage required to use the classifier
High	10557	10903	96.82%
Medium	3112	4772	65.21%
Low	286	692	41.33%

Group Agent

In the last phase of the Processing Cascade, the Group Agent is responsible for the processing and storage of historical data of the system. The goal is to communicate with other agents who need some historical data, and the creation of social role models for studying the behavior of the targets in each scenario using data mining techniques.

4   Results and conclusions

Below, the proposed improvements at each stage will come together to see what the average computational savings in each case would be through the use of intelligent agents that lead to a non-linear optimized system.

Table 4. Improved system efficiency of each phase

Training Data: 12,000 Frames in the initial video				
Congestion charge	Computation in tracking phase.	Computation in detection phase.	Computation in classiffication phase.	Percentage relative to unop-timized system
High	47.37 %	90.85%	96.82%	**78.34 %**
Medium	39.43 %	39.76 %	65.21%	**48.13 %**
Low	27.62 %	5.76 %	41.33%	**24.90 %**

The main conclusion drawn is that the improvement of the system compared to algorithmically correct system in the treatment of the images but without any parametric regulation depends heavily on the "Congestion Charge", although in all cases significant improvements occur with regard to computational effort. We should also note that when cascaded stages, we would be leading to a greater reduction in computational effort because the improvements would chaining.

In short, although the general increase in computer performance allow us to run the system without improvements, with the architecture and the behavior of the agents proposed the system will stop committing inconsistencies as to process frames in which there are no eligible targets by size, try to classify all the frames when all objects have already been definitively classified or process a number of FPS, much larger than needed to obtain a reliable result. With these improvements, it has been possible to run the system satisfactorily in a Single-Board Computer (SBC) with a 700MHz processor and 512Mb of Random Access Memory (RAM), capable of running a Linux distribution.

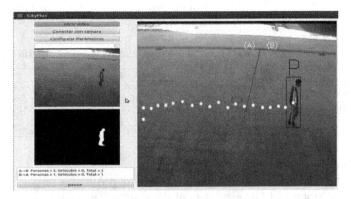

Fig. 7. Screenshot of the developed software running in a SBC

Acknowlegements. This work has been carried out by the project Sociedades Humano-Agente en entornos Cloud Computing (Soha+C). SA213U13. Project co-financed with Junta Castilla y León funds.

References

1. Baker, S., Roth, S., Scharstein, D., Black, M.J., Lewis, J.P., Szeliski, R.: A Database and Evaluation Methodology for Optical Flow. In: IEEE 11th International Conference on Computer Vision, ICCV 2007, pp. 1–8 (2007), doi:10.1109/ICCV.2007.4408903, ISSN 1550-5499
2. Bellifemine, F.L., Poggi, A., Rimassa, G.: Developing multi-agent systems with JADE. In: Castelfranchi, C., Lespérance, Y. (eds.) ATAL 2000. LNCS (LNAI), vol. 1986, pp. 89–103. Springer, Heidelberg (2001)
3. Benezeth, Y., Jodoin, P.-M., Emile, B., Haurent, H., Rosenberg, C.: Comparative Study of Background Subtraction Algorithms. Journal of Electronic Imaging 19 (2012)
4. Boissier, O., Demazeau, Y.A.: An architecture for social and invidivual control and its application to Computer Vision. In: European Workshop on Modeling Autonomous Agents In a Multiagent World, pp. 107–118 (1994)
5. Brooks, R.A.: Intelligence without representation. Artificial Intelligence 47, 139–159 (1991)
6. Liu, Y.H., Wang, S.Z., Du, X.M.: A multi-agent information fusion model for ship collision avoidance. In: Proceedings of the International Conference on Machine Learning and Cybernetics, pp. 6–11 (2008)
7. Lucas, B.D., Kanade, T.: An iterative image registration technique with an application to stereo vision. In: Proc. of the 7th IJCAI, Vancouver, Canada, pp. 674–679 (1981)
8. Luo, H., Yang, S., Hu, X.: Agent oriented intelligent fault diagnosis system using evidence theory. Expert Systems with Applications 39(3), 2524–2531 (2012)
9. Sundaram, N.: Making computer vision computationally efficient. EECS Department University of California, Berkeley Technical Report No. UCB/EECS-2012-106 (May 11, 2012)
10. Piccardi, M.: Background subtraction techniques: a review. In: 2004 IEEE International Conference on Systems, Man and Cybernetics (2004)
11. Zhu, Q., Avidan, S., Yeh, M.-C., Cheng, K.-T.: Fast Human Detection Using a Cascade of Histograms of Oriented Gradients. In: 2006 IEEE Computer Society Conference on Computer Vision and Pattern Recognition, vol. 2, pp. 1491–1498 (2006), doi:10.1109/CVPR.2006.119, ISSN:1063-6919
12. Rodríguez, S., De Paz, Y., Bajo, J., Corchado, J.M.: Social-based planning model for multiagent systems. Expert Systems with Applications 38(10), 13005–13023 (2011)
13. Rodríguez, S., De la Prieta, F., García, E., Zato, C., Bajo, J., Corchado, J.M.: Virtual Organizations in Information Fusion. In: 9th International Conference on Practical Applications of Agents and Multiagent Systems - Special Session on Adaptive Multiagent Systems, Salamanca, Spain, pp. 195–202 (2011)
14. Shi, J., Tomasi, C.: Good features to track. In: IEEE Conference on Computer Vision and Pattern Recognition (CVPR 1994), Seattle (June 1994)
15. Stauffer, C., Grimson, W.E.L.: Adaptive background mixture models for real-time tracking. In: IEEE Computer Society Conference on Computer Vision and Pattern Recognition, vol. 2 (1999), doi:10.1109/CVPR.1999.784637, ISSN:1063-6919

16. Tapia, D.I., de la Prieta, F., Rodríguez González, S., Bajo, J., Corchado, J.M.: Organizations of Agents in Information Fusion Environments. In: Antunes, L., Pinto, H.S. (eds.) EPIA 2011. LNCS, vol. 7026, pp. 59–70. Springer, Heidelberg (2011)
17. Viola, P., Jones, M., Snow, D.: Detecting pedestrians using patterns of motion and appearance (2003)
18. Wang, X.: Intelligent multi-camera video surveillance: A review. Pattern Recognition Letters 34(1), 3–19 (2013), doi:10.1016/j.patrec.2012.07.005
19. Zato, C., Sanchez, A., Villarrubia, G., Rodriguez, S., Corchado, J.M., Bajo, J.: Platform for building large-scale agent-based systems. In: 2012 IEEE Conference on Evolving and Adaptive Intelligent Systems (EAIS), pp. 17–18 (May 2012)

Improved Local Weather Forecasts Using Artificial Neural Networks

Morten Gill Wollsen and Bo Nørregaard Jørgensen

Centre for Energy Informatics
The Maersk Mc-Kinney Moller Institute
University of Southern Denmark, Denmark
{mgw,bnj}@mmmi.sdu.dk

Abstract. Solar irradiance and temperature forecasts are used in many different control systems. Such as intelligent climate control systems in commercial greenhouses, where the solar irradiance affects the use of supplemental lighting. This paper proposes a novel method to predict the forthcoming weather using an artificial neural network. The neural network used is a NARX network, which is known to model non-linear systems well. The predictions are compared to both a design reference year as well as commercial weather forecasts based upon numerical modelling. The results presented in this paper show that the network outperforms the commercial forecast for lower step aheads (< 5). For larger step aheads the network's performance is in the range of the commercial forecast. However, the neural network approach is fast, fairly precise and allows for further expansion with higher resolution.

1 Introduction

Solar irradiation and temperature forecasts are in today's world used in a variety of different control systems. One of such systems is intelligent climate control, where solar irradiance affects the planning of supplemental artificial lighting. Typically, weather forecasts for a local area are purchased from commercial weather forecast agencies. Unfortunately, forecasts are never as precise as one might hope. The Danish Meteorological Institute (DMI) states for the last running year, that they have had a 97% accuracy rate within $2°C$ [2]. The accuracy is very good, but the interval is also fairly wide. The commercial weather forecasts are typically generated in a 10 by 10km grid. If it was possible to predict the local weather based on the previous days' weather and sufficiently fast, smaller and more precise local weather forecasts could be distributed across an area. A local forecast using previous data will learn about local variation in the weather, something that cannot be captured by a 10 by 10km grid size.

This paper proposes a method that uses previous measurements to create short term ($<= 25$ hours) weather forecasts. The predictions will be made using an artificial neural network. To be useful, the network will have to perform at the same level of a weather forecast agency or even better.

© Springer International Publishing Switzerland 2015   75
S. Omatu et al. (eds.), *Distributed Computing & Artificial Intelligence, 12th Int. Conference*,
Advances in Intelligent Systems and Computing 373, DOI: 10.1007/978-3-319-19638-1_9

Artificial neural networks have previously been shown successful in weather forecasting. The work by Shank et al. used an artificial neural network with the backpropagation training algorithm to predict dew point temperatures from 1 to 12 hours ahead [11]. They included other inputs such as relative humidity, solar irradiation and air temperature to increase their performance. They also used a total of three different activations functions in their hidden layer. Smith et al. also used an artificial neural networks with the backpropagation algorithm to predict air temperature for an early warning system in agriculture [12]. Their network also predicted from 1 to 12 hours ahead. They experimented with network ensembles in linear and parallel settings, but neither improved their results. Multiple types of neural networks (multilayered perceptron, Elman recurrent, radial basis and Hopfield model) were used in an ensemble to predict weather parameters like temperature, wind speed and humidity for 1 to 24 hours ahead in [7]. With both a weighted average and a winner-takes-all approach, their architecture outperforms the individual network types.

The background information on artificial neural networks will not be reviewed. The reader is refered to [4, 5, 10].

The rest of this paper is structured as follows. Section 2 describes the methodology used for the experiments in this paper. Section 3 presents the results of the experiments and relevant discussions. And finally section 4 concludes the results found in this paper.

2   Method

Artificial Neural Network. The artificial neural network used in this paper is a NARX network which has been demonstrated to be well suited for the modelling of non-linear systems [9]. NARX stands for Nonlinear AutoRegressive with eXternal input. The NARX model can algebraically be described as

$$y(t) = f\left(u(t - n_u), \ldots, u(t - 1), u(t), y(t - n_y), \ldots, y(t - 1)\right) \qquad (1)$$

where the function f can be approximated with an artificial neural network. y is the parameter of interest and u is the external input. n_y and n_u are the numbers of previous values to include for y and u respectively. The number of previous values to include is called delay size. The network is created using the EnCog framework for Java [3]. The network contains a single input layer, a hidden layer with bias node and an output layer, also with a bias node. There are 20 nodes in the hidden layer. All nodes are using the *tanh* activation function. All data presented to the network has been normalized between -1 and 1. The EnCog framework uses an adaptive learning rate that depends on the gradient error.

The input to the network is created by the use of a delay block. Given an input u and a delay block with delay size 3, the delay block will generate $u(t)$, $u(t - 1)$, $u(t - 2)$ and $u(t - 3)$. For every new sample, the output is shifted so that the delay block always outputs the correct amount of elements. For solar irradiance prediction the delay size is 12. This size will allow the network to see how much time has passed since the sun went up. For temperature prediction, the delay size is 2 which was selected through a parameter search. These values are suited for the weather patterns in Denmark, but might be different for other countries.

Additionally, the network has a boolean input describing if the sun is up or down. This input is calculated based on the day and time of the year [8]. The boolean input is the only external input used in the network, and is only used for solar irradiance prediction. This input is also used to force the network's output to be 0 at night.

The structure of the network with its delay block is illustrated in Fig. 1. The network will only predict 1-step ahead, meaning at time t, the calculated parameter is the predicted value at time $t+1$. The closed loop allows the network to run continously, which enables the network to predict further steps ahead. For further steps ahead, the network uses its own output as input, instead of the actual weather.

The artificial neural network will be demonstrated through two different experiments. Common for both experiments is the use of a sliding window for the training data. The training data is 14 days long (336 samples). The network is trained using the backpropagation training algorithm [10]. The network is trained until achieving an error of 0.05%. For every new sample the network's weights are kept, which will allow the network to be trained faster. This approach will not influence the output even after many samples, because the network is trained until a sufficient margin of error has been achieved.

No steps will be taken towards ensuring that the network generalizes well. The target of this paper is to show that the raw NARX network is able to predict the weather parameters without generalization measures.

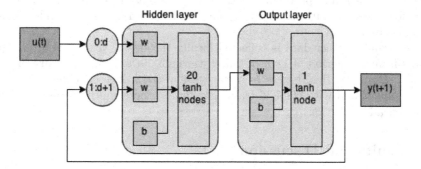

Fig. 1. The setup of the artificial neural network. The green delay blocks have different numbers for input and previous output. For delay size $d = 1$ the inputs will be $u(t)$, $u(t-1)$, $y(t)$ and $y(t-1)$. The bias node and the trained weights are b and w respectively.

Statistics. The results of the predictions are compared visually, and with the root mean square error (RMSE) measurement:

$$RMSE = \sqrt{\frac{\sum_{t=1}^{n}(\hat{y}_t - y_t)^2}{n}} \qquad (2)$$

where $\hat{y}_t$ is the prediction and y_t is the actual value. n is the amount of predictions in the series.

2.1   Experiment 1 - Baseline Measurement of Precision

The first experiment will demonstrate that the proposed network architecture is able to predict solar irradiance and temperature. The network will predict from 1 through 25 hours ahead, and will be compared to actual data. The data is a design reference year (DRY) from Copenhagen, Denmark of 1995 [6]. A period is picked out (11th of March through 20th of May) for comparison with the data available in experiment 2. This period is the longest available period of continuous data.

A DRY is a dataset for one year, arranged as 8760 hourly weather parameters. DRY's are often used as climate input data for system modelling and simulation. The DRY will contain the daily, weekly and monthly patterns which are common throughout the year. This makes a DRY a good referencing point for any predictions.

2.2   Experiment 2 - Comparing Forecasts with ConWX Forecast

The second experiment will use actual weather data from a greenhouse research facility in Aarslev, Denmark. The network's predictions will be compared both against the actual measurements as well as ConWX's weather forecasts for the same period. ConWX is a commercial weather forecast agency that provides local forecasts [1]. ConWX produces weather forecasts based on numerical models. Their forecasts are initiated every 6 hours, and takes 4 hours to complete. The forecasts span is 144 hours (= 6 days).

The actual weather data is collected locally through a weather station. The data is from the period of 11th of March through 20th of May, just like experiment 1. However, this data is from the year of 2013. These 71 days provides 1704 samples, and with the 336 training samples, there will be a total of 1368 1-step ahead predictions.

3   Results and Discussion

In the visualizations of the predictions, only a section containing 5 days will be shown. This improves the inspection possibility and allows for smaller details to be seen. The chosen section is from hours 925 through 1045. This section has been chosen randomly, and will be used for all results. The error values however, are calculated from the entire 1368 predictions.

3.1   Baseline Comparison

Solar Irradiance. The output of the network's prediction of solar irradiance can be seen in Fig. 2. The figure shows both the prediction for 1-step ahead and 25-step ahead. The actual measurements are from the 1-step ahead. The 25-step ahead prediction has been shifted 24 hours, to match the time period.

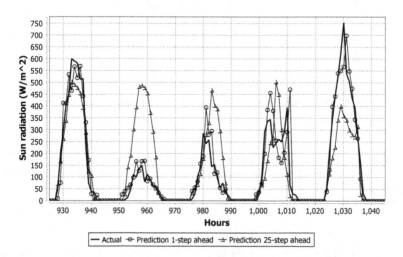

Fig. 2. Prediction output from 1-step and 25-step ahead prediction of solar irradiance. The RMSE of the 1-step ahead is 59.7 W/m^2 and 134.1 W/m^2 for the 25-step ahead.

The 1-step ahead prediction is very precise with only minor mishaps during sudden changes. The 25-step ahead however, is not very precise. It has correctly learned the sunrise and sunset pattern, but the values during the day are not correct. Even though the actual measurement in the section has both cloudy and sunny days, the 25-step ahead prediction's max values are almost equal.

For any timestep the 1-step ahead prediction will be trained with the actual data in the next timestamp. At this point, it will learn if its prediction was right or wrong. Because of this, it can predict both shiny and cloudy days very precisely.

The 25-step ahead prediction does not learn from its mistakes at the next hour. Instead it needs to trust its own output, and use its output to predict the next one. This means that any error will be magnified for every hour it predicts ahead. This is reflected in the output, where it can be seen that the network has a harder time differencing between a shiny and a cloudy day. Despite this lack of precision, the network has still learned to predict the daily pattern of sunrise and sunset.

Temperature. The prediction of temperature of the design reference year can be seen in Fig. 3. Again, the 25-step ahead prediction has been shifted to match the time period.

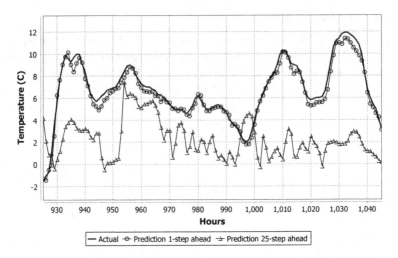

Fig. 3. The network output from 1-step and 25-step ahead prediction of outdoor temperature. The predictions are compared to the design reference year. The RMSE is measured to $0.86°C$ for the 1-step ahead prediction. For the 25-step ahead prediction, the RMSE is $4.92°C$.

Just like the 1-step ahead prediction for solar irradiance, the 1-step ahead prediction for temperature is very precise which is confirmed by its RMSE. With temperature there is no strong daily pattern of sunrise and sunset like with solar irradiance. However, a hint of this pattern can be seen from $t = 1000$ through $t = 1045$. Just as with the solar irradiance, the network will learn about its errors in the next hour, and can therefore quickly adjust.

The 25-step ahead prediction is not very good, just as it was shown for solar irradiance. It has a negative bias in mean values, and is fairly stable. The RMSE is more than five times higher than the 1-step prediction for temperatures. For comparison, the solar irradiance 25-step ahead prediction was only showing RMSE values twice of those found for the 1-step ahead prediction. This indicates that because the 25-step ahead prediction for solar irradiance had learned the daily pattern of sunrise and sunset, it had a better performance. Despite the very poor precision, it might be possible to use the prediction for a rough estimate. The means of the temperatures in the entire period are $y = 9.025$ and $\hat{y} = 9.198, \sigma(\hat{y}) = 3.617$ for actual and prediction respectively. This is even within the interval of two degrees used by DMI for verification and is definately useable for a rough estimate.

3.2   Comparison with Commercial Weather Forecast

Solar Irradiance. The prediction of solar irradiance of the 2013 dataset can be seen in Fig. 4 for 1-step ahead. Just like with the design reference year, the network is able to predict the precise values as well as learning the daily pattern.

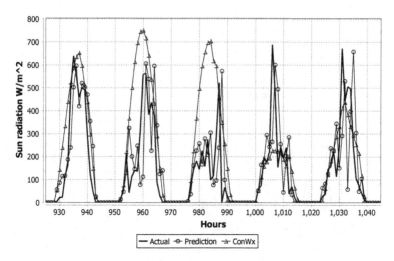

Fig. 4. The predictions of the artificial neural network and ConWX for 1-step ahead. The RMSEs are 85.8 W/m^2 and 148.4 W/m^2 respectively.

For the 25-step ahead prediction, only the daily pattern is learned. Compared to the reference year, the actual measurements are more fluctuating. This is most likely because the data measurements are from a single weather station, where as the reference year is comprised of multiple stations which will even the readings.

These fluctutations clearly have an influence on the precision of the 1-step ahead prediction. Because the network will be trained with this fluctation in the data at the next hour, its influence can then be seen at the network's output.

A zero-phase low pass filter was applied to the network output to smooth any fluctuations, but the overall RMSE dropped because of the filtering, and was therefore discarded.

The ConWX prediction only predicts the daily pattern of sunrise and sunset. The shape of the ConWX prediction is very round and shows only very little changes during the day. At best the ConWX prediction can be used as an estimate.

The 25-step ahead prediction in Fig. 5 shows once more that the closed loop operation influences the precision. Generally the same problem witnessed in experiment 1 is present here as well.

The network's prediction is not very precise, but it correctly predicts the daily pattern. Additionaly, it predicts some fluctuations during the day, but these fluctuations do not match with the actual data.

ConWX's prediction is almost identical to their 1-step prediction. The shape is again round with no fluctuation, but with a correct prediction of the daily pattern.

Fig 6 shows how the error of the prediction relates to the amount of step aheads predicted. ConWX's errors are pretty constant where as the network's predictions are much more dependent on the amount of step aheads that are

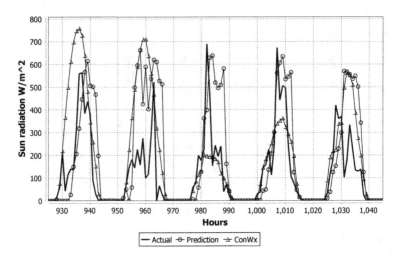

Fig. 5. The predictions for 25-step ahead. The prediction of the network is less precise, whereas the ConWX forecasts stays about the same as for 1-step, precisionwise. The RMSEs are 136.5 W/m^2 and 157.5 W/m^2 for the network prediction and ConWX forecasts respectively.

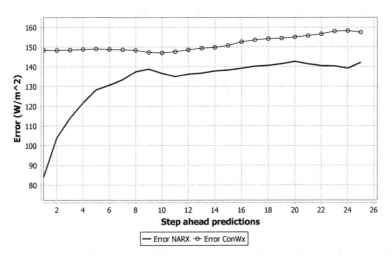

Fig. 6. The error of the network prediction increase drastically after the first few steps and then smooths out. Interestingly the error of the ConWX prediction is pretty constant through out.

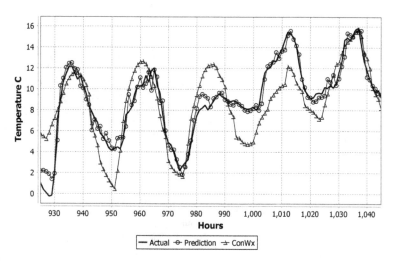

Fig. 7. Temperature prediction by the neural network and ConWX. The network's prediction is very precise and the ConWX prediction lacks behind. Both predictions have correctly predicted the daily pattern. The RMSEs are $0.95°C$ and $3.01°C$ of the network and ConWX respectively.

being predicted. However, overall the network is more precise in the sense of RMSE. This indicates that the approach is a good alternative to the ConWX numerical model.

Temperature. The prediction of temperature in the same 2013 dataset can be seen in Fig. 7 for 1-step ahead. Just like the reference year, the network is able to predict very precisely as confirmed with a RMSE of only 0.95. In this dataset, the daily patterns are much more visible which might help the network's prediction.

Just like the solar irradiance, the actual measurements are more fluctuating, but the network copes with this very well for the 1-step ahead prediction.

The ConWX prediction shows much of the same pattern as with the solar irradiance with a round shape. The day beginning at $t = 1000$ is predicted almost spot on, besides a negative bias.

Fig 8 shows the 25-step ahead prediction. The network's prediction is not very good, but it has correctly predicted the daily pattern. The more visible daily pattern in the actual data did help the network, but the fluctuations during the day are not matched. Because there is no other input to the network than previous values, the bad prediction can only be explained by the training period. The training period is 14 days (336 values), and if these 14 days are not a good generalization of the coming period, then the prediction will be bad. In this case there are several generalization measures that could be examined, but this is out of scope for this paper. One of such measures is training a base model with data from an entire year, which will be examined in future work.

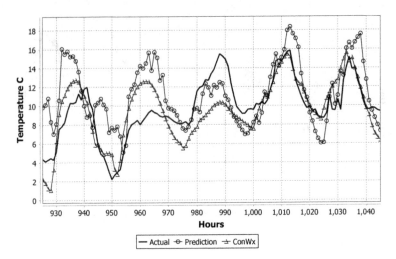

Fig. 8. 25-step ahead temperature prediction. Both predictions lack precision, but have predicted the daily pattern. RMSEs are $3.59°C$ for the network and $3.3°C$ ConWX.

The prediction by ConWX is pretty good, even though it does not contain the fluctuations during the day. The daily patterns are predicted very well for this cutout.

In Fig. 9 it can be seen how the error changes according to the amount of step aheads that is being predicted. As with the solar irradiance, the ConWX error is very stable. The network prediction climbs in the beginning and then flattens out. Compared to the solar irradiance, ConWX is the better choice for prediction beyond 5-steps ahead. At lower step aheads prediction, the network clearly outperforms ConWX.

Despite the fact that the network's prediction is not the optimal choice for longer step aheads prediction, it should still be considered an alternative. The time it takes the network to calculate these predictions gives the network a strong advantage for some cases. The network completes all calculations in a matter of minutes on a laptop from 2013. The laptop contains a 3 GHz Intel Core i7 processor, 8 GB memory and a SSD harddrive.

3.3 Comparison with Related Work

In [12], Smith et al. succeeds to achieve a MAE of $0.52°C$ for 1-step ahead prediction and $1.87°C$ for 12-step ahead prediction of temperature. It is not possible to compare these numbers directly, because they are calculated differently. With mean absolute error (MAE) negative and possitive errors will even out each other. With RMSE which is used in this paper, both negative and possitive errors will have an increasing effect on the RMSE.

In both the 1-step ahead and the 12-step ahead prediction, Smith et al. shows a better performance, $0.52°C$ vs $0.95°C$ and $1.87°C$ vs $3.93°C$ respectively.

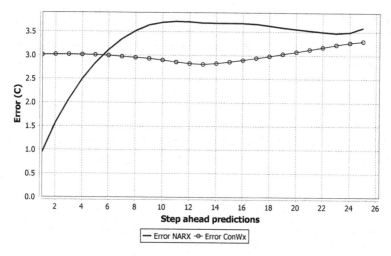

Fig. 9. The error compared to the amount of step ahead predicted. The ConWX error is pretty stable throughout the step ahead predictions. The network error increases fast in the beginning and then stays fairly constant.

This may be because of the different error measurement, but the difference is fairly small, especially for 1-step. They included many input variables such as previous measurements, solar irradiation, wind speed, rainfall, relative humidity and a fuzzy encoded time-of-day variable. These additional input variables would be interesting to test with the NARX network type as well.

Maqsood et al. [7] shows even better results for 24-step ahead prediction of temperature. In a matching period, their ensemble approach has a RMSE of only $0.0053°C$ vs $3.59°C$ in this paper. Their network ensemble gives a better result than the included networks individually. This indicates that the ensemble approach is definately worth investigating for the NARX network type. The main reason for their better performance is, that they only predict 24 steps ahead. This is very opposite of how the prediction is done in this paper, where the 25-step ahead prediction is calculated by running in a closed loop. In the closed loop operation, any error in the prediction will magnify for every iteration. A possible future ensemble approach could utilize this, and have 24 networks in the ensemble, one for each step ahead.

4   Conclusion

An approach for predicting weather parameters using a NARX artificial neural network has been proposed. The parameters of the network have been explained and the approach has been tested through two experiments. One of the experiments compared the results to a commercial weather forecast from ConWX that uses numerical models.

The parameters predicted were solar irradiance and outdoor temperature. The parameters were predicted from 1-step ahead through 25-step ahead. At lower step aheads the artificial neural network clearly outperformed the commercial forecast. The network was also better at higher step aheads for solar irradiance, but this was not the case for temperature predictions.

The network approach is a good alternative to the commercial forecast that takes four hours to complete, since it can complete its computation in minutes. This allows the artificial neural network to be used in higher resolutions or even embedded in a weather station. The network only requires previous measurements to work, and these may already be available in many cases.

References

1. ConWX: Conwx, http://www.conwx.com/ (accessed: January 5, 2015)
2. DMI: Dmi verification (in Danish),
   http://www.dmi.dk/vejr/til-lands/regionaludsigten/verifikation/
   (accessed: January 8, 2015)
3. Heaton, J.: Encog machine learning framework,
   http://www.heatonresearch.com/encog (accessed: December 23, 2014)
4. Hornik, K., Stinchcombe, M., White, H.: Multilayer feedforward networks are universal approximators. Neural Networks 2(5), 359–366 (1989),
   http://linkinghub.elsevier.com/retrieve/pii/0893608089900208
5. Lee, Y., Member, S., Cha, Y.T., Member, S., Park, H.: Short-term load forecasting using an artificial neural network. Transactions on Power Systems 7(1), 124–132 (1992)
6. Lund, H.: The Design Reference Year User Manual, A Report of Task 9: Solar Radiation and Pyranometer Studies (1995)
7. Maqsood, I., Khan, M., Abraham, A.: An ensemble of neural networks for weather forecasting. Neural Computing and Applications 13(2), 112–122 (2004),
   http://link.springer.com/10.1007/s00521-004-0413-4
8. Office, N.A.: Almanac for Computers 1990. U.S. Government Printing Office, 1990 (1989)
9. Powell, K.M., Sriprasad, A., Cole, W.J., Edgar, T.F.: Heating, cooling, and electrical load forecasting for a large-scale district energy system. Energy 74, 877–885 (2014), http://linkinghub.elsevier.com/retrieve/pii/S036054421400886X
10. Rumelhart, D.E., Hinton, G.E., Williams, R.J.: Learning representations by back-propagating errors. Cognitive Modeling (1988)
11. Shank, D.B., Hoogenboom, G., McClendon, R.: Dewpoint temperature prediction using artificial neural networks. Journal of Applied Meteorology and Climatology 47(6), 1757–1769 (2008)
12. Smith, B.A., Hoogenboom, G., McClendon, R.W.: Artificial neural networks for automated year-round temperature prediction. Computers and Electronics in Agriculture 68(1), 52–61 (2009)

On the Bias of the SIR Filter in Parameter Estimation of the Dynamics Process of State Space Models

Tiancheng Li[1,2], Sara Rodríguez[1], Javier Bajo[3],
Juan M. Corchado[1], and Shudong Sun[2]

[1] BISITE group, Faculty of Science, University of Salamanca, Salamanca 37008, Spain
[2] School of Mechanical Engineering, Northwestern Polytechnical University, 710072, China
[3] Department of Artificial Intelligence, Technical University of Madrid, 28660, Spain
{t.c.li,srg,corchado}@usal.es, jbajo@fi.upm.es, sdsun@nwpu.edu.cn

Abstract. As a popular nonlinear estimation tool, the sampling importance re-sampling (SIR) filter has been applied with the expectation–maximization (EM) principle, including the typical maximum a posteriori (MAP) estimation and maximum likelihood (ML) estimation, for estimating the parameters of the state space model (SSM). This paper concentrates on an inevitable bias existing in the EM-SIR filter for estimating the dynamics process of the SSM. It is analyzed that the root reason for the bias is the sample impoverishment caused by the resampling procedure employed in the filter. A process noise simulated for the particle propagation that is larger than the real noise involved with the true state will be helpful to counteract sample impoverishment, thereby providing better filtering result. Correspondingly, the EM-SIR filter tends to yield a biased (larger-than-the-truth) estimate of the process noise if it is unknown and needs to be estimated. The bias is elaborated via a straightforward roughening approach by means of both qualitative logical deduction and quantitative numerical simulation. However, it seems hard to fully remove this bias in practice.

Keywords: Particle filter, expectation–maximization, parameter estimation.

1 Introduction

Nonlinear state space models (SSMs) widely exist in the field of control and signal processing, often in which the observed data may be used to estimate the parameter(s) of the model (in the process of estimating the state) if it is unknown. The goal of parameter estimation is to compute an estimate of the true parameter that can provide best match to the observations [1, 2, 3], on a specific principle such as the most typical expectation–maximization (EM) estimation including maximum a posteriori (MAP) estimation and maximum likelihood (ML) estimation. The EM could be implemented in both batch and online manners [4]. This is also the main content of system identification for the state estimation, for which the Bayes filter provides a useful solution. A powerful approximation of the Bayes filter for nonlinear SSMs is based on random sampling, namely Sequential Monte Carlo, which is often known as the particle filter. A variety of particle filters have been developed in the last two decades.

© Springer International Publishing Switzerland 2015
S. Omatu et al. (eds.), *Distributed Computing & Artificial Intelligence, 12th Int. Conference*,
Advances in Intelligent Systems and Computing 373, DOI: 10.1007/978-3-319-19638-1_10

This paper concerns with estimating the state process noise by using the sampling importance resampling (SIR, also referred to as sequential importance sampling (SIS) and resampling, SISR) filter, which is the most commonly used type of particle filters and is the basis of most of the other types of particle filters. Based on SIS and resampling, the SIR filter and its further extension have been applied for various parameter estimation for the state space model since they relaxes linearity and Gaussian assumptions; see some very recent work e.g. [5-7].

However, unwanted bias has been observed in experiments e.g. [1, 3, 8, 9] which could not be ruled out completely; see also [10]. This bias was abstractly attributed to the sample degeneracy [9, 10, 11] of the PF simulation for on-line Bayesian estimation. That is, most of the particles will have negligible weights after a few iterations and, as a result, only few particles really work. The sample degeneracy is an inherent drawback of SIS and to combat it, resampling is often applied that is apparently replicating high-weighted particles to replace low-weighted particles. However, this often comes at the price of generating a large number of particles of the same state (as they are replicated from the same patent particles) and small-weighted particles are discarded, reducing particle diversity and causing sample impoverishment. The more serious the sample degeneracy, the more serious the sample impoverishment after resampling; see further explanation given in [12, 13]. It is fairer to attribute the bias of the SIR filter for parameter estimation to sample impoverishment rather than degeneracy. This is a critical problem for the SIR filter's application for parameter estimation, in addition to the state estimation, but a specific discussion on this problem seems still missing.

In this paper, we clarify the problem from a new perspective based on an in-depth analysis of the side-effect of resampling, with particularly regards to parameter estimation of the dynamics noise of the SSM. It is formally demonstrated that the EM parameter estimation of the state process noise implemented by the SIR filter [3, 8, 9, 14, 15, 16] suffer from a bias, i.e. the estimation of the process noise given by the EM-SIR filter is larger, statistically, than the truth. It is not the intention of this paper to go into details of any specific EM estimator or even the content of the SIR filter but instead, we focus on the statistical property of the SIR filter.

The rest of the paper is organized as follows. Section 2 describes the qualitative logical deduction model and Section 3 presents the quantitative simulation evidences model. The simulation results and discussion are shown in Section 4. Finally, Section 5 concludes the paper.

2   Problem Statement

In the context of the general non-linear SSM, it is often of great interest and significance to estimate the parameter(s) of the model that can be time-varying during the process of estimating the state. The SSM consists of two recursive equations as shown in the following:

$$x_k = f_{k|k-1}\left(x_{k-1}, \theta_k\right) \quad \text{(state transition equation)} \tag{1}$$

$$y_k = g_{k|k}\left(x_k, \beta_k\right) \quad \text{(observation equation)} \tag{2}$$

where k indicates time, x_k denotes the state, y_k denotes the observation, θ_k and β_k denote noises affecting the state Markov process equation $f_{k|k-1}(\cdot)$, and observation equation $g_{k|k}(\cdot)$, respectively and this paper focuses particularly on estimating the static process noise θ_k by the SIR filter, i.e. for $t=1, 2,..., k$, $\theta_t = \theta^*$, where θ^* is a static parameter.

Available solutions for the parameter estimation can be classified into two groups: MAP estimation (also called Bayesian estimation or approximate Bayesian computation [5]) and ML estimation. There are several well-established MAP estimation filters for sequentially learning both x_t and θ, such as [1, 6, 17, 18], see the discussion and comparison in [19]. Bias issue has been pointed out in the online MAP estimation and is attributed to the degeneracy problem of the particle filter [9-11]. On the contrary, a different group of work [3, 5-7, 14] particularly focuses on the ML estimation and points out that the bias issue also exists in ML parameter estimation by using the SIR filter. It is demonstrated that this bias is independent of the Bayesian recursion path but instead it can occur at the very first step.

Before we go into the logical deduction and simulation demonstration of the bias, the basic content of ML and MAP estimation is provided in the following context for clarification. Assuming that true noise θ^* is involved with the state, generating observations $y_{1:k}$ and its value is unknown, our goal is to compute point estimates of θ^* from the observations. In general ML principles, the estimate of θ^* is the maximizing argument of the marginal likelihood of the observed data, which can be termed as:

$$\hat{\theta} = \arg \max_{\theta \in \Omega} p\left(y_{1:k} | \theta\right) \tag{3}$$

where Ω is a specified parameter space, $y_{1:k} \triangleq (y_1, y_2, ..., y_k)$ denotes the history path of the observation process. For the detail of various implementations of the ML estimation, the reader is refer to [3, 8, 9, 14] and the references therein.

Given a prior distribution of θ^*, the MAP estimate is

$$\hat{\theta} = \arg \max_{\theta \in \Omega} \frac{p\left(y_{1:k} | \theta\right) g(\theta)}{\int_{\vartheta} p\left(y_{1:k} | \vartheta\right) g(\vartheta)} = \arg \max_{\theta \in \Omega} p\left(y_{1:k} | \theta\right) g(\theta) \tag{4}$$

As shown, the difference between MAP and the ML estimates is the use of a prior distribution. However, the choice of prior will heavily influence quality of result of MAP as well as the Bayes posterior.

3   Logical Deduction

3.1   EM Estimation of the State Dynamics Noise

Remark 1. $\forall$ two parameters θ_1 and θ_2 that are close-enough to each other in the monotonic domain of Eq. (3): if a better particle approximation of the posterior is obtained by using θ_1 as the particle propagation noise parameter in the SIR filter than by using θ_2, then the estimate of the true state dynamics noise parameter θ^* obtained by the SIR filter will be more likely closer to θ_1 than to θ_2, i.e. parameters θ_1 is more likely to match the real observations than θ_2 in the sense of likelihood or the posterior

$$p\left(y_{1:k}\,|\theta_1\right) > p\left(y_{1:k}\,|\theta_2\right), \quad \text{or}$$
$$p\left(y_{1:k}\,|\theta_1\right)g(\theta) > p\left(y_{1:k}\,|\theta_2\right)g(\theta) \tag{5}$$

This Remark is just for the content of the ML or MAP principle and proof seems unnecessary here. By saying a better particle approximation of the posterior, it means that the underlying particle approximation of the state is closer to the real state distribution and is therefore more likely to match the real observation. θ_1 and θ_2 are limited to be close enough with each other so as to eliminate any local maximum point between them, for monotonicity.

3.2   Direct Roughening

A critical step for the SIR filter is resampling [13] that is designed to reduce the weight variance. As a result of this, many particles may constitute the very similar or the same state i.e. they are replications of the same particle, leading to the so-called sample impoverishment problem. To counteract this problem, one effective solution is to spread the replicated particles by introducing additional noise, namely the roughening. This can be realized in two equivalent ways. One way is to increase the dynamics noise for particle propagation directly which is known as direct roughening [20], and the other way is to apply roughening separately after resampling similar to the separate roughening scheme proposed in [21]; with similar idea called the 'move step' in [22].

In contrast to the state dynamics given in (1), the Markov process (called propagation) of the ith particle that is perturbed by a roughening noise r in the direct roughening approach, i.e. the proposal function, can be written as:

$$x_k^{(i)} = f_{k|k-1}\left(x_{k-1}^{(i)}, \theta^* + r\right) \tag{6}$$

where θ^* is the dynamics noise involved with the state and the roughening noise r is normally a zero-mean Gaussian $N(0, \Sigma r)$ distribution. In the case of sample impoverishment, the direct roughening helps to improve the approximation quality of the posterior by spreading particles in the state space. It is vital to note that, over roughening (too significant r) however will lead to very dispersive distribution of particles and will conversely reduce the estimation accuracy. As long as the SIR filter suffers from sample impoverishment, we have

Remark 2. $\forall \theta$ that is 'slightly' larger than θ^*: the SIR filter that uses θ as the particle propagation noise will obtain better approximation of the posterior than uses θ^*.

This Remark is no more than a re-statement of the validity of the direct roughening strategy. It is worth nothing that θ is limited to be 'slightly' larger than θ^* to eliminate any local peak between them, for monotonicity.

The theory suggests that the particle filter benefits from a sampling proposal function that that has a 'heavier' tail so that the filter is sensitive to the outliers [12]. This also indicates that the SIR filter can benefit from a comparably large state dynamics noise. Combining Remark 1 and 2, we can arrive at the assertion that the SIR filter will tend to yield a larger-than-the-truth estimate of the state dynamics noise in the forward-only filtering when sample impoverishment occurs. In fact, existing experiments e.g. [1, 8, 9] have observed the bias but attributed the reason to the sample degeneracy without providing clear explanation. Furthermore, the effect of roughening has been verified in e.g. [21, 22]. In the following, further simulations are provided to demonstrate the bias of the EM-SIR filter for estimating the state Markov process noise.

4   Simulation Demonstration

As indicated by remark 1 and 2, a sufficient condition for the occurrence of the bias of the EM-SIR filter in estimation of the dynamics noise is that the filter benefits from the direct roughening approach, which will be demonstrated quantitatively below. Without the loss of generality, we consider estimating the static dynamics noise in a classical 1-dimensional SSM as follows:

$$x_k = 0.5x_{k-1} + \frac{25x_{k-1}}{\left(1+x_{k-1}^2\right)} + 8\cos\left(1.2\left(k-1\right)\right) + \theta \qquad (7)$$

$$y_k = 0.05x_k^2 + \beta \qquad (8)$$

where Gaussian noise $\theta \sim N(0, Q)$, $\beta \sim N(0, 1)$, Q is the unknown variance of the zero-mean Gaussian Markov process noise to be estimated. Without the loss of generality, the state x_k evolves with the dynamics noise with variance $Q^*=1$.

In order to evaluate the estimation accuracy, the RMSE (root mean square error) is used for evaluation, which is defined as follows:

$$\text{RMSE}=\left(\frac{1}{T}\sum_{k=1}^{T}\left(x_k - \hat{x}_k\right)^2\right)^{1/2} \qquad (9)$$

where $\hat{x}_k$ is the estimate of the state x_k which is the mean state of all particles. Nevertheless, the RMSE is unavailable in practice since the true state x_k is unknown. Hence, the RMSD (root mean square discrepancy) between the estimated observation $\hat{y}_k$ and the real observation y_k are defined as a measurement of the likelihood for batch parameter estimation as:

$$\text{RMSD}=\left(\frac{1}{T}\sum_{k=1}^{T}\left(y_k - \hat{y}_k\right)^2\right)^{1/2} \qquad (10)$$

where $\hat{y}_k = 0.05\hat{x}_k$.

In order to capture the average performance, the simulation length is set to T=1000 steps and each simulation runs 500 trials.

In the first simulation, four bootstrap SIR filters are designed that apply Q=1, 1.1, 1.2 and 1.5 respectively for particle propagation. Q=Q^*=1 is the basic SIR filter and Q=1.1, 1.2 and 1.5 are roughening-enhanced. Their average RMSEs are given in Fig.1, which show that roughening-enhanced SIR filters perform better than the basic SIR filters. Especially when the number of particles is small (e.g. between 20~60), the sample impoverishment is getting more serious and therefore roughening is more helpful. Just as the impoverishment is case specific, the effectiveness of the roughening approach for the SIR filter also varies on case by case basis see the discussion given in [20] therein, which is a multi-dimensional SSM. This indicates that the bias of the SIR filter is also case specific.

In the second simulation, the SIR filter uses different parameters Q (from 0.5 to 4 with interval 0.1) and the same 50 particles. The average RMSE and RMSD results are plotted in Fig.2, which provide more details of the bias of the SIR filter in terms of estimating Q. As indicated, the RMSE result compared with the red dotted line, the SIR filter benefits from a state dynamics noise that is larger, but not too much to prevent overshooting, than that involves with the true state.

It is worth noting that, RSMD is not monotonically proportional with RMSE in the whole domain but instead, the larger the Q used for the particle propagation, the smaller the RMSD. Now that RMSD can be used as a measure for forward-only offline/batch estimation of Q^*, while the maximum likelihood estimate of Q shall constitute a value that is larger than the real Q^*=1 in the most basic sense. This directly demonstrates that the SIR filter can yield a larger-than-the-truth estimate of (the variance of) the state dynamics noise, although here we did not specify the gradient used to search for the optimal parameter.

As stated, more experimental evidences of online parameter estimation can be found in e.g. [1, 8, 9] which are consistent with our assertion. As the importance sampling theory suggest, the simulation result can be interpreted as that the particle filter benefits from a sampling proposal with a heavy tail that is insensitive to the outliers. This in turn can be taken as that a smoother distribution enjoying better particle diversity to alleviate the sample impoverishment. Based on this, we conclude that the sample impoverishment is the primary cause of the bias of the SIR-like filter in parameter estimation of the SSM. Therefore, attentions shall be paid to the use of the SIR-like filter for estimating the dynamics parameters of the SSM, which includes not only the process noise but also the state transition function.

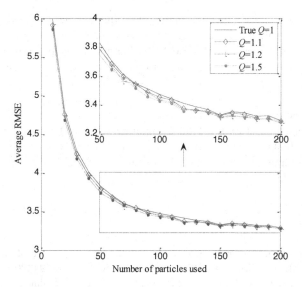

Fig. 1. RMSE against different number of particles

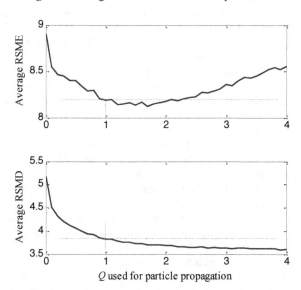

Fig. 2. RMSE and RMSD against different Q used in the filter

5 Conclusion

Based on the findings that a comparably large noise used for the particle propagation helps to alleviate the sample impoverishment caused by resampling and will therefore produce better approximation of the posterior, we assert that the EM-SIR filter tends to yield a biased (larger than the truth) estimate of the state dynamics noise. This has been elaborated by means of both qualitative deduction and quantitative simulations. However, the sample degeneracy and impoverishment can hardly be ruled out completely and so is the bias of the EM-SIR filter for the estimation of the dynamics noise. Therefore, one must be careful with the use of the SIR filter for parameter estimation. More concerns on the use and misuse of Bayes filters can be found in [23].

Acknowledgments. This work has been partly supported by the project Sociedades Humano-Agente en entornos Cloud Computing (Soha+C) SA213U13. Project cofinanced with Junta Castilla y León funds. Tiancheng Li's work is supported by the Excellent Doctorate Foundation of Northwestern Polytechnical University and the Postdoctoral Fellowship of the University of Salamanca.

References

1. Storvik, G.: Particle filter for state-space models with the presence of unknown static parameters. IEEE Transactions on Signal Processing 50(2), 281–289 (2002)
2. Schön, T.B., Wills, A., Ninness, System, B.: identification of nonlinear state-space models. Automatica 47(1), 39–49 (2011)
3. Doucet, A., Tadic, V.B.: Parameter estimation in general state-space models using particle methods. Annals of the Institute of Statistical Mathematics 55, 409–422 (2003)
4. Yildirim, S., Jiang, L., Singh, S.S., Dean, T.A.: Calibrating the Gaussian multi-target tracking model. Statistics and Computing, 1–14 (2014)
5. Dean, T.A., Singh, S.S., Jasra, A., Peters, G.W.: Parameter estimation for hidden Markov models with intractable likelihoods. Scand. J. Statist. 41, 970–987 (2014)
6. Lundquist, C., Karlsson, R., Ozkan, E., Gustafsson, F.: Tire radii estimation using a marginalized particle filter. IEEE Transactions on Intelligent Transportation Systems 15(2), 663–672 (2014)
7. Zikmundová, M., Helisová, K.S., Beneš, V.: On the use of particle Markov Chain Monte Carlo in parameter estimation of space-time interacting discs. Methodology and Computing in Applied Probability 16(2), 451–463 (2014)
8. Andrieu, C., Doucet, A., Tadic, V.B.: Online parameter estimation in general state-space models. In: Proc. IEEE CDC/ECC, Seville, Spain, pp. 332–337 (2005)
9. Poyiadjis, G., Doucet, A., Singh, S.S.: Parameter approximations of the score and observed information matrix in state space models with application to parameter estimation. Biometrika 98, 65–80 (2011)
10. Kantas, N., Doucet, A., Singh, S.S., Maciejowski, J.M., Chopin, N.: On particle methods for parameter estimation in state-space models. Statistical Science (2015)
11. Olsson, J., Cappe, O., Douc, R., Moulines, E.: SMC Smoothing with application to parameter estimation in nonlinear state-space models. Bernoulli 14(1), 155–179 (2008)
12. Li, T., Sun, S., Sattar, T.P., Corchado, J.M.: Fight sample degeneracy and impoverishment in particle filters: A review of intelligent approaches. Expert Systems with Applications 41(8), 3944–3954 (2014)
13. Li, T., Bolic, M., Djuric, P.: Resampling methods for particle filtering, IEEE Signal Processing Magazine (to appear, May 2015), doi:10.1109/MSP.2014.2330626
14. Poyiadjis, G., Doucet, A., Singh, S.S.: Maximum likelihood parameter estimation in general state-space models using particle methods. In: Proc. American Stat. Assoc. (2005)

15. Hong, M., Bugallo, M.F., Djurić, P.M.: Joint model selection and parameter estimation by population Monte Carlo simulation. Journal of Selected Topics in Signal Processing 4(3), 526–539 (2010)
16. Míguez, J., Crisan, D., Djurić, P.M.: On the use of sequential Monte Carlo methods for maximum a posteriori sequence estimation and stochastic global optimization. Statistics and Computing 23(1), 91–107 (2013)
17. Liu, J., West, M.: Combined parameters and state estimation in simulation based filtering. In: Doucet, A., de Freitas, N., Gordon, N. (eds.) Sequential Monte Carlo Methods in Practice, pp. 197–223. Springer, New York (2001)
18. Carvalho, C.M., Johannes, M., Lopes, H.F., Polson, N.: Particle learning and smoothing. Statistical Science 25(1), 88–106 (2010)
19. Lopes, H.F., Tsay, R.S.: Particle filters and Bayesian inference in financial econometrics. J. Forecasting 30, 168–209 (2011)
20. Li, T., Sattar, T.P., Han, Q., Sun, S.: Roughening methods to prevent sample impoverishment in the particle PHD filter. In: 16th International Conference on Information Fusion, Istanbul, Turkey, July 10-12 (2013)
21. Gordon, N., Salmond, D., Smith, A.: Novel approach to nonlinear/non-Gaussian Bayesian state estimation. IEE Proc. F Radar Signal Processing 140(2), 107–113 (1993)
22. Fulop, A., Li, J.: Efficient Learning via Simulation: A marginalized resample-move approach. Journal of Econometrics 176, 146–161 (2013)
23. Li, T., Corchado, J.M., Bajo, J., Sun, S., De Paz, J.F.: Do we always need a filter, arXiv:1408.4636 [stat.AP]

Hierarchical Multi-label Classification Problems: An LCS Approach

Luiz Melo Romão[1] and Julio César Nievola[2]

[1] Universidade da Região de Joinville, Departamento de Informática, Joinville, Brasil
`luizmromao@gmail.com`
[2] Pontifícia Universidade Católica do Paraná, PPGIA, Curitiba, Brasil
`nievola@ppgia.pucpr.br`

Abstract. Traditional classification tasks deal with assigning instances to a single label. However, some real world databases classes are structured in a hierarchy and its instances can have their classes associated with two or more paths in the hierarchical structure. In this case, such situations are referred as hierarchical multi-label classification problems. The purpose of this paper is to explore the concept of hierarchical multi-label classification problems and present a solution based on Learning Classifier Systems (LCS) to solve this kind of problem. The Hierarchical Learning Classifier System Multi-label (HLCS-Multi) proposed, presents a comprehensive solution to hierarchical multi-label classification problems building a global classifier to predict all classes in the application domain.

Keywords: Hierarchical Multi-label Classification Problems, Learning Classifier Systems, Protein Function.

1 Introduction

According to [1], traditional computational approach to automated classification assumes that each object should be assigned to only one out of two or more classes. However, some real world applications digress from this generic scenario in two important ways. First, each example can belong to several classes simultaneously (multi-label classification). Second, the classes can be hierarchically ordered in the sense that some are more specific versions of others (hierarchical classification). In this case, such situations are referred as hierarchical multi-label classification problems.

The task has recently received considerable attention, databases in various fields, including text categorization, web content searches, image annotation, digital libraries, or functional genomics (focus this work), are known to be organized as hierarchies.

The purpose of this paper is to explore the concept of hierarchical multi-label classification problems and present a solution based on Learning Classifier Systems (LCS) to solve this kind of problem. Conceived in 1975 by John Holland [2], the Learning Classifier System (LCS) consists of a set of rules called classifiers. As defined in [3], the LCS develops a model of intelligent decision-making,

© Springer International Publishing Switzerland 2015
S. Omatu et al. (eds.), *Distributed Computing & Artificial Intelligence, 12th Int. Conference,*
Advances in Intelligent Systems and Computing 373, DOI: 10.1007/978-3-319-19638-1_11

using two biological metaphors, evolution and learning, where learning guides the evolutionary component to move in the direction of the best rules.

The proposed approach, called HLCS-Multi (Hierarchical Learning Classifier System Multi-label) will be used in this work for predicting protein functions.

The remainder of this paper is organized as follows: Section 2 discusses the hierarchical multi-label classification concept and how to distinguish hierarchical problems. The HLCS-Multi architecture is described in Section 3. Section 4 demonstrates the computational results achieved and Section 5 presents the conclusions of this study and the possible directions for future research.

2   Hierarchical Multi-label Classification

According to [4], traditional classification tasks deal with assigning instances to a single label. In multi-label classification, the task is to find the set of labels that an instance can belong to rather than assigning a single label to a given instance. For example, a medical patient may suffer from more than one health condition: diabetes, high blood pressure, high cholesterol.

Hierarchical classification is a variant a traditional classification where the task is to assign instances to a set of labels where the labels are related through a hierarchical classification scheme. In this case, when an instance is labeled with a certain class, it should also be labeled with all of its superclasses.

The hierarchical classification problems can be differentiated according to the organization of its structure (tree or DAG), and according to type of algorithmic approach used (Local or Global).

In the local approach a model trains a binary classifier for each node of the class hierarchy. In this case, it is necessary to use N independent local classifiers, one for each class except the root node. Therefore, the number of classifiers to be trained can be very large in situations where there are many classes. Moreover, in using the local approach, the technique can provide inconsistent results, because there is no guarantee that the class hierarchy will be respected.

In the global approach, a single classification model is built from the training set, taking into account the hierarchy of classes as a whole during a single execution of the classifier algorithm. In the global approach, the fact that the algorithm maintains hierarchical relationships between classes during the phases of training and testing makes the outcome of the prediction easier to understand.

Therefore, when a problem is said hierarchical multi-label besides classes are structured in a hierarchy, instances can have their classes associated with two or more paths in the hierarchical structure. Thus when a solution is said hierarchical multi-label indicates that the algorithm could potentially predict multiple data paths in the hierarchy of classes.

3   HLCS-Multi

The Hierarchical Learning Classifier System Multi-label (HLCS-Multi) algorithm proposed in this paper uses as a development model the Learning Classifier

System (LCS). Solutions to hierarchical problems using LCS are already being exploited in other algorithm previously proposed by the authors. In [5] is presented the HLCS-DAG, which also uses a global approach with the capacity to work with databases structured in DAGs.

The HLCS-Multi is an evolution of these work and presents a comprehensive solution to hierarchical multi-label classification problems building a global classifier to predict all classes in the application domain.

In order to work with the class hierarchy, the HLCS-Multi presents a specific component for this task which is the evaluation component of the classifiers. This component has the task of analyzing the predictions of classifiers considering the class hierarchy. In addition to this, the HLCS-Multi architecture consists of the following modules: population of classifiers, performance component, credit assignment component and GA component, which interacts internally. The details of the main differences of HLCS-Multi components, in relation other version follow below.

The HLCS-Multi starts its execution by analyzing the data hierarchy at the training base. Reading of data is performed at the training base of dimension m. This base consists of samples represented by $B_{train} = [v_1, v_2, v_3, ..., v_N]$, where N is the total number of instance. Each instance of B_{train} characterized by attributes, $v_i = [a_1, a_2, ..., a_j, a_{jk}]$, wherein a_j is the attribute j-th, $1 \leq j$ and a_{jk} represents the attribute class. As this is a multi-label problem the attribute a_{jk} can be formed by one or more classes.

Reading data is also performed the learning process of the hierarchy of classes H which will be used by the algorithm HLCS-Multi, represented in the database as follows: $H = (root/des_1, root/des_2, root/des_x, des_1/des_3, des_x des_y)$ where, $root$ represents the root node, des represents the descendent node, A/B indicates that A is the parent of B, x and y the amount of relations A/B.

Following we have the decomposition of the instances at the training base. This process makes the multi-label instances are transformed into a set of simple-labels instances as shown in Table 1.

Table 1. Decomposition Multi-label

Instance	$Attrib_1$	...	$Attrib_n$	Class
1	1	...	3	GO0003674 @ GO0005624
1.1	1	...	3	GO0003674
1.2	1	...	3	GO0005624

In the example, the instance 1 is composed of two classes, after decomposition this instance is replaced by two new instances (1.1, 1.2). In this case, the attribute values of the instances simple labels generated are kept equal to the attributes of multi-label instances, ensuring hierarchical knowledge.

With the new set of training defined starts creating the initial population of classifiers. In the HLCS-Multi, the size of population ($SizePop$) is determined in the initial settings by a percentage ($Perc\_Pop$) in relation to number of instances in the training base ($Total\_Instance$), according to Equation 1.

$$SizePop = Total\_Instance * Perc\_Pop/100 \tag{1}$$

In the initial population only exclusive classifiers are added, the set of all classifiers forms the prediction model. Each classifier C_i ($0 < i \leq SizePop$) of the HLCS comprises: a t set of conditions (where t=number of attributes of an training instance), the class value and the classifier quality measure, according to Equation 2.

$$C_i = (([Cond_0]and[Cond_t])(VClass)(Q_{classifier})) \tag{2}$$

Each condition has three parameters: OP, VL, A/I, where: OP: operator relation (= or !=), VL: condition value and A/I: the choice of an active or inactive attribute, which determines whether the condition will be used in the classifier or not.

In order to form each classifier, the HLCS-Multi randomly chooses an instance $Instance_i$ ($0 < i \leq N$) of the training base as a model. For each attribute of an training instance, a condition in the classifier is created. At the beginning, the conditions start with the operator relation (OP) "=". The condition value (VL) receives the value attribute of the training instance and whether the condition will be active (A) or inactive (I) is randomly determined.

The HLCS-Multi runs only on databases with nominal attributes, in the case of databases with continuous attributes, it is necessary to use some method of discretization.

Each classifier created, presents in its structure, the definition of a rule (IF-THEN) that will be used for prediction. After the definition of conditions of the classifier is necessary to determine the class of prediction that is assigned to the classifier. For this, the classifier is analyzed with all instances of the training base. Each instance covered by the classifier has scored his class. After the analysis, the class scored more times is then chosen as the class of the classifier.

The last step in the creation of the initial population of classifiers is defining the quality of the classifier. To calculate the quality of the classifier two factors are considered: the percentage of positive classes predicted and the hierarchical control evaluation of the classifier.

The hierarchical control evaluation represents the predictive ability of the classifier, considering not only the class in question, but all the class antecedents in the hierarchy. This process is performed by the evaluation component which is essential for the HLCS-Multi to solve problems with hierarchical structures.

The steps to calculate the quality of the classifier and the functions of evoluation component, of component performance, of credit assignment component and of GA component, follow the same definitions shown in [5].

After the step of credit assignment a new population of classifiers is generated. This population, called the final population, constitutes the final model of prediction of HLCS-Multi. It contains the winners of each competition who predicted correctly or partially correct the class of instance chosen at the beginning of the competition. This final population does not have a number of pre-defined classifiers, this value will depend on the power of learning model.

To define the number of executions of learning, each classifier inserted into the final population is compared with all instances of the training base. All instances that are covered by the classifier and the real class of the instance is correct or partially correct with the predicted class classifier, are excluded from the training base. After this, the learning process is restarted until a minimum percentage of instances are covered by the classifiers of final population.

This process maintains consistency with respect to classifiers sent to the final population and the hierarchical characteristic of instances of the training base.

Every classifier inserted into the final population is excluded from the initial population. When this occurs, a new classifier is generated and included in the initial population to keep the size of the population.

4 Computational Results

In order to demonstrate the results obtained with the HLCS-Multi, the algorithm was compared with another version called HLCS-DAG proposed by the authors in [5]. The main difference between these versions is that HLCS-DAG does not present a complete solution for databases multi-label. The results of the HLCS-Multi was also compared with the Clus-HMC approach proposed in [6].

Table 2. Summary of the data sets used in our experiments. The ('data set') gives the data set name, the ('training') gives the number of training examples, the ('test') gives the number of test examples, the ('attributes') gives the number of attributes and the ('classes') gives the number of classes in the class hierarchy.

Data Set	Training	Test	Attributes	Classes
Cellcycle	2473	1278	77	4126
Church	1627	1278	27	4126
Derisi	2447	1272	63	4120
Expr	2485	1288	551	4132
Pheno	1005	581	69	3128
Spo	2434	1263	80	4120

We have selected six bioinformatics data sets from [6] described in the Gene Ontology (GO) and structured in DAG. The different data sets describe different aspects of the genes in the yeast genome. They include five types of bioinformatic data: sequence statistics, phenotype, secondary structure, homology, and expression. For the evaluations in this paper, the algorithms were executed using the training and test sets, retaining the same examples available at (http://dtai.cs.kuleuven.be/clus/hmcdatasets). Table 2 presents the details of the bases used in this experiment.

In order to evaluate the algorithms we have used the metrics of hierarchical precision (hP), hierarchical recall (hR) and hierarchical F-measure proposed by [7]. These measures are, in fact, extended versions of the known measures like precision, recall and F-measure, tailored to the scenario of hierarchical classification.

According to the results in the first test, the HLCS-Multi showed significantly better results than HLCS-DAG in all bases analyzed. GO bases are very complex, some examples have up to 22 classes and the total classes exceeds by more than 4100 at most bases. However, the HLCS-Multi through its hierarchical and multi-label solutions, proved to be more suitable for this type of problem. The results of the comparison between the model HLCS-Multi and HLCS-DAG are shown in Table 3.

Table 3. Hierarchical measures of precision (hP), recall (hR) and F-measure (hF) values (mean ± standard deviation) calculated over 10 runs. An entry in the 'hF' column is shown in bold if the hierarchical F-measure value obtained by one of the methods was significantly greater than the other method - according to a Wilcoxon test with 95% confidence.

	hP	hR	hF
HLCS-DAG			
Cellcyle	0.2207 ± 0.03	0.1463 ± 0.04	0.1759 ± 0.03
Church	0.2848 ± 0.05	0.1469 ± 0.06	0.1939 ± 0.04
Derisi	0.2208 ± 0.02	0.1029 ± 0.03	0.1404 ± 0.02
Expr	0.1564 ± 0.04	0.0889 ± 0.03	0.1134 ± 0.03
Pheno	0.2303 ± 0.05	0.0730 ± 0.04	0.1109 ± 0.05
Spo	0.1789 ± 0.03	0.1645 ± 0.03	0.1714 ± 0.04
HLCS-Multi			
Cellcyle	0.2611 ± 0.04	0.4209 ± 0.02	**0.3223 ± 0.03**
Church	0.2259 ± 0.06	0.6183 ± 0.03	**0.3309 ± 0.04**
Derisi	0.2948 ± 0.03	0.5655 ± 0.03	**0.3873 ± 0.03**
Expr	0.3254 ± 0.05	0.3568 ± 0.03	**0.3109 ± 0.04**
Pheno	0.2962 ± 0.04	0.7444 ± 0.04	**0.4281 ± 0.05**
Spo	0.2975 ± 0.03	0.3992 ± 0.04	**0.3410 ± 0.04**

The second test was conducted against Clus-HMC algorithm. The Clus-HMC consists of a hierarchical classification model based on decision tree method. In the test performed, the data evaluation had adverse effect by fact that the authors of it, publish and make available through the algorithm only results based in binary measure. As seen above, the HLCS-Multi is a hierarchical multi-label global algorithm, which promotes their classifiers that have the ability to predict at least some kind of antecedent class from the real class.

This fact makes the HLCS-Multi have a low percentage accuracy when compared to traditional binary evaluation methods, precision and recall. However, taking into account the hierarchy, HLCS-Multi presents some satisfactory results.

However, to evaluate the HLCS-Multi against Clus-HMC, tests were performed using the GO database. With precision and recall obtained by Clus-HMC using standard threshold settings and run as material provided by the authors, the graphics of the PR-curve for all databases were designed. The PR-curve uses the relation between the precision measure plotted on the Y axis, and recall measure plotted on the X axis.

As the precision and recall values of HLCS-Multi do not change much with different parameter settings, the results of the HLCS-Multi were marked as a point on the graph. Thus, as the comparison is being made between curve and point, you can define a better performance HLCS-Multi when the curve is below the point and better performance Clus-HMC when the curve is above the point. Also, another factor that can be considered is the fact that, these bases, the number of negative examples for each class far outweighs the number of positive examples. This suggests that, in this case, the value obtained by the recall measure is more significant than the precision.

Thus, as shown in Figure 1, the HLCS-Multi algorithm had recall values above 0.4 points and also in some bases a favorable position of the point relative to Clus-HMC curve.

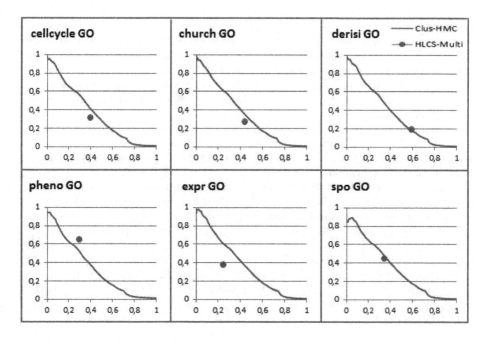

Fig. 1. Grfico Preciso Revocao - Bases GO

5   Conclusions

The paper introduced a new global model approach for hierarchical multi-label classification problems using LCS and applied it to the classification of biological dataset. The proposed HLCS-Multi unveils a global classification model in the form of an ordered list of IF-THEN classification rules which can predict terms at all levels of the hierarchy, satisfying the parent-child relationships between terms. The advantage of HLCS-Multi in contrast to other approaches is their adaptability. Based on the LCS model, the HLCS-Multi makes constant iterations of environmental samples to create their classification rules, making it a more flexible classification model.

The purpose of this paper was to show a underexploited topic, but with a lot of applications in the world. The LCS has been used with great success in several areas like robotics, environment navigation, function approximation and others. However, the topic of this work, hierarchical multi-label classification problems like protein function, has not been approached by neither proposal based on LCS.

The computational results with HLCS-Multi prove that the use of LCS models can be an alternative to the hierarchical multi-label. As future research, we intend to evaluate this method on a larger number of datasets and compare it against other global hierarchical classification approaches.

References

1. Vateekul, P.: Hierarchical Multi-Label Classification: Going Beyond Generalization Trees. Open Access Dissertations, Paper 723 (2012)
2. Holland, J.H.: Adaptation in Natural and Artificial Systems: An Introductory Analysis with Applications to Biology, Control and Artificial Intelligence. MIT Press, Cambridge (1992)
3. Urbanowicz, R.J., Moore, J.H.: Learning classifier systems: a complete introduction, review, and roadmap. Journal Artif. Evol. App., 1:1–1:25 (2009)
4. Alaydie, N., Reddy, C.K., Fotouhi, F.: Exploiting Label Dependency for Hierarchical Multi-label Classification. In: Tan, P.-N., Chawla, S., Ho, C.K., Bailey, J. (eds.) PAKDD 2012, Part I. LNCS, vol. 7301, pp. 294–305. Springer, Heidelberg (2012)
5. Romão, L.M., Nievola, J.C.: Hierarchical Classification of Gene Ontology with Learning Classifier Systems. In: Pavón, J., Duque-Méndez, N.D., Fuentes-Fernández, R. (eds.) IBERAMIA 2012. LNCS, vol. 7637, pp. 120–129. Springer, Heidelberg (2012)
6. Vens, C., Struyf, J., Schietgat, L., Džeroski, S., Blockeel, H.: Decision trees for hierarchical multi-label classification. Mach. Learn. 73(2), 185–214 (2008)
7. Kiritchenko, S., Matwin, S., Fazel, A.F.: Functional Annotation of Genes Using Hierarchical Text Categorization. In: Proceedings of BioLINK SIG: Linking Literature, Information and Knowledge for Biology (2005)

Analysis of Web Objects Distribution

Manuel Gómez Zotano[1], Jorge Gómez Sanz[2], and Juan Pavón[2]

[1] Corporación Radio Televisión Española,
Alcalde Sainz de Baranda 92, 28007 Madrid, Spain
manuel.gomez@rtve.es

[2] Universidad Complutense de Madrid, Facultad de Informática, 28040 Madrid, Spain
{jjgomez,jpavon}@fdi.ucm.es

Abstract. Understanding how the web objects of a website are demanded is relevant for the design and implementation of techniques that assure a good quality of service. Several authors have studied generic profiles for web access, concluding that they resemble a Zipf distribution, but further evidences were missing. This paper contributes with additional empirical evidences that confirm that a Zipf distribution is present in different domains and that its form has changed from past studies. More specifically, the α parameter has become higher than one, as a consequence that the popularity factor has become more critical than before. This analysis also considers the impact of web technologies on the characterization of web traffic.

Keywords: Web traffic analysis, Web technology, Zipf distribution, Cache policy.

1 Introduction

Patterns of web site access have changed in the last few years. There are much more users, accessing from mobile devices as well as from the desktop, and demanding more personalized contents. One consequence of this evolution is that contents are still consumed as *HTML* pages but also as structured documents published via *APIs*. It is common finding that the information is obtained as *JSON* documents in mobile applications or in web pages, using technologies such as *AJAX*. Moreover, multimedia content appears in almost every site in the form of videos, audios, but mainly as images.

In this context, it is worth analyzing whether websites workload has also changed, in terms of web objects distribution. How web resources are demanded and whether there is a distribution that can characterize the incoming requests requires some analysis in order to foresee demand trends. Several works have analyzed traces from websites and proxies (see section 3) and have found that the Zipf distribution can well characterize incoming requests. However, the consistence of the collected data that supports this distribution function may require some revision because it was very domain specific and the method to obtain part of the data could have biased the results to some extent.

© Springer International Publishing Switzerland 2015   105
S. Omatu et al. (eds.), *Distributed Computing & Artificial Intelligence, 12th Int. Conference,*
Advances in Intelligent Systems and Computing 373, DOI: 10.1007/978-3-319-19638-1_12

Distribution of requests enables to identify user's behavior properties [2,15] that can be inferred from users' consumption patterns in an application domain. Among these properties, the maximum theoretical hit ratio that can be reached is the most relevant to determine the best policy for caching web contents [13]. After checking the bibliography, Zipf appears as the most general distribution found that captures the demand behavior.

This paper presents new evidences obtained in a diversity of websites to support the existence of the Zipf distribution function. It also analyses whether this function depends on the web server technology and the domain of application. The experimentation comprises 14 websites using different technologies. They belong to different sectors and objectives: some are media and news sites, others are used for e-commerce, and the rest show public information to citizens. Additionally, a ticketing tool and a wiki site (both made by Atlassian) have been included as examples of login based sites.

A Zipf distribution is characterized by α, a parameter that determines the shape of the cumulative probability function that web objects show in the distribution. Our analysis has confirmed that web requests follow a Zipf distribution. Furthermore, a relevant result has appeared: α is greater than in previous published works. According to previous articles α varies between 0,6 and 1, but our analysis shows that current web traffic has made an evolution of α to values higher than one in all the traces, regardless the domain and the technology. This means that the majority of the requests concentrate on a small set of objects, and therefore the percentage of web objects needed to reach a certain hit ratio is smaller than in previous analysis. This result is relevant for the design and configuration of web cache policies and mechanisms, for instance.

The rest of the paper is structured as follows. Section 2 explains how to define a function that represents the distribution of incoming web objects requests. In concrete, it shows how this distribution function can be characterized as a Zipf. Once Zipf is properly presented, section 3 presents and discusses previous works that have already shown evidences supporting the Zipf distribution. Section 4 presents the evidences derived from our recent analysis of 14 web sites of different domains and technologies. Section 5 elaborates more on this knowledge, and suggests a change in the function that allows to determine the theoretical hit ratio that can be reached given a cache size, when the website's requests follow a Zipf behavior. Finally, section 6 presents the conclusions and raises some issues for further study.

2   Introduction to Zipf and Power-Law Distributions

Zipf is a power-law distribution, which was first used in linguistics, for analyzing words occurrence. *Zipf is a discrete distribution defined in the rank-frequency domain, which states that when items are ranked (R) in descending order of their popularity, then the frequency (F) of the item is inversely proportional to the rank item* [11]. The Zipf formula can be written as:

$$f(r) \sim k r^{-\alpha} \tag{1}$$

where k and α are constants and α is close to 1. By applying a logarithm in both sides, the formula is transformed to:

$$log(f(r)) \sim -\alpha log(x) + log(k) \qquad (2)$$

As it can be seen, the formula 2 corresponds to a straight line shape with slope $-\alpha$. Therefore a power-law function, once transformed to log-log, is plot as a line.

Depending on the value of α, a Zipf distribution is called strict when α is almost equals to 1, and it is called Zipf-like when α is different to one.

In the web domain, Zipf is interpreted as the distribution of the user requests accessing the servers. As the requests are distributed among a finite number of objects, and the client requests are considered as independent events, the popularity of a web object is characterized by its probability. Therefore, the web object popularity model [15] is established based on a redefinition of a Zipf in the form of $P_i = \frac{C}{i^\alpha}$, where C is a constant, i the i_{th} most popular object and α the Zipf parameter seen before.

Although power laws are defined in the domain of real numbers, natural numbers are used for web sites. Also the straight line is only an approximation where values at the bottom (e.g., those objects requested only once) and values in the firsts ranks (e.g., the most popular web objects) can appear outside the straight line. These two features can be found in previous traces as it will be shown in section 3. There is a rank from where the distribution behaves as a Zipf. This value is noted as x_{min} [4].

In the next section previous evidences of Zipf presence are shown.

3   Evidences for Zipf Distribution Function

The existence of the Zipf distribution for web requests is well supported by empirical results, which are summarized in table 1. These results are built over data sets gathered with two techniques: analyzing directly the website *access.log* or using proxies that register every request between the servers and the clients during a period of time. Differences have been noticed in the results when using the different methods as mentioned in [2].

It is important to note that the majority of the traces analyzed up today, are based on proxies, gathering user requests in a facility or mixing requests from different websites. In the process of collecting data for a website, it is not accurate to use proxies from facilities because many of the requests are lost. Therefore, in order to find the user behavior it is better to obtain data from the users. Most of the evidences are not from final users, so some bias can be found related to a website.

Checking the bibliography and from the evidences, Zipf distribution appears in almost every case, what means that web objects' popularity is skewed toward a small set of them. The α value varies between 0,6 to 1 in most works [2,15,12] although some exceptions can be found.

Table 1. Summary of traces found

Authors	α	Year	Number Traces	Technique
Breslau et al. [2]	0,64 - 0,83	1999	6	Proxy
Roadknight et al. [14]	0,5 - 0,9	1999	5	Proxy
Artlitt and Jin [1]	1,16	2000	1	Web Server
Mahanti et al. [10]	0,74 - 0,84	2000	3	Proxy
Challenger et al. [3]	1	2004	1	Web Server
Krashakov et al. [9]	$1,02 \pm 0,05$	2006	5	Proxy
Gill et al. [6]	0,56	2007	1	Proxy
Urdaneta et al. [17]	0,64	2009	1	Web Server
Traverso et al. [16]	0,7 - 0,85	2012	4	Proxy
Huang et al. [7]	1 in client layer	2013	1	Web Server
Imbreda et al. [8]	0,83	2014	1	Proxy

However, a deeper analysis on these works raises some issues. More specifically, the number or requests that are considered in these studies is far from the volume of requests that is common nowadays, so it seems necessary to review the measurement of α by increasing the number of requests in at least two orders of magnitude. Additionally, the way of using the web has also changed, and some authors such as Krashakov et al. pointed that α varies through years [9].

These issues imply the need for new analysis of web objects requests in diverse types of web domains and technologies. The next section shows a wider analysis with access logs of current web sites.

4   Data Analysis

We have collected data from different web sites as a set of Apache-like access logs. These show information for every single request that servers receive, including the response code, user-agent, size, *URL* and time-stamp. For legal restrictions, the *IP* addresses and other personal data are omitted from the logs. But not all the collected requests are needed for this analysis: only those with response code 200, 203, 206, 300, 301 or 410 are used [2,12].

Obtaining a wide variety of logs has been a difficult task due to commercial and privacy issues. After contacting to some webmasters we have obtained 14 logs. They are listed in table 2 together with the number of requests, the collecting period, the sector, and finally the technology the server uses. These logs were properly anonymized by the webmasters themselves. It can be seen that the volume of the web traces is higher than in most of the previous works. We pay special attention to Spanish Television web site (*www.rtve.es*) because it is a very demanding website, with more than 100 million requests per day of news and multimedia. Another website, *eldiario.es* offers news and media too and its data is useful to compare conclusions with respect the media sector.

The rest of sites are from other sectors. Among them, *bodaclick.com* and *www.ucm.es* are interesting for the huge volume of requests that they receive; *jira.rtve.int* for being a login based ticketing tool used in Spanish Television

Table 2. Logs analyzed

Sector	Site	Period	Requests	Technology	α
Media	www.rtve.es	22/06/2014	130 millions	J2EE	1,41
Media	www.eldiario.es	10/10/2013	8.961.959	PHP	1,16
B2B and B2C	www.bodaclick.com	26/06/2013	13.987.028	Symphony 2	1,30
Sport Event B2B	www.ebuops.com	1-30 Jun 2014	330.043	J2EE	1,61
University	www.ucm.es	1-30 Oct 2013	8.379.230	PHP	1,09
University	www.fdi.ucm.es	Jul - Sep 2013	3.080.476	ASP.Net	1,4
Research	grasia.fdi.ucm.es	Jul - Sep 2013	74.023	Drupal	1,53
Gaming	www.otogami.com	May - Jul 2013	2.547.172	J2EE	1,13
Public Sector	www.pedro-munoz.es	Jul - Oct 2013	4.302.641	Drupal	1,49
Public Sector	pdmmadridejos.es	Sep 2013	221.062	Drupal	1,53
Public Sector	cobisa.es	Ago - Sep 2013	42.753	Drupal	1,41
Sport facility	deportesmota.es	May - Oct 2013	384.416	Drupal	1,50
Ticketing tool	jira.rtve.int	Oct 2013	1.246.808	J2EE	1,66
Internal wiki	wiki.rtve.int	Oct 2013	3.469.508	J2EE	1,17

and *wiki.rtve.int*, a login base tool for sharing documents used in the Spanish Television as well. Finally, other sites with a not so high number of requests are included to enrich the analysis.

For determining α, the estimator by Clauset et al. [4] is used, which is available for downloading at [5]. It estimates x_{min} and α according to the goodness-of-fit method described in [4] using the Kolmogorov-Smirnov goodness-of-fit statistic D for each possible choice of Smirnov x_{min}, and after the estimator selects the x_{min} that gives the minimum value D over all values of x_{min}.

5   Analysis of the Results

After filtering the logs as discussed in section 4, their analysis produces different demand frequency rank graphics for every web sites. This drives to a form of power law distribution, with the characteristic straight line.

Figure 1 shows the diagrams of 6 of the most demanded websites from the 14 traces. For them, the estimated straight-line and the distribution is shown. As it can be seen, the left hand side of the distribution varies, but from certain point the Zipf form can be recognized. The straight-line appears during most the trace, and finally a tail is found as expected. This behavior is similar to the evidences found in previous works (see section 3).

After observing that the graphics follow the Zipf-like behavior, the α value has been calculated for all the traces. This has shown that all of them follow a Zipf distribution, with α ranging from 1,09 to 1,87, as it can be seen in the last column of the table 2. This result differs from those shown in section 3. Note that α in the previous works ranged between 0,6 to 1,0 and the new values obtained are higher. It must be noted also that this fact happens regardless the underneath technology. Therefore, it can be concluded that the web objects' popularity in the analyzed websites shows a bigger skewness, and therefore, most

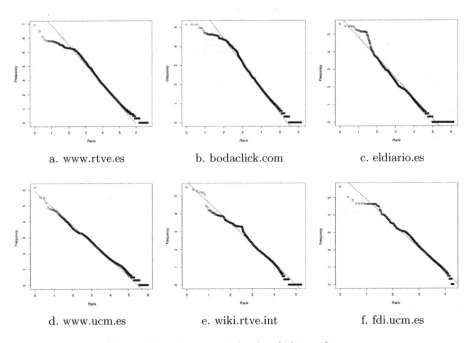

a. www.rtve.es   b. bodaclick.com   c. eldiario.es

d. www.ucm.es   e. wiki.rtve.int   f. fdi.ucm.es

Fig. 1. Zipf diagrams in log-log fashion of traces

of the workload falls in only a few highly demanded web objects. This fact can be used to improve the quality of the services: for example, when it comes to caching, storing such objects can drive to a high hit ratio.

Other interesting results can be seen: those sites implemented using *Drupal* show an α between 1,41 to 1,53. In fact, with the exception of *cobisa.es*, the α values are very close among the traces from 1,49 to 1,51. This result is interesting although a huge number of requests are not involved in these traces. Other combinations of technologies do not show similar results: Java based websites range from 1,13 to 1,66, or *PHP* based websites range from 1,09 to 1,87. Moreover, the four traces with the smaller α are implemented with four different technologies, and something similar happens with the technology of the traces with the higher α.

The analysis by sectors shows some conclusions. First of all, media sites' traces include a huge amount of requests, showing a very different α value: in *www.rtve.es* α is equals to 1,41 while in *www.eldiario.es* the value reaches 1,16. This difference does not occur in the public institutions, where all traces show a similar α value, but it is the consequence of using the same technology and not a specificity derived from the sector. Regarding *B2B* and e-commerce sites, α is not enclosed in similar values. Finally, as a final conclusion, when comparing login based websites with free access ones, no correlation appears since α varies from 1,17 to 1,66 in login based traces, and between 1,09 and 1,87 in free access sites.

When analyzing the existence of correlation between the number of requests received by a site and the α value obtained, the result is that no correlation exists, given the Pearson correlation a value equals to 0,5.

Finally, the main result from this analysis is that no matter the domain or the technology, the Zipf distribution appears for all the websites. This implies that popularity is inherent to a web based solution. Furthermore, the only correlation found regarding the relation between the technology and α are found in the *Drupal* implementations, but a general correlation between the technology and the technology used has not been found.

6 Conclusions and Future Work

We have gathered new evidences supporting that the frequency of web objects demand follows a Zipf distribution. A new analysis of the traffic in web sites was necessary because the volume of information that is exchanged today has considerably increased. This analysis also shows that the α of Zipf is greater today than in the previous studies. In fact, its value in some cases is almost 60% higher than the strict Zipf distribution.

The consequence is the high skewness that shows the web objects' popularity: only a few objects receive a high volume of requests. For some techniques like web caching, this result is interesting, because considering only the most demanding objects, a high hit-ratio can be obtained. This result is also relevant when designing a cache policy, and when designing self-managed caches, since it is possible to determine the optimal cache size and policy for the traces with minimal costs. In the case of the *RTVE* web site, with millions of requests per day, we have calculated that just storing around 1.000 objects can result in a hit ratio higher than 0,8.

The second conclusion is that the Zipf distribution seems to be inherent to current web traffic patterns, because, as it can be seen in the traces, no matter the technology, the sector or the business usage of the site, a Zipf distribution always appears.

A deeper study must be done to determine further consequences from these results, as well as the reasons behind these observations. The analysis of the different web objects and their relations is needed to understand the nature of the obtained distribution.

Another issue to be analyzed is the existence of differences in α when using only dynamic or only static contents. The consumption of dynamic structured information for mobiles and devices using *JSON* or *XML* is higher than some years ago, and perhaps such analysis can bring a new perspective to be considered in the design of services.

Acknowledgement. This work is partially supported by the Programa de Financiación de Grupos de Investigación UCM-Banco Santander with reference GR3/14.

References

1. Arlitt, M., Jin, T.: A workload characterization study of the 1998 world cup web site. IEEE Network 14(3), 30–37 (2000)
2. Breslau, L., Cao, P., Fan, L., Phillips, G., Shenker, S.: Web caching and Zipf-like distributions: Evidence and implications. In: Proceedings of the IEEE INFOCOM 1999, pp. 126–134. IEEE (March 1999)
3. Challenger, J.R., Dantzig, P., Iyengar, A., Squillante, M.S., Zhang, L.: Efficiently serving dynamic data at highly accessed web sites. IEEE/ACM Transactions on Networking 12(2), 233–246 (2004)
4. Clauset, A., Shalizi, C.R., Newman, M.E.: Power-law distributions in empirical data. SIAM Review 51(4), 661–703 (2009)
5. Clauset, A., Shalizi, C.R., Newman, M.E.: Power-law Distributions, http://tuvalu.santafe.edu/aaronc/powerlaws/
6. Gill, P., Arlitt, M., Li, Z., Mahanti, A.: Youtube traffic characterization: a view from the edge. In: Proceedings of the 7th ACM SIGCOMM Conference on Internet Measurement, pp. 15–28. ACM (October 2007)
7. Huang, Q., Birman, K., van Renesse, R., Lloyd, W., Kumar, S., Li, H.C.: An analysis of Facebook photo caching. In: Proceedings of the Twenty-Fourth ACM Symposium on Operating Systems Principles (pp, pp. 167–181. ACM (November 2013)
8. Imbrenda, C., Muscariello, L., Rossi, D.: Analyzing cacheable traffic in isp access networks for micro cdn applications via content-centric networking. In: Proceedings of the 1st International Conference on Information-Centric Networking, pp. 57–66. ACM (September 2014)
9. Krashakov, S.A., Teslyuk, A.B., Shchur, L.N.: On the universality of rank distributions of website popularity. Computer Networks 50(11), 1769–1780 (2006)
10. Mahanti, A., Williamson, C., Eager, D.: Traffic analysis of a web proxy caching hierarchy. IEEE Network 14(3), 16–23 (2000)
11. Mahanti, A., Carlsson, N., Arlitt, M., Williamson, C.: A tale of the tails: Power-laws in internet measurements. IEEE Network 27(1), 59–64 (2013)
12. Nair, T.R., Jayarekha, P.: A rank based replacement policy for multimedia server cache using zipf-like law. arXiv preprint arXiv:1003.4062 (2010)
13. Podlipnig, S., Böszörmenyi, L.: A survey of web cache replacement strategies. ACM Computing Surveys (CSUR) 35(4), 374–398 (2003)
14. Roadknight, C., Marshall, I., Vearer, D.: File Popularity Characterisation. In: Proceedings of the 2nd Workshop on Internet Server Performance (WISP 1999), Atlanta, GA (May 1999)
15. Shi, L., Gu, Z.-M., Wei, L., Shi, Y.: Quantitative analysis of zipf's law on web cache. In: Pan, Y., Chen, D.-X., Guo, M., Cao, J., Dongarra, J. (eds.) ISPA 2005. LNCS, vol. 3758, pp. 845–852. Springer, Heidelberg (2005)
16. Traverso, S., Ahmed, M., Garetto, M., Giaccone, P., Leonardi, E., Niccolini, S.: Temporal locality in today's content caching: why it matters and how to model it. ACM SIGCOMM Computer Communication Review 43(5), 5–12 (2013)
17. Urdaneta, G., Pierre, G., Van Steen, M.: Wikipedia workload analysis for decentralized hosting. Computer Networks 53(11), 1830–1845 (2009)

Investors or Givers ?
The Case of a Portuguese Crowdfunding Site

Paulo Mourao[1] and Catarina Costa[2]

[1] Department of Economics/ University of Minho, 4700 Braga – Portugal
paulom@eeg.uminho.pt
[2] MSc Social Economics / University of Minho, 4700 Braga – Portugal
catapcosta@hotmail.com

Abstract. This is the first paper studying the determinants behind the different values raised by project and by investor using data collected from Portuguese crowdfunding platforms. We concluded that higher prize-moneys, expanded ad duration, and stating that the investment will be used for multiple ends tend to maximize the value that investors send to projects. However, there is a downtrend of the raised values observed at the most recent years that generates additional challenges for all interested on building, managing and promoting web-crowdfunding platforms.

Keywords: Electronic Business, Data Analysis, Crowdfunding.

1   Introduction

Crowdfunding is not a new strategy for getting funds for a project. Until few years ago, the pages of newspapers usually had the function that the supporters of projects ask now to the internet, specifically to web platforms (Mourao [1]). Several papers discussed in international scientific meetings, like International Symposium on Distributed Computing and Artificial Intelligence (DCAI), have also addressed this issue (Miglieta and Parisi [17] or Rao et al. [18]).

The common structure of crowdfunding platforms is simple for explaining - a supporter promotes a project using a web-platform, he/she uses an attractive message that clarifies the ends for the money that will be sent; then, the investors send money using digital channels and investors get a reward if certain objectives are reached (the most common objective is the minimum funding for making the project works).

However, this recent phenomenon (the web platforms) has also a scarce literature on this topic. We are trying to fill this gap with our studies. In this study, we focus on the main Portuguese crowdfunding platform – the site *ppl.com.pt*. Our motivation is related to the intention of understanding the statistically significant determinants that can explain why certain projects get substantial amounts of money invested and other projects do not get so expressive values.

The structure of this paper is the following. At section 2, we are going to review the literature on giving/crowdfunding, and on the business structure of the common

© Springer International Publishing Switzerland 2015
S. Omatu et al. (eds.), *Distributed Computing & Artificial Intelligence, 12th Int. Conference*,
Advances in Intelligent Systems and Computing 373, DOI: 10.1007/978-3-319-19638-1_13

crowdfunding platforms. At section 3, we are going to discuss our empirical efforts on analyzing the determinants that can be statistically identified as influencing the different values raised by project and by investor. Finally, we conclude at section 4.

2   Literature Review

Since Marcel Mauss's gift essay in 1925, the subject of the gift and reciprocity has been discussed in academic domains. In particular, the crowdfunding subtheme has been increasing in terms of popularity. In 2012, the turnover of crowdfunding was almost three billions dollars worldwide (Crowdfunding Report [2]) and the number of platforms emerging through the last years is also a proof of the popularity. However, as crowdfunding is a recent subject the existence of related literature is still scarce.

The term Crowdfunding appeared shortly after crowdsourcing, initially used by Jeff Howe in 2006 (Howe [3]). The concept of crowdfunding finds its roots in the broader concept of crowdsourcing. The word Crowdsourcing is a compound contraction of Crowd and Outsourcing, thus it means outsourcing to the crowd (Guittard & Schenk [4]). Crowdsourcing is therefore the ability to use the public as a way to get ideas, feedback and solutions in order to develop corporate activities (Howe [3]; Kleemann et al [5]).

Basically, crowdfunding's main goal is to raise money to invest, in other words, small amounts of money are collected to a project. This project can be the edition of a book or of a music record, the improvement of the house of a non-profit organization, etc.

According to Belleflamme et al. [6], "Crowdfunding involves an open call, essentially through the Internet, for the provision of financial resources either in form of donations (without rewards) or in ex-change for some form of reward and/or voting rights in order to support initiatives for specific purposes".

Kappel [7] describes crowdfunding as an informal act usually, via the Internet, in order to generate and distribute funds by a group of people. These funds can have several purposes typically related with social, personal or entertainment activities.

Crowdfunding might be seen as financial contributions from investors, sponsors or donors to finance initiatives or certain companies with or without profit. It is a different approach to raise capital and support new projects, soliciting contributions from a large number of stakeholders.

Some authors defend that crowdfunding became a variant of classic fundraising for nonprofit organizations. Crowdfunding also seeks inspiration on microcredit concepts (Morduck [8]) and crowdsourcing (Poetz and Schreier [9]), but it represents a unique category of fundraising.

According to De Buysere et al [10], crowdfunding is used, by organizations, to help with market research, financing and marketing. Depending on the project need, different business models can be used. The main advantage of this type of financing is that lenders are also project ambassadors/supporters and may even promote the product.

Crowdfunding's success is intrinsically linked with each project quality but also with the personal network of the founder of the project, since family and friends are the first to contribute. Another aspect worth mentioning regards the expansion of the boundaries between founders of the projects and the investors. According to Agrawal et al.[11], start-ups investments tend to be local due to the personal relations and the monitoring process and the costs of such activities are sensitive to distance.

These authors (Morduck, [8]; Poetz and Schreier, [9]; De Buysere et al, [10]) emphasize three characteristics that can diminish the effect of the geographical distance: a standardized format to disclosure projects to anyone who has Internet access; the low cost of investment and the provision of information on investment and project itself and followed up (Agrawal et al [11]).

2.1   Crowdfunding Models

As the provision of capital may take the form of donations, sponsorships, pre-order or pre-sale, loans and private equity, the complexity of the processes greatly varies (Hemer et al., [12]) .According to Nesta [13], there are four types of crowdfunding models: donation, reward, loan, and equity. The donation model comprises donations with nothing in return as well as intangible or material rewards (merchandising). The loan model includes three subcategories: traditional loan agreement, where there is refund by rate and the terms are standardized; forgivable loan when there is only refund in the case of one of the following two conditions is fulfilled, a) if and when the project generate revenue or b) if and when the project generate profit; pre-sale, in which the individual receives the product or service before anyone else. The equity model is divided in two subcategories: the financial investments model, in which the investors buy company shares; profit model or revenue sharing, in which a part of the revenue or profit goes to the investors.

2.2   Crowdfunding Platforms

Platforms work as intermediaries and, despite emerging only a few years ago, they have started to deeply transform the traditional ways to raise money. Due to the increase of popularity, the number of platforms has been quickly increasing in the last years over different continents (Holzer, [14]). The crowdfunding platforms vary in the way they are structured. Therefore, the three types of platforms are: Specialized platforms (mainly used at specific industries such as music, games or television); Platforms of a specific activity (mainly used at several industries focused on some sort of projects, like creative, technological or hardware development), and Platforms of general purposes (they have no restrictions). Theses platforms feature a diversity of campaigns which can be created by individuals who need help to pay a surgery, to pay a wedding or even by artists who need money for their projects. Crowdfunding platforms usually structured the allocation of funds in two ways. The most usual way is named "all-or-nothing model" in which the project founder only gets the money if his goal is reached or exceeded, during the defined period. In the other case, all the money raised over the financing period is transferred to the founder, even if the project is not totally financed. These platforms also have different fees for the services they provide. The most common fees are through commissions, receiving the web platforms' managers a percentage of the raised funds (usually between 2 and 5%).

Although crowdfunding is not limited to web platforms, none doubts that the presence of this kind of private and social funding owes its great recent growth and popularity to the potentialities of world wide web (Schenk and Guittard, [4]). Actually, Holzer[14] argues that there are dozens of web platforms for crowdfunding purposes. Around the world there is a great heterogeneity of crowdfunding sites. For the trade volume, the history track, and the number of funded projects, some of the most relevant sites are kickstarter.com, indiegogo.com, crowdfunder.com, rockethub.com, and crowdrise.com (Barnett, [15]). In Portugal, the crowdfunding has a recent history. Considering the records from the two most notorious Portuguese managed sites (ppl.com.pt and Novo Banco Crowdfunding), we can claim that the Portuguese electronic crowdfunding is not more aged than five years-old. However, year after year, Portuguese has become more fans of crowdfunding for getting resources for so different purposes as creating the own project, editing a book or organizing a trip. At next section, we are going to test determinants able to explain total raised value per project, the percentage of the target, and the raised value per investor.

3   Empirical Section – Determinants for the Value Raised in a Portuguese Web-Platform for Crowdfunding

In this section, we will comment the main results we got when testing the set of potential determinants of the three dependent variables. We recall these dependent variables are: "total raised value per project", "percentage of target", and "raised value per investor". The independent variables follow the review of literature. Therefore, we will use variables focused on the project's characteristics (Poetz and Schreier, [9]), on the number of investors (Schenk and Guittard [4]) and on the prizes available to the investors (Kappel, [7]). The independent variables are: the ad duration, the number of investors, the type of promoters, the existence of multiple ends attributed to the raised money, the existence of a well-defined validity for the project, the minimum invested value with prize, and the minimum invested value with the maximum prize. Table 1 exhibits the main descriptive statistics of our variables. We observe that the median project promoted by http://ppl.com.pt has the following characteristics: it received 2000 euros, being excessively funded 6% above the required value, and each one of the 45 investors gave around 40 euros to this representative project. Most likely it was funded in 2014, announced for 1 month, and promoted by an informal group of promoters. This representative project had no validity restraints and it did not involve buildings repairing. The minimum invested value that granted a prize to the investor was 5 euros and the minimum invested valued that granted the maximum prize to the investor had the median value of 100 euros.

Table 1. Descriptive statistics

	Number of cases	Mean	Standard Deviation	Maximum	Minimum	Median
Total raised value per project (euros)	253	2346.9	1965.0	21917	500	2000
Percentage of target (= Funded Value / Required Value), euros	253	1.136	0.194	2.06	0.94	1.06
Raised value per investor (euros)	253	56.479	76.680	1000	4.24	39.79
Year	253	2013.4	0.773	2014	2011	2014
Ad duration (months)	253	1.296	0.756	3	0	1
Number of investors	253	64.269	88.163	1122	1	45
Type of promoters*	253	1.783	0.778	5	1	2
Multiple functions (1, yes; 0, no)	253	0.031	0.175	1	0	0
Validity of the project (1, yes; 0, no)	253	0.352	0.478	1	0	0
Minimum invested value with prize (euros)	253	5.535	6.862	100	1	5
Minimum invested value with the maximum prize (euros)	253	282.48	525.96	5000	1	100

* Type of promoters: 1, a single promoter; 2, an informal group of promoters; 3, a non-profit organization; 4, a public entity; 5, a private company/firm.

The correlation matrix of our variables (revealed if requested) did not exhibit coefficients with a value higher than 0.60 (which minimizes the threats of endogeneous regressors). Anyway, we run the Durbin-Hu-Hausman tests against the endogeneity of the variables more likely to introduce endogeneity troubles and we were not able to reject the hypotheses of exogenous regressors; full details available under request. Therefore, OLS is more appropriate than Two Stages Least Squares (Wooldridge, [16]). Table 2 exhibits these results, estimated by ordinary least squares with heteroscedasticity robust standard errors.

Table 2. Estimations for total funded value, excess of funding, and value per investor (site: http://ppl.com.pt , 2011-2014)

	Total raised value per project	Percentage of target	Raised value per investor
Constant	370369.7***	-14.99	-476.0
	(213503.9)	(32.77)	(12346.8)
Year	-183.46*	0.008	0.267
	(106.0)	(0.016)	(6.131)
Ad duration	267.1**	-0.014	-5.897
	(119.8)	(0.017)	(6.487)
Number of investors	8.858***	0.0005***	omitted
	(2.120)	(0.0001)	
Type of promoters	-108.7	0.021	-0.713
	(89.7)	(0.017)	(6.365)
Multiple functions/ends for the invested money	1081.6**	0.067	12.709
	(444.9)	(0.070)	(26.566)
Validity of the project	-157.5	-0.027	12.167
	(175.5)	(0.027)	(10.113)
Minimum invested value with prize	26.8**	0.0001	-0.001
	(11.8)	(0.002)	(0.002)
Minimum invested value with the maximum prize	1.82**	-0.0001	-0.118
	(0.844)	(0.0018)	(0.861)
R2 / Number of cases	0.496 / 247	0.065 / 247	0.012 / 247
F-stat	5.94***	2.07*	0.41

Note: Robust heteroskedasticity standard errors between parentheses. Significance levels: *, 10%; **, 5%; ***, 1%.

4   Conclusion

This research is the first to test determinants for the values raised for the most relevant Portuguese web platform of crowdfunding. Web-crowdfunding has been characterized by a rising interest and by a rising volume of trade for the last years. Usually, web platforms are used to promote the projects of individuals, but there is a rising number of informal groups of citizens or of non-profit organizations recurring to this way of getting funds for their own projects.We focused on the data collected from the main Portuguese web platform of crowdfunding – the site ppl.com.pt . We have observed all of the 247 well-succeeded projects funded through this platform until December of 2014. We concluded that the most recent years exhibit lower values related to the total raised value per project. A higher ad duration and a higher number of investors tend to increase the total raised value per project. The existence of multiple ends for the invested money and higher values for the prizes available to the investors also tend to increase the collected money per project. It is very interesting to conclude there are no statistically significant dimensions for "value per investor" – this must be read as a further challenge for researching, possible contributed by extensive surveys distributed to the sample of investors focused on each investor's motivations, idiosyncrasies and characteristics. So, let recover our title question: "Investors or givers?" Based on our results, we can claim that most of the people who send money to the analyzed projects react to the expected reward, as a typical investor. Our results generate two challenges and two implications. The first challenge relates to the need of collecting additional data from other Portuguese/European platforms in order to compare and update these estimates. The second challenge relates to the advantage of conducting surveys responded by crowdfunding investors and promoters, in order to get additional explanations for dimensions still unsolved, like the value per investor. The two implications are related to the importance of these kind of studies for trade and electronic business analysis.

Actually, this study proved that web-crowdfunding/crowdsourcing can use some dimensions in order to raise the value per project by improving the ad duration or the ad quality, by attracting an enlarged number of investors and by increasing the prize-money for the investment. The second implication relates to the high significance of the presence of multiple ends for the invested money (for instance, for improving the infra-structures of a daycare center and for empowering the local population). Our estimates revealed that crowdfunding investors are sensitive to the complex of needs stated at the ad – therefore, a special emphasis must also be put on the quality of the messages posted in the web.

Acknowledgements. The authors acknowledge the reviews of two anonymous reviewers of DCAI Scientific Committee on a previous version. Remaining errors are authors' exclusive ones.

References

1. Mourao, P.: Dar olhando a quem - estudo sobre o Projeto 'Todo o Homem é Meu Irmão'. Innovar 23(49), 131–139 (2013)
2. Crowdfunding Report (2013), http://www.crowdsourcing.org/editorial/2013cf-the-crowdfunding-industry-report/25107
3. Howe, J.: Crowdsourcing: Why the Power of the Crowd is Driving the Future of Business (2008), http://www.bizbriefings.com/Samples/IntInst%20-%20Crowdsourcing.PDF

4. Schenk, E., Guittard, C.: Crowdsourcing: What can be Outsourced to the Crowd, and Why?.<halshs-00439256v1> (2009)
5. Kleemann, F., Vob, G.G., Rieder, K.: Un(der)paid Innovators: The Commercial Utilization of Cosumer Work through Crowdsourcing. Science, Technology and Innovation Studies 4(1), 5–26 (2008)
6. Belleflamme, P., Lambert, T., Schwienbacher, A.: Crowdfunding: An Industrial Organization Perspective. Working Paper. Université Catholique de Louvain (2010)
7. Kappel, T.: Ex Ante Crowdfunding and the Recording Industry: A Model for the US. Loyola of Los Angeles. Entertainment Law Review 29, 375 (2008)
8. Morduch, J.: The microfinance promise. Journal of Economic Literature 37, 1569 (1999)
9. Poetz, M., Schreier, M.: The value of crowdsourcing: can users really compete with professionals in generating new product ideas? Journal of Product Innovation Management 29(2), 245–256 (2012)
10. De Buysere, K., Gajda, O., Kleverlaan, R., Marom, D.: A framework for European Crowdfunding (2012),
    http://www.europecrowdfunding.org/files/2013/06/
    FRAMEWORK_EU_CROWDFUNDING.pdf
11. Agrawal, A., Catalini, C., Goldfarb, A.: Friends, Family, and the Flat World: The Geography of Crowdfunding (2011),
    https://www.law.northwestern.edu/
    research-faculty/searlecenter/workingpapers/documents/
    AgrawalCataliniGoldfarb.pdf
12. Hemer, J.: A snapshot on crowdfunding, Working papers firms and region, No. R2/2011 (2011), http://hdl.handle.net/10419/52302
13. Nesta, An Introduction to Crowdfunding (2012), http://www.nesta.org.uk
14. Holzer, J.: SEC Is Reviewing Small Internet Sales of Stock. The Wall Street Journal (2011),
    http://online.wsj.com/article/SB10001424052748704843404576251160999848924.html
15. Barnett, C.: Top 10 Crowdfunding sites for fundraising. Forbes (August 5, 2013),
    http://www.forbes.com/sites/chancebarnett/2013/05/08/
    top-10-crowdfunding-sites-for-fundraising/
16. Woodridge, J.: Econometric Analysis of Cross Section and Panel Data. MIT Press, Cambridge (2002)
17. Miglietta, A., Parisi, E.: Means and Roles of Crowdsourcing Vis-À-Vis CrowdFunding for the Creation of Stakeholders Collective Benefits. In: Aiello, L.M., McFarland, D. (eds.) SocInfo 2014 Workshops. LNCS, vol. 8852, pp. 438–447. Springer, Heidelberg (2015)
18. Rao, H., Xu, A., Yang, X., Fu, W.-T.: Emerging dynamics in crowdfunding campaigns. In: Kennedy, W.G., Agarwal, N., Yang, S.J. (eds.) SBP 2014. LNCS, vol. 8393, pp. 333–340. Springer, Heidelberg (2014)

Effects of Organizational Dynamics in Adaptive Distributed Search Processes

Friederike Wall

Alpen-Adria-Universitaet Klagenfurt, 9020 Klagenfurt, Austria
friederike.wall@aau.at
http://www.aau.at/csu

Abstract. In this paper, the effects of alternating the organizational setting of distributed adaptive search processes in the course of search processes are investigated. The organizational properties under change include, for example, the coordination mechanisms among agents. We analyze different temporal change modes using an agent-based simulation. Results suggest that inducing organizational dynamics has the potential to increase the effectiveness of search, but that the mode of change in conjunction with the complexity of the search problem considerably affects the order of magnitude of this beneficial effect.

Keywords: agent-based simulation, complexity, coordination.

1 Introduction

Adaptive distributed search processes occur in a large variety of real-world systems like, for example, networks of roboters, "swarms" of unmanned aerial vehicles or managers of firms collaboratively searching for higher levels of business performance. The question of how to organize these search processes, and, in particular, to coordinate among distributed agents is of fundamental relevance and has been investigated in various disciplines, e.g. complex systems science, robotics or computational organization theory (for reviews [3], [10], [2]).

In this paper, we focus on the temporality of the organizational set-up for adaptive search and, in particular, we are interested in the effects of alternating the organizational setting on the effectiveness of distributed adaptive search processes. With this, our research effort relates to broad field of adaptive evolutionary networks. However, it is worth mentioning that the paper mainly focusses on "dynamics of networks" rather than on "dynamics on networks" [3]. With respect to distributed multi-agent coordination, research has been categorized into streams on consensus, formation control, optimization, task assignment, and estimation [10]: The research effort presented here is directed towards the latter aspect, meaning that - since, for example, global information on the search problem is not available - a distributed search and assessment of partial solutions to the search problem is reqired. Hence, in the center of our research is not to figure out the performance of, for example, certain consensus mechanisms

S. Omatu et al. (eds.), *Distributed Computing & Artificial Intelligence, 12th Int. Conference,*
Advances in Intelligent Systems and Computing 373, DOI: 10.1007/978-3-319-19638-1_14

in dynamic task environments; rather, the effects of different modes of changing the organizational set-up in the course of adaptive search processes is focussed.

For this, we employ an agent-based simulation and, in particular, observe adaptive search processes conducted collaboratively by the agents of a search system. Our agents operate on fitness landscapes and seek to find superior levels of fitness or, in other terms, of the search system's overall performance. The fitness landscapes are set-up following the idea of NK fitness landscapes [4], [5]. The distinctive feature of our model is that we employ different modes of mid-term dynamics imposed on the organizational setting of the search process. Moreover, not only the coordination mechanism [1] is altered, but also other properties of the distributed search system.

2   Outline of the Simulation Model

2.1   Search Problem and Complexity

In line with the NK model, in each time step t of the observation period, our search systems under investigation face an N-dimensional binary search problem, i.e., they have to find a configuration $d_{it} \in \{0,1\}$, $i = 1, ...10$. Hence, at each time step, the search space consists of 2^N different binary vectors $\boldsymbol{d_t} \equiv (d_{1t}, ..., d_{Nt})$ possible. Each of the two states $d_{it} \in \{0,1\}$ makes a certain contribution C_{it} to overall fitness $V(\boldsymbol{d_t})$ of the organization. C_{it} is randomly drawn from a uniform distribution with $0 \leq C_{it} \leq 1$.

The NK framework allows for representing interactions among choices with level K. K reflects the number of other choices d_{jt}, $j \neq i$ which also affect the fitness contribution C_{it} of choice d_{it}. K can take values from 0 (no interactions) to $N-1$ (maximum interactions). With this, fitness contribution C_{it} might not only depend on the single choice d_{it} but also on K other choices d_{jt} where $j \in \{1, ...N\}$ and $j \neq i$:

$$C_{it} = f_i(d_{it}, d_{jt, j \in \{1, ...N\}, j \neq i}) \tag{1}$$

The overall fitness (performance) V_t achieved in period t results as normalized sum of contributions C_{it} from

$$V_t = V(\boldsymbol{d_t}) = \frac{1}{N} \sum_{i=1}^{N} C_{it} \tag{2}$$

2.2   Organizational Set-up

Search Agents and Central Unit. The search processes are conducted in a distributed way. In particular, the N-dimensional search problem is segmented into M disjoint partial problems where each of these sub-problems is delegated to one search agent subscripted by r, $r = 1, ..., M$ correspondingly. Each search agent has primary control of its "own" subset of the N choices. Hence, from the perspective of search agent r the search problem is partitioned into a partial search vector d_t^r for those choices which are in the "own" responsibility and into $d_t^{r,res}$ for the residual choices that the other search agents $q \neq r$ are in charge of. However, in case of cross-segment interactions, choices of agent r might affect the contributions of the other agents' choices and vice versa.

Additionally, our simulation model captures a kind of "central agent" whose particular role within the search process depends on the mode of coordination. We go more into detail of the mode of coordination below.

Shaping the Search Agents' Perspective. In each time step t, a search agent seeks to identify the best configuration for the "own" subset of N^r choices d_t^r assuming that the other agents do not alter their prior choices. In particular, an agent randomly discovers two alternatives of partial configurations assigned to this agent: an alternative configuration that differs in one choice ($a1$) and another alternative ($a2$) where two bits are flipped compared to the current configuration. Hence, together with the status quo d_{t-1}^{r*} and the two alternatives $d_t^{r,a1}$ and $d_t^{r,a2}$ agent r has three options to choose from.

However, which option a search agent favors, depends on the "perspective" P_t^r the agent has on the search problem or, in other words, how the agent's *objective* in evaluating options is shaped. An agent might focus only on the "own" partial problem or may also take the rest of the search problem into consideration as controlled by parameter α^r in Eq. (3):

$$P_t^r(\mathbf{d_t}) = P_t^{r,own}(d_t^r) + \alpha^r \cdot P_t^{r,res} \tag{3}$$

$$P_t^{r,own}(d_t^r) = \frac{1}{N} \sum_{i=1+p}^{N^r} C_{it} \tag{4}$$

with $p = \sum_{s=1}^{r-1} N^s$ for $r > 1$ and $p = 0$ for $r = 1$,

$$P_t^{r,res} = \sum_{q=1,q \neq r}^{M} P_t^{q,own} \tag{5}$$

Coordination among Search Agents. The search system can employ three modes of coordination among the search agents: (1) In a fairly decentralized mode each search agent decides on the "own" partial choices d_t^r autonomously, and the overall configuration $\mathbf{d_t}$ results as a combination of these choices without any intervention by the central agent. (2) As a type of horizontal coordination our search agents inform each other about their preferences, and are allowed to veto against each other mutually. (3) In a rather central mode of coordination each search agent transfers a list with the two most preferred options from d_{t-1}^{r*}, $d_t^{r,a1}$ and $d_t^{r,a2}$ to the central agent which than chooses that combination of the r lists of proposals that promises the highest overall fitness V.

Increasing Diversity of Search. There is some evidence that imperfect information on the outcome (fitness, performance) of options could have beneficial effects on search processes (e.g. [6], [8], [9]): In particular, the more rugged a fitness landscape, the more likely it becomes that adaptive search processes employing local search stick to local maxima. False-positve evaluations of options provide the opportunity to leave a local peak and, by that, to eventually find higher levels of fitness. In order to make use of these findings, we endow our agents, eventually, with slighty distorted information about the fitness of options. In particular, $P_t^{r,own}(d_t^r)$ and $P_t^{r,res}(d_t^r)$ (search agents) and V_t (central unit) each are distorted by adding an error term as exemplarily given in Eq. (6):

$$\tilde{P}_t^{r,own}(d_t^r) = P_t^{r,own}(d_t^r) + e^{r,own}(d_t^r) \qquad (6)$$

We reflect distortions as relative errors imputed to the true performance (for other functions see [6]), and, for simplicity, the error terms follow a Gaussian distribution $N(0; \sigma)$ with expected value 0 and standard deviation σ. In particular, standard deviations $\sigma^{r,own}$ and $\sigma^{r,res}$, for the sake of simplicity, are assumed to be the same for search agents r and stable in time. The latter also holds for σ^{cent} which is relevant for the central unit. Errors are assumed to be independent from each other.

2.3   Organizational Dynamics

In the center of our research effort is the question whether altering the organizational set-up of the search processes could increase the fitness of solutions achieved. The model captures three temporal modes of change:

- *"Once"*: The organizational set-up is modified once in period T^*.
- *"Periodical"*: The set-up is altered periodically after T^{**} periods.
- *"Fitness-driven"*: The set-up is changed depending on the fitness increase, i.e., in every T^{***}-th time step the fitness change $\Delta V = V_t - V_{t-T^{***}}$ is assessed and, if ΔV is below a certain threshold v, then the set-up is altered.

Alterations are put forward along three dimensions: (1) The *objective* (or perspective) of the distributed search agents as given in Eq. (3) can be switched between focusing overall fitness V_t, fitness achieved with respect to the partial search problem d_t^r only, or the latter plus a certain ratio of residual performance as controlled by parameter α^r in Eq. (3). (2) The *mode of coordination* can be changed between the three modes as introduced above. (3) The *precision of ex ante-evaluation* as given by $\sigma^{r,own}$, $\sigma^{r,res}$ and σ^{cent} for the information errors.

In all of the three temporal change modes, alternations in two randomly selected dimensions out of three (i.e., objective, coordination mode, precision of ex ante-evaluations) can take place. Moreover, within each of these dimensions the alternative set-up is also chosen randomly. This also applies to the initial organizational set-up of the distributed search processes.

3   Simulation Experiments and Parameter Settings

For simulating the search processes under the regime of different temporal modes of change, after a fitness landscape is generated, the initially organizational setting of the search system is determined randomly. Then the systems are placed

Table 1. Parameter Settings

Parameter	Values / Types
Observation period	$T = 200$
Number of choices	$N = 10$
Interaction structures	self-contained ($K = 4$); full interdependent ($K = 9$)
Number of search agents	$M = 2$, agent 1: $d^1 = (d_1, ...d_5)$, agent 2: $d^2 = (d_6, ...d_{10})$
Perspective/Objective	three levels: $\alpha^r \in \{0; 0.5; 1\}$
Precision of evaluation $\sigma^{r,own}$, $\sigma^{r,res}$ and σ^{cent}	$(0.1; 0.15; 0.125)$; $(0.05; 0, 2; 0.125)$; $(0; 0; 0)$
Coordination mode	decentralized, lateral veto, central
Change mode	no change; once: $T^* = 25$; periodical: $T^{**} = 25$; fitness-driven: $v = 0.01; T^{***} = 25$

randomly in the fitness landscape and observed while searching for higher levels of overall fitness.

The simulations are conducted for two interaction structures of the search problem which, in a way, represent two extremes (for these and other structures see [7]): in the *self-contained* structure the overall search problem consists of two disjoint subproblems with maximal intense intra-subproblem interactions while no cross-subproblem interactions exist. In contrast, in the *full interdependent* case all single options d_i affect the fitness contributions of all other choices, i.e., the intensity of interactions is raised to maximum. Table 1 summarizes the parameter settings.

4   Results

Table 2 reports the condensed results of our simulations where each row represents results of 5,000 adaptive walks: 1,000 distinct fitness landscapes with 5 adaptive walks on each. In particular, the table displays the final performance achieved in the end of the observation period ($V_{t=200}$) and the average performance $\bar{V}_{\{0;200\}}$ over the observation time, i.e., achieved by average in each of the 200 periods. Both measures can serve as an indicators for the effectiveness of the search process as well as the frequency of how often the global maximum is found in the final period. Figures 1 and 2 display the adaptive walks for the different change modes and interaction structures under investigation.

In general, results suggest that alternations of the organizational set-up in the course of search may have beneficial effects on the performance levels achieved; however, results also suggest that the effects of alternations depend on the change mode in conjunction with the complexity of the underlying search problem: In the self-contained structure, the three modes involving organizational dynamics clearly outperform the organizationally stable scenario. The same applies to the "once" and the "periodical" mode in the full-interdependent interaction stucture;

Table 2. Condensed Results

Change Mode	Final Perf. $V_{t=200}$	CI^* of $V_{t=200}$	Avg. Perf. $\bar{V}_{\{0;200\}}$	CI^* of $\bar{V}_{\{0;200\}}$	Frequ. Glob.Max. in $t = 200$	Ratio Alt.Conf. of d	Number of Changes
self-contained structure							
no change	0.9460	±0.0027	0.9395	±0.0026	24.60%	2.88%	0
once	0.9592	±0.0022	0.9505	±0.0021	31.40%	4.80%	1
periodical	0.9674	±0.0020	0.9532	±0.0015	38.10%	7.30%	7
fitness-dr.	0.9612	±0.0024	0.9482	±0.0017	34.38%	8.07%	4.32
full interdependent structure							
no change	0.8722	±0.0034	0.8630	±0.00307	3.96%	4.76%	0
once	0.8861	±0.0032	0.8738	±0.00290	5.72%	5.49%	1
periodical	0.8914	±0.0037	0.8736	±0.00243	7.22%	8.54%	7
fitness-dr.	0.8769	±0.0047	0.8645	±0.00282	7.36%	12.58%	4.31

* Confidence intervals at a confidence level of 0.001

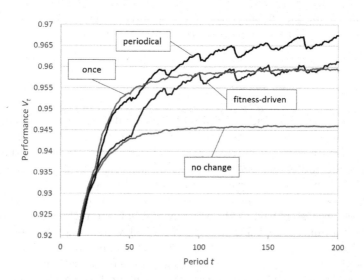

Fig. 1. Adaptive search processes in the self-contained interaction structure. Parameter settings see Table 1.

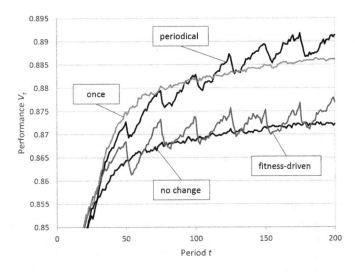

Fig. 2. Adaptive search processes in the full interdependent interaction structure. Parameter settings see Table 1.

however, with maximum complexity the "fitness-driven" mode provides final and average performance levels that are nearly the same as achieved without any change.

A closer analysis reveals that organizational dynamics apparently increase the diversity of search as can be seen in the second column from the right in Table 2. Moreover, we find that in the final period the global maximum is discovered more often in all of the scenarios employing organizational dynamics (third column from the right). In sum, this lets us hypothesize that *inducing organizational dynamics in the search processes increases the diversity of search and, by that, reduces the peril of sticking to an inferior local maximum.*

However, the periodical and the fitness-driven mode induce *oscillations* with respect to the performance levels as is obvious in Figures 1 and 2. Reasonably, this is due to a trade-off between exploration of new areas in the fitness landscape and exploitation of the current position in the landscape. In particular, each alternation brings along the chance to discover new options but, by that, also might disrupt an incremental improvement in the neighborhood of the status quo.

Having said that, the question remains why the "fitness-driven" mode shows such a rather low performance in the highly complex search problem. We argue that in this mode, additionally to the trade-off between exploration and exploitation, at least, two further effects are at work: First, the number of alterations is reduced compared to the periodical mode (most right column in Table 2). Second, with the fitness-driven mode organizational alternations, in tendency, are shifted to later periods (as becomes also apparent in the self-contained structure in Fig. 1). This is since in the early periods fitness gains usually are rather high (i.e., above threshold v) and, thus, alternations are shifted. Apparently, search for complex problems is rather sensitive to these effects.

5 Conclusion

The major finding of our study is that modifying the organizational set-up of distributed adaptive search processes has the potential to increase the effectiveness of search, but that the temporal mode of change together with the complexity of the search problem considerably affects the effectiveness of organizational dynamics.

Our analysis is subject to some limitations which should be overcome in further research efforts. For example, further studies should investigate more into detail the role of the change parameters like, frequency and thresholds triggering alternations or the dimensionality of change - which all were fixed in the simulation presented in this paper.

References

1. Baumann, O.: Distributed Problem Solving in Modular Systems: the Benefit of Temporary Coordination Neglect. Systems Research and Behavioral Science (2013), doi:10.1002/sres.2218
2. Carley, K.M., Gasser, L.: Computational organization theory. In: Weiss, G. (ed.) Multiagent Systems: A Modern Approach to Distributed Artificial Intelligence, pp. 299–330. MIT Press, Cambridge (1999)
3. Gross, T., Blasius, B.: Adaptive coevolutionary networks: a review. J. R. Soc. Interface 20, 259–271 (2008)
4. Kauffman, S.A., Levin, S.: Towards a general theory of adaptive walks on rugged landscapes. J. Theor. Biol. 128, 11–45 (1993)
5. Kauffman, S.A.: The origins of order: Self-organization and selection in evolution. Oxford University Press, Oxford (1993)
6. Levitan, B., Kauffman, S.A.: Adaptive walks with noisy fitness measurements. Mol. Divers 1, 53–68 (1995)
7. Rivkin, R.W., Siggelkow, N.: Patterned interactions in complex systems: Implications for exploration. Manage Sci. 53, 1068–1085 (2007)
8. Wall, F.: The (Beneficial) Role of Informational Imperfections in Enhancing Organisational Performance. Lecture Notes in Economics and Mathematical Systems, vol. 645, pp. 115–126. Springer, Berlin (2010)
9. Wall, F.: Comparing basic design options for management accounting systems with an agent-based simulation. In: Omatu, S., Neves, J., Rodriguez, J.M.C., Paz Santana, J.F., Gonzalez, S.R. (eds.) Distrib. Computing & Artificial Intelligence. AISC, vol. 217, pp. 409–418. Springer, Heidelberg (2013)
10. Yongcan, C., Wenwu, Y., Wei, R., Guanrong, C.: An Overview of Recent Progress in the Study of Distributed Multi-Agent Coordination. IEEE Trans. on Industrial Informatics 9, 427–438 (2013)

A Multi-agent Framework for Research Supervision Management

Omar Abdullatif Jassim, Moamin A. Mahmoud, and Mohd Sharifuddin Ahmad

Centre for Agent Technology, College of Information Technology,
Universiti Tenaga Nasional, Kajang, Selangor, Malaysia
omar4049@yahoo.com, {moamin,sharif}@uniten.edu.my

Abstract. In this paper, we propose an agent-based framework to enhance, control and manage the research supervision process. The proposed framework consists of three phases which are Research Development Activities, Performance and Completion Measurement, and Tracking Activities. The Research Development Activities phase proposes a number of activities to develop a research. These activities consist of two layers, abstract and detail. Performance and Completion Measurement phase works on measuring a student performance and expected completion date. The Tracking Activities phase presents the proposed activities to track and trigger a student's tasks. We discuss the components of the framework as possible implementation for a general application of research supervision management.

Keywords: Task Management, Supervision Management, Research Development Activities, Intelligent Software Agents.

1   Introduction

The rate of students' enrolment for postgraduate studies is rapidly increasing [27, 28, 29]. According to Patterns and Trends in UK higher education [27], there is a percentage increase of 32% between 2002–03 and 2010–11 for students registering for postgraduate study. Indirectly, this increasing trend raises some management concerns about the challenges in research supervision and development activities affecting supervisors and students. Some of these challenges include miscommunication between supervisors and students, ambiguities of research development activities, lack of effective tracking processes for status of different research activities, and last but not least, lack to effective methods to measure students' performance that reflect their real progress.

From the literature, we have not discovered any comprehensive research supervision system that formally manages research activities except some segments of processes that implement research supervision management activities [1, 2, 3] and some software that monitor students' progress [4, 5, 6]. To fill this gap, we attempt to investigate and develop a system that handles comprehensive processes of research supervision management involving supervisors and students with the following capabilities:

© Springer International Publishing Switzerland 2015
S. Omatu et al. (eds.), *Distributed Computing & Artificial Intelligence, 12th Int. Conference*,
Advances in Intelligent Systems and Computing 373, DOI: 10.1007/978-3-319-19638-1_15

- Track the Stages of Research Activities: Tracking the various research activities such as Literature Review, Modelling, Designing, etc., that are undertaken by students. Such tracking reduces the burden of having to remember many research issues (cognitive load).
- Measure Students' Performance: An outcome of this performance measure is a student's progress within a certain period.
- Gauge Students Completion: This estimates when a student would finish his/her research work.
- Housekeeping activities: Such as alert, acknowledge, remind, declare, warn etc., directed to both the supervisors and students.

We postulate that adaptive IT techniques could deliver effective solutions to supplement and enhance the performance of research supervision management activities. Specifically, based on the desired system abilities mentioned above, we propose to utilize intelligent software agents to assist supervisors in managing and monitoring those activities. This paper presents our initial findings of the work-in-progress of this research.

The objective of this project is two-fold: (i) To analyse the most efficient standard of research supervision activities, (ii) To propose a research supervision management framework based on multi-agent systems (MAS). The outcome of this paper is a model that enables software agents to assist supervisors in managing and monitoring students' research progress. The significance of this outcome contributes to a more efficient supervision and more qualified researchers.

2   Related Work

Many researchers have employed agent based-systems as effective tools to improve task management [7, 8, 9, 10, 11, 12]. However, in the domain of research supervision management, the literature do not seem to provide ample support for agent technology application in such domain. Some fragmented attempts have been made on using electronic versions of the domain process but with limited capabilities.

Pearson and Kayrooz [13] argued that research supervision is a facilitative process requiring support that involves providing educational tasks and activities which include: progressing the candidature, mentoring, coaching the research project and sponsoring students' participation in academic practices. A defining question which draws the line between facilitation and enculturation model is: "how much responsibility should a student or a supervisor take for arriving at the destination?" Mentoring is a powerful concept in this arena [13, 14].

AlBar [4] proposed an Electronic Supervision System Architecture to build an educational collaborative environment between supervisors and teachers which include several kinds of tasks to perform such as skills development, experience sharing, group meeting and tasks, and discussing about teaching and administrative strategies. Romdhani et al. [5] presented a student project performance management system developed as an integrated and collaborative online supervision system for Bachelor final year and dissertation projects. The system activities are:

- Development of project proposal.
- Development of the problem description.
- Following the objectives.
- Presenting and analysing the data.
- Drawing conclusions and identifying future work.
- Presenting and defending the work orally.
- Development of the final version of the report.

Yew et al. [1] proposed a conceptual framework that integrates supervision process, knowledge management (KM) activities, and enabling information technology (IT) for designing such research supervision as a KM System. Swanson and Watt [3] conducted theoretical reviews on supervisors and postgraduate responsibilities in research in order to provide efficient monitoring of supervision activities. Lubega and Niyitegeka [6] presented a pedagogical model for E-supervision that is facilitated by the available technology instead of traditional supervision activities such as e-mails, discussion boards, forums, telephony, chat rooms, Wiki, blogs and e-research group.

We discovered that only fragments of efforts have been made that address the issues of research supervision management, as reviewed above. While we do not claim that the review is exhaustive, there seems to be a serious lack of comprehensive models or systems for research supervision management. Consequently, this paper attempts to tease out a comprehensive model for research supervision management based on the multi-agent approach.

Software agents have been widely used to assist humans in complying with the schedules of a collaborative work process [9, 10] and task management applications [24, 25, 26]. Consequently, in this research, we exploit the software agent technology due to its autonomous, reactive, proactive and social ability characteristics [9, 10].

3   A Conceptual Multi-agent Research Supervision Management Framework

Initial investigation from the literature reveals six main stages that constitute research development processes which are Basement Stage [15,16]; Review Stage [17]; Data Collecting Methods [18], Data Analysis [19, 20], Development Stage [21], and Testing Stage [22]. Based on our analysis of these six stages, we synthesize a conceptual framework for an agent-based research supervision management that espouses the six stages.

Figure 1 shows each of these stages that involves numerous steps which are applicable to diverse research scopes. Based on the nature and characteristics of a research, the supervision team selects some of these steps which are applicable to the particular research scope.

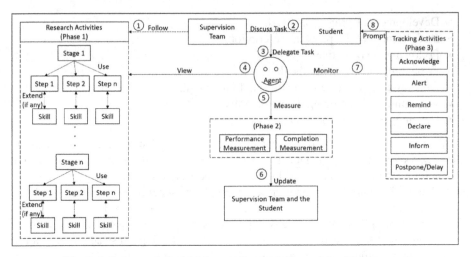

Fig. 1. A Framework for Multi-agent Research Supervision Management

As shown in Figure 1, a supervisor or a supervision team (1) follows the given stages, (2) discusses a new task with a student, and (3) delegates the tasks to the student's agent, which communicates with the student. The agent then performs several tasks; it (4) views the research processes' contents and specifies the given task to a particular stage and step. It also (5) measures the performance and the completion of the research work and (6) updates the student and the supervision team. In addition, the agent (7) monitors the student's achievement and performs some activities to (8) prompt the student to meet the tasks' deadlines.

In general, the presented framework consists of three main phases which are Research Development Activities, Performance and Completion Measurement, and Tracking Activities.

3.1   Research Development Activities

The literature reveal several activities for research development. We propose that these activities can be divided into two layers; abstract and detail. As shown in Figure 2, the abstract layer consists of six stages, and the detail layer consists of numerous steps. The stages are basement stage, review stage, data collecting stage, data analysis stage, development stage, and testing and validation stage [15, 17, 18, 19, 23].

A supervision team must absolutely follow the abstract layer stages. But it is able to pick up appropriate steps (and not all suggested steps in Figure 2) from the detail layer since the complexity varies from one research to another.

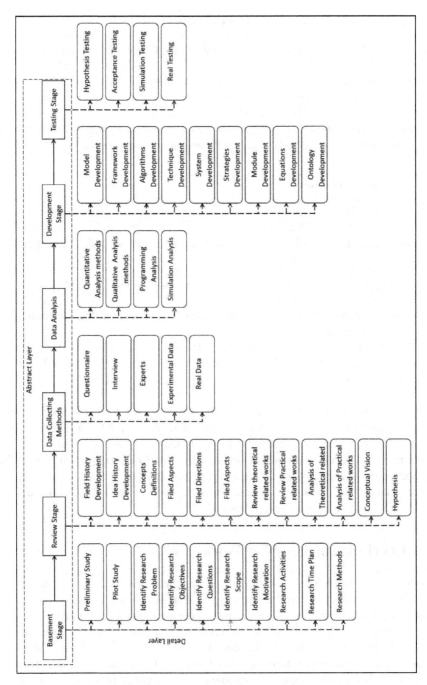

Fig. 2. Research Development Activities

3.2   Performance and Completion Measurement

The second phase entails measuring a student's performance of past tasks and measuring the expected completion date. The completion is influenced by performance, if the performance is high, the completion is imminent and vice versa. The performance is measured by dividing the given time for a task by the real time taken to achieve that task. It is gauged as Meet Expectation (ME) if the result equals 1, Exceed Expectation (EE) if it is greater than 1, and Low Expectation (LE) if it is less than 1. However, the details of measurements are beyond the scope of this paper.

3.3   Tracking Activities

This phase involves two activities. The first records different tasks and messages between a student and his/her supervisor. The second activity activates and monitors the student's efforts to achieve his/her task within a given deadline. We suggest six actions for this activity as follows:

- Acknowledge: The agent notifies a message sender that the message is sent successfully and received by a recipient.
- Remind: The agent reminds the student regarding a task and the remaining time before the deadline.
- Alert: The agent alerts the student when a deadline is imminent. A penalty token is attached with an alerting message. For example, *"please be informed that you have to submit your progress report in one hour, otherwise the meeting will be cancelled and this will affect your performance."*
- Declare: The agent declares a message to the student and his/her supervisor when the student fails to comply with a given deadline. For example, the agent declares that *"the meeting is cancelled due to failure in submitting the assignment report."*
- Inform: This function provides communication between the student and his/her supervisor to share information about a particular matter.
- Postpone/Delay: The student or the supervisor may request to postpone/delay a meeting due to some emergency issues.

4   Conclusion and Further Work

In this paper, we present our initial findings on our research to develop a comprehensive framework for research supervision management utilizing the multi-agent system's approach to enhance the supervision quality, accurately reveal a student's progress status, and track the activities of a student and his/her supervisor.

Accordingly, the proposed framework introduces three phases; Research Development Activities, Performance and Completion Measurement, and Tracking Activities. Research Development Activities phase proposes two layers; an abstract layer that all supervisors must follow, and a detail layer from which supervisors may select some or all of the activities according to a particular project's needs. The second phase entails measuring a student performance and eventually the expected completion date. There are three possible results, Exceed Expectation (EE), Meet Expectation (ME), or Low Expectation (LE). The last phase involves tracking different activities and produce appropriate housekeeping messages. We suggest six actions which are Acknowledge, Remind, Alert, Declare, Inform, and Postpone/ Delay.

In our future work, we shall study the required MAS aspects that animate tasks, e.g. interaction, collaboration, delegation, learning, etc. Subsequently, we shall build an agent-based system simulation model to test the proposed framework and validate its efficiency.

Acknowledgements. This project is sponsored by Universiti Tenaga Nasional (UNITEN) under the Internal Research Grant Scheme No. J510050546.

References

1. Yew, K.: A framework for designing postgraduate research supervision knowledge management systems. In: National Postgraduate Conference (NPC), September 19-20, pp. 1–6 (2011)
2. Ismail, A., Abiddin, N.Z., Hasan, A.: Improving the Development of Postgraduates' Research and Supervision. International Education Studies 4(1), 78–89 (2011)
3. Swanson, C.C., Watt, S.: Good Practice in the Supervision & Mentoring of Postgraduate Students. McMaster University (2011)
4. AlBar, A.M.: An Electronic Supervision System Architecture in Education Environments. European Journal of Business and Management 4(8) (2012)
5. Romdhani, I., Tawse, M., Habibullah, S.: Student Project Performance Management System for Effective Final Year and Dissertation Projects Supervision. In: Infonomics Society (ed.) London International Conference on Education, LICE 2011. Infonomics Society, London (2011)
6. Lubega, J., Niyitegeka, M.: Integrating E-Supervision in Higher Educational Learning. In: Aisbett, J., Gibbon, G., Rodrigues, A.J., Kizza, M.J., Nath, R., Renardel, G.R. (eds.) Strengthening the Role of ICT in Development, vol. IV, pp. 351–358 (2008)
7. Hsieh, F., Lin, J.: Project Scheduling Based on Multi-Agent Systems. Journal of Advanced Management Science 3(3) (September 2015)
8. Itaiwi, A.K., Ahmad, M.S., Hamid, N.H.A., Jaafar, N.H., Mahmoud, M.A.: A Framework for Resolving Task Overload Problems Using Intelligent Software Agents. In: 2011 IEEE International Conference on Control System, Computing and Engineering, ICCSCE 2011 (2011)
9. Ahmed, M., Ahmad, M.S., Yusoff, M.Z.M.: A Collaborative Framework for Multiagent Systems. International Journal of Agent Technologies and Systems (IJATS) 3(4), 1–18 (2011)
10. Ahmed, M., Ahmad, M.S., Yusoff, M.Z.M.: Mitigating Human-Human Collaboration Problems using Software Agents. In: The 4th International KES Symposium on Agents and Multi-Agent Systems – Technologies and Application (AMSTA 2010), Gdynia, Poland, June 23-25, pp. 203–212 (2010) ISBN:3-642-13479-3 978-3-642-13479-1
11. Sánchez, E., Lama, M., Amorim, R., Riera, A., Vila, X., Barro, S.: The EUME Project: Modelling and Design of an Intelligent Learning Management System. In: Proceedings of the AIED-Workshop on Intelligent Learning Management Systems, Sydney, Australia, pp. 183–191 (2005)

12. Decker, K., Sycara, K., Zeng, D.: Designing a Multi-Agent Portfolio Management System. In: AAAI 1996 Workshop on Internet-Based Information Systems, Portland (1996)
13. Pearson, M., Kayrooz, C.: Enabling Critical Reflection on Research Supervisory Practice. International Journal for Academic Development 9(1), 99–116 (2004)
14. Brew, A.: Conceptions of Research: a phenomenographic study. Studies in Higher Education 26(3), 271–285 (2001)
15. Blasius, J., Walliman, N., Baiche, B.: Your research project. A step-by-step guide for the first-time researcher. KZfSS Kölner Zeitschrift für Soziologie und Sozialpsychologie 53(3), 607–608 (2001)
16. Mmuya, M.: Developing and writing a research proposal: instruction manual (2007)
17. Krauss, S.E., Hamzah, A., Nor, Z.M., Omar, Z., Suandi, T., Ismail, I.A., Zahari, M.Z.: Preliminary investigation and interview guide development for studying how Malaysian farmers' form their mental models of farming. The Qualitative Report 14(2), 245–260 (2009)
18. Marshall, C., Rossman, G.B.: Data Collection Methods. In: Designing Qualitative Research, ch. 4 (2006) (retrieved June 7, 2006)
19. Waters, E.: The Goodness of Attachment Assessment: There Is A Gold Standard But It Isn't As Simple As That (2002) (retrieved October 22, 2013)
20. Wögerer, W.: A survey of static program analysis techniques. Vienna University of Technology (2005)
21. French, S.D., Green, S.E., O'Connor, D.A., McKenzie, J.E., Francis, J.J., Michie, S., Grimshaw, J.M.: Developing theory-informed behaviour change interventions to implement evidence into practice: a systematic approach using the Theoretical Domains Framework. Implementation Science 7(1), 38 (2012)
22. Wilcox, R.R.: Introduction to robust estimation and hypothesis testing. Academic Press (2012)
23. Thabane, L., Ma, J., Chu, R., Cheng, J., Ismaila, A., Rios, L.P., Goldsmith, C.H.: A tutorial on pilot studies: the what, why and how. BMC Medical Research Methodology 10(1), 1 (2010)
24. Lavendelis, E., et al.: Multi-agent Robotic System Architecture for Effective Task Allocation and Management. In: Recent Researches in Communications, Electronics, Signal Processing & Automatic: Proceedings of the 11th WSEAS International Conference on Signal Processing, Robotics and Automation (ISPRA 2012), United Kingdom, Cambridge, February 22-24, pp. 167–174 (2012)
25. Arauzo, A.J., Pavon, J.: Agent-based modeling and simulation of multiproject scheduling. Inteligencia Artificial (42), 12–20 (2009)
26. Lacouture, J., Gascueña, J.M., Gleizes, M.-P., Glize, P., Garijo, F.J., Fernández-Caballero, A.: ROSACE: Agent-based systems for dynamic task allocation in crisis management. In: Demazeau, Y., Müller, J.P., Rodríguez, J.M.C., Pérez, J.B. (eds.) Advances on PAAMS. AISC, vol. 155, pp. 255–260. Springer, Heidelberg (2012)
27. Patterns and trends in UK higher education (2012), http://www.universitiesuk.ac.uk/highereducation/Documents/2012/PatternsAndTrendsinUKHigherEducation2012.pdf
28. The rise of graduate education and university research, Higher Education in Asia (2014), http://www.uis.unesco.org/Library/Documents/higher-education-asia-graduate-university-research-2014-en.pdf
29. Trends in Higher Education, The Association of Universities and Colleges of Canada (2011), http://www.aucc.ca/wp-content/uploads/2011/05/trends-2011-vol1-enrolment-e.pdf

A Formal Machines as a Player of a Game

Paulo Vieira[1] and Juan Corchado[2]

[1] Polythecnic Institute of Guarda,
UDI – Research Unit for Inland Development of Guarda, Portugal
[2] University of Salamanca, Salamanca, Spain

Abstract. The Four In Line game is a known game and is played in a game board. A lot of references of this game can be found in internet. This game is a game played between two players. For play this game, against a good player, is necessary a lot of skills. We are working in a new computational system, that we call a Formal Machine, and we developed a library where we implemented this formalism. This library is in version 0.03. We are choosing to implement this game to show; that is possible to implement the new formalism as a player of that game, that the new formalism can have considerable Artificial Intelligence, how we can implement the new formalism for solve an engineering problem. In this paper we described an implementation of a Formal Machine that is a Four In Line player. For measure the Artificial of Intelligence, of the new formalism implemented, we present a statistical study whose conclusions are obtained from a collected sample of 100 different plays/players of the game. The Formal Machine implemented is a computational model whose abstract methods are implemented to make it an efficient player. This implementation of the Four In Line game is played between a human being and a Formal Machine. Thus the Formal Machine plays against an intelligent player. The game board consists in one matrix that can be chosen and setting among a range of matrices that are between a 4 x 4 matrix and a 10 x 10 matrix. Each one of the elements of this matrix is called by us square. It wins the player that completes 4 sequential and consecutive squares in horizontal, vertical or oblique way.

Keywords: Formal Machines, Computational systems, Artificial Intelligence, and Inference Statistic.

1 Introduction

The Four In Line (FIL) Game is a widely game found in internet. The FIL game is constituted by a game board that, is a $n \times m$ matrix, n rows and m columns, a matrix of $n \times m$ squares. In this implementation[1] the game board can be chosen[2] among a matrix 4×4 and a matrix 10×10. This game is a game that is played between two players, in this implementation one of the players is a Formal Machine (FM). The versatility of the FM allows to that it is a good computational

---

[1] http://www.ipg.pt/user/~pavieira/private/SW/FILGame5x5.jar
[2] We only implemented the game boards 4 x 4, 4 x 5, 5 x 4, 5 x 5.

© Springer International Publishing Switzerland 2015   137
S. Omatu et al. (eds.), *Distributed Computing & Artificial Intelligence, 12th Int. Conference*,
Advances in Intelligent Systems and Computing 373, DOI: 10.1007/978-3-319-19638-1_16

model to write and solve engineering problem with formal methods. We have two kinds of algorithms for implement Formal Machines (FMs). One of it is a serial FM implementation and the other is a parallel FM implementation. In this paper we, only, implemented the serial FM implementation.

This article has the following structure. It starts with the 1-*Introduction* where, we talk a little about the game, about the players, about how we implemented it and how the article is structured. The second section is called *Playing the Game*, in it are described the algorithms that allow to implement the FM as a FIL game player, and is made a description of how the FM is playing this game. In section 3, called *The Formal Machine, the Player*, is defined the new computational model that we are working, called Formal Machine (FM), and its Computational Structure (shorting, CSFM)[3]. CSFM denoted either the Computational Structure of a FM and its implementation. The CSFM is a set of variables, abstract classes and methods. When someone wants to implement a FM it should download the CSFM classes and overwrite its abstract methods. Next is the section 4, *Results*. In it we present a statistical study about several complete games and we extrapolated the results obtained through of confidence intervals. In section 5, *Conclusion and Future Work*, we present a conclusion and we talk about the work that is ongoing. Finally we have the section 6, *Appendix*. In the appendix can be seen; figures about the interfaces of the game, a table about the moves of one complete game, and the description of the CSFM.

2   Playing the Game

In this section we present the algorithm that is implemented for to do the FM a FIL game player. The FM is a formalism created by us, this formalism is a computational system, and it intended to be a formalism where, in the sense of the Category Theory ([Mac10]), we can rewrite any computational system without losing its structure. The aim is that a computational system ([Eil74]), when rewritten as a FM, should to maintain its original structure in the FM[4]. Our goal when we started to work, in all of this, was to find concepts, define them and quantify them, that are associated with idea of behavior intelligent and skills in computational systems. In this work process was then that we built this model to do this, since we have proved, not here, that lot of computational systems can be rewritten in this model without losing its structure. The problem become, thus, in defining and quantifying these concepts and skills in the new model. But with the construction of this model the need arose and curiosity to show that we could use it directly in solving a problem that requires intelligence and skills. It is in this context that the idea of implementing a game, choosing the FIL game, arises and appears this article. As the FM is a computational system

---

[3] CSFM is an acronym of Computational Structure of a Formal Machine, is in version 0.03, and can be download from `http://www.ipg.pt/user/~pavieira/private/SW/FM_OpComp_v003.jar` documentation `http://www.ipg.pt/user/~pavieira/private/SW/javadocFM_OpComp/index.html`

[4] Is not our pretension to show, in this paper, how it is done.

it has a computational structure and we implement it, in Software, through of a set of abstract methods and classes called the CSFM. When someone wants to implement a FM it must overwrite the CSFM methods.

Now let's go to talk about the game. The way as the FM sees the game depends of the initial environment. This environment is a game board that is set as matrix where the game is running. If the game board is a matrix 4×4, 4×5, 5×4 or 5×5, then the FM implemented does serial processing, and we say that the FM has a serial procedure. If the game board implemented is from those until a matrix 10×10, then the FM does parallel processing. The option to make parallel processing to this boards game was because the serial implementation originated, in the drafts implementation, long response times of the FM player which made us give up to implement them in serial mode. The two distinct ways to work of the FM are illustrated in the figure 1, 2.

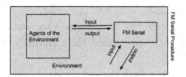

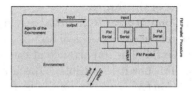

Fig. 1. FM serial procedure   **Fig. 2.** FM parallel procedure

In a FM serial procedure we obtain the input from the environment and/or from an Agent. The input, in the machine, is translated to the language of the machine, the configuration c. c is put in a set of configuration $c_P \in ConfM$. In this implementation the $c_P = \{c\}$. After to introduce the input the FM runs the CSFM, that was written, and from c_P is processed the output set of configurations c'_P. The c'_P is sent to the output. In this implementation $c'_P = \{c'\}$. Then c'_P is translated in the environment language and is sent for the environment. Thus, we describe a, new, move of the FM.

The FM with a parallel procedure receives the input from the environment or from agents and is translated in the language of the machine, c_P. After that the machine, the FM, inside it, launches several threads (suppose k threads) and c_P is translated in each one of the threads. The translation of the c_P in thread i is c_P^i. Thus, $c_P^1, c_P^2, ..., c_P^k$ are the inputs respectively of these threads $1, 2, ..., k$. Then each thread produces one output, respectively, $c_P'^1, c_P'^2, ..., c_P'^k$. The $c_P'^i$ is obtained from the thread i c_P^i. After this is necessary to produce only one output, c'_P, that is obtained from the output's threads. For last the c'_P is translated in the environment language and we have a move of the FM.

3 The Formal Machine, the Player

In this section we are going to present; the definition of the new computational model, the FM, the definition of its computational structure, the CSFM in version 0.03, and we refer too how to conceive a problem in a FM. Therefore, let's start. We start to present the definition of a Formal Machine.

3.1 The Formal Machine

A FM, fM, is a 7-tuple with a computational operator, $\vdash$. This is a short definition

$$\text{fM} = (\underbrace{CompM_B, CompM_R,}_{components}\ \underbrace{ConfM}_{relations\ configuration}\ ,\underbrace{ConfM_i,}_{initial}\underbrace{ConfM_f,}_{final}\ \underbrace{InstM}_{instructions}\ ,\ \underbrace{VN_{Alg}}_{VonNeumann}\ )$$

wherein:

i) $CompM_B = \{C_1, ..., C_n\}$ is a finite set. $CompM_B$ is called the set of *components* of the fM,

ii) $CompM_R = \{R_1, ..., R_m\}$ is a finite set. $CompM_R$ is called the *relations* of the fM. These elements have not directly importance in the Structural Computation of the fM. The importance of the set $CompM_R$ is in to build the instructions of the fM, but it is not in the discussion here.

iii) $ConfM \subseteq C_1 \times ... \times C_n$ is a set. $ConfM$ is called the set of *configurations* of the fM,

iv) $ConfM_i \subseteq ConfM$ is a subset of the $ConfM$. $ConfM_i$ is called the set of *initial configurations* of the fM,

v) $ConfM_f \subseteq ConfM$ is a subset of the $ConfM$. $ConfM_f$ is called the set of *final configurations* of the fM,

vi) The $InstM$ set has two kinds of elements. A kind of elements that is called instructions and is denoted by inst[5] and the other denoted a special operator NOP, that is an acronym of, No OPeration, and

vii) VN_{Alg} is an algorithm that describes how the Machine works.

The elements of the sets $CompM_B$, $CompM_R$ and $InstM$ should verify additional conditions. Here we do not talk about it, it is out of the article's scope.

3.2 Computational Structure of a Formal Machine

The CSFM is a computational implementation of a FM. Here we are describing the CSFM in is version 0.03. A CSFM is a 4-tuple CSFM=$(VN_{Alg}$, psm, $\mathcal{A}$, $\vdash$) where: VN_{Alg} is the Von Neumann's Algorithm (describe how the machine work), psm is the Physical State of the Machine (describe the current state of the machine), $\mathcal{A}$ is a Finite Automaton (describe how work the instructions of the machine), and $\vdash$ is the Computational Operator and it is defined by an algorithm called Computational Operator Algorithm (COA)[6](describe how is doing a step of computation). A more detailed presentation of the definition of the CSFM of a FM can be seen in appendix subsection A.2 page 144.

---

[5] Inst is a partial function, inst:$(ConfM - ConfM_f)^k \to \mathcal{P}(ConfM)$.

[6] Download `http://www.ipg.pt/user/~pavieira/private/SW/FM_OpComp_v003.jar`

3.3 How to Conceive a Problem in FMs

For write a problem in FM Technology it should be had in counts the following; i) The components of the FM are the components of the problem, ii) The instructions of the FM are the actions carried out for solve the problem iii) In the psm should be possible to appear any possible state of the problem, iv) NOP is an operator that serves for restrict the number of active configurations of the Machine or to provoke delays in the system. In the delays the Machine not does anything, And v) In the definition of the operator ⊢ is necessary to do some choices through of the implementation of the motives, the motives are abstract methods of the CSFM software. The implementation, in FM Technology, should have in counts the problem that we are solving.

Now we can ask, but how to transport the idea referred above for the FIL game. In the FIL game first is necessary to setting the game. To set the game consist in to choose the size of the game board where the game is running and to choose "whom" starts the game. After the game have been setting, the moves consist in to click in buttons (the squares) of the game board. Thus, the constitution of the FM is generically the following:

- the components of the FM are three, C_1, C_2 and C_3:
i) C_1 - describes the moves, in the game, made by the two players
ii) C_2 - describes the moves, in the game, made by the Human Being (HB)
iii) C_3 - describes the moves, in the game, made by the FM.
- the set of the instructions of the FM. The only action is to use the buttons, clicking in it, for play the game. Thus the FM only has one instruction.
- the operator ⊢. Is defined through of the COA

4 Results

In this section we present a statistical study ([LaW04]) about this game, we use inference statistic. Thus, we collect a set of data that are the results of to play the game. We collected a sample of 100 results of the game and this collects was made through of a random choice of players that is a way to guarantee the representation of the sample.

Table 1. Set of data collected for a game board 4×4

Matrix table, 4×4																			
FM, First player										HB, First player									
FM	FM	D	D	D	FM	D	D	D	D	FM	D	D	FM	D	D	D	D	D	FM
D	D	D	D	D	FM	FM	FM	FM	FM	D	D	FM	D	FM	D	FM	D	FM	D
D	FM	D	FM	FM	FM	D	D	FM	FM	D	D	D	D	D	D	FM	FM	FM	D
D	FM	FM	FM	D	D	D	FM	FM	FM	D	FM	D	D	FM	D	FM	FM	FM	D
D	D	D	FM	D	FM	D	FM	FM	D	D	FM	D	D	FM	D	D	D	FM	FM

D-draw, FM - wins the FM, HB-wins the Human Being. We define a random variable $X_{P_{FM}}$ (respectively, $X_{P_{HB}}$, X_{P_D}) as "the quantity of victories that the FM obtains" (respectively, "the quantity of victories that the HB obtains", "the quantity of draws in the game"). We know that $X_{P_{FM}}$ has Bernoulli distribution with parameter p_{FM}, $X_{P_{FM}} \sim B(p_{FM})$ (respectively, $X_{P_{HB}} \sim B(p_{HB})$, $X_{P_D} \sim B(p_D)$). This random variable possesses Bernoulli distribution and, as the sample, $n = 100$, has more than 30 elements the theory and practical say to us that it can be approximated through of a normal distribution with average p_{FM} and variance $\frac{p_{FM}\, q_{FM}}{n}$. Thus, $\frac{X_{P_{FM}} - p_{FM}}{\sqrt{\frac{p_{FM}\, q_{FM}}{n}}} \sim N(0,1)$ (respectively, $\frac{X_{P_{HB}} - p_{HB}}{\sqrt{\frac{p_{HB}\, q_{HB}}{n}}} \sim N(0,1)$, $\frac{X_{P_D} - p_D}{\sqrt{\frac{p_D\, q_D}{n}}} \sim N(0,1)$) with $q_{FM} = 1 - p_{FM}$ (respectively, $q_{HB} = 1 - p_{HB}$, $q_D = 1 - p_D$). From this we can extrapolated the results for the population using a confidence interval with 95 percent of confidence. By the theory, $p_{FM} \in [\hat{p}_{FM} - z_{\alpha/2}K, \hat{p}_{FM} + z_{\alpha/2}K]$ where $K = \sqrt{\frac{p_{FM}\, q_{FM}}{n}}$ with confidence of $100\,(1 - \alpha)\,\%$ (respectively and in similar way, for p_{HB} and p_D). Analyzing data we can say with a confidence of 95% that the FM wins between 33% to 52% of the games played, that there are draw between 47% to 68% of the games played and that the wins of the Human Being are residual and have not expression. This shows that the algorithm implemented, the FM player, is a good algorithm and possesses a considerable Artificial Intelligence.

Table 2. values calculated from the sample

Matrix table, 4 × 4								
symbols	α	$\frac{\alpha}{2}$	$z_{\frac{\alpha}{2}}$	$\hat{p}_{FM}$	$\hat{q}_{FM}$	$K = \sqrt{\frac{\hat{p}_{FM}\,\hat{q}_{FM}}{n}}$	$\hat{p}_{FM} - z_{\alpha/2}K$	$\hat{p}_{FM} + z_{\alpha/2}K$	$\hat{p}_{HB}$
values	0,05	0,025	1,959963985	0,43	0,57	0,049507575	0,332966936	0,527033064	0
symbols	$\hat{q}_{HB}$	$K = \sqrt{\frac{\hat{p}_{HB}\,\hat{q}_{HB}}{n}}$	$\hat{p}_{HB} - z_{\alpha/2}K$	$\hat{p}_{HB} + z_{\alpha/2}K$	$\hat{p}_D$	$\hat{q}_D$	$K = \sqrt{\frac{\hat{p}_D\,\hat{q}_D}{n}}$	$\hat{p}_D - z_{\alpha/2}K$	$\hat{p}_D + z_{\alpha/2}K$
values	1	0	0	0	0,57	0,43	0,049507575	0,472966936	0,667033064

5   Conclusion and Future Work

The FIL game was chosen for implementing a FM, as a player of the game, for three reasons. One of it is because the game is played between two players, the second because one of the players is an human being and the third because for play it is necessary some admixture of properties that normally are associated with the intelligence and presupposes the exercise of some skills. Thus, the FM player is in competition against, an intelligent player, a Human Being. At the moment that we write this paper we have implemented the FM serial procedure not the parallel FM procedure (this is ongoing). Thus, now we can play the game in game boards that are matrices of 4×4, 4×5, 5×4 and 5×5. The implementation of game boards, that are matrices from 5×6, 6×5 until 10×10, are ongoing and following the algorithm illustrated in figure 2, the parallel FM procedure (justification in section 2). For measure the skills of the FM player we did a statistical study when the game board is a matrix of 4×4. The results measured in section 4 show to us that the FM player has a considerable skill to play this game. In the future we going to do similar analysis for the others possibles environments.

References

LaW04. Wasserman, L.: All of Statistics: A concise course in statistical inference. Springer Science + Business Media Inc. (2004)

Eil74. Eilenberg, S.: Automata, languages, and machines, vol. A. Academic Press (1974)

Hop08. Hopcraft, J.E.: Introduction to Automata Theory, Languages, And Computation, 3/E. Pearson Education (2008)

Mac10. Mac Lane, S.: Categories for the Working Mathematician. Graduate Texts in Mathematics. Springer, Berlin (2010)

Ma98. Mac Lane, S.: Categories for the Working Mathematician. Graduate Texts in Mathematics. Springer (1998) ISBN 0-387-98403-8

A Appendix

A.1 Appendix: Playing the Game

In this subsection, of the appendix, we are going to describe one complete play of the game. We present two figures, figures 3 and 4, and one table, table 3. In the figure 3 we can see the interface of the FIL game with its different possibles choices for the game board and for the player that start the game. In figure 4 we can see all the moves of one complete game and its result. In the table 3 is showed step by step all the moves of one complete game.

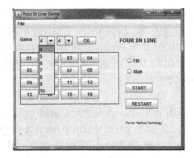
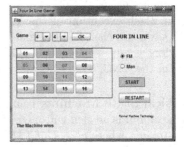

Fig. 3. FIL Game Interface   **Fig. 4.** Square 14, FM wins

Table 3. All moves of one complete game

Moves	FM: FM	Human Being	comment
First player, FM	10		random move of the FM
HB	10	4	obstruction to the HB's play
FM	10,3	4	
HB	10,3	4,7	
FM	10,3,2	4,7	tentative to do a vertical alignment of the FM
HB	10,3,2	4,7,5	
FM	10,3,2,6	4,7,5	FM's obstruction of the HB's tentative to win
HB	10,3,2,6	4,7,5,11	in-consequent move of the HB
FM	10,3,2,6,14	4,7,5,11	the FM is the winner

A.2 The Computational Structure of a FM

In this subsection we present each one of the constituents of the CSFM. We start for remember the definition of the CSFM of a FM. The CSFM of a FM is a 4-tuple CSFM=$(VN_{Alg}$, psm, $\mathcal{A}$, $\vdash)$ where: VN_{Alg} is the Von Neumann's Algorithm of the FM, psm is the Physical State of the Machine, $\mathcal{A}$ is a Finite Automaton, and $\vdash$ is the Computational Operator of the FM and it is defined by an algorithm called Computational Operator Algorithm (COA). The Finite Automaton referred in the CSFM definition is the computational model where the $InstM$ is represented. The COA Algorithm describe how the operator $\vdash$ works.

Von Neumann's Algorithm, VN_{Alg}. The Von Neumann's Algorithm describes how the machine works. The VN_{Alg} is written in agree with the problem that we have to solve and should allow to the machine to work for solve the problems for which it was created. The VN_{Alg} is an algorithm that obeys to a structure, it is divided in three spaces, three zones. A *definition zone*, a *setting zone* and an *execution zone*. The definition zone is the place in the algorithm where the constants, the variables and all kinds of objects that are used in the algorithm are defined. In the setting zone, is a zone where the initial configuration of the machine is set and, from out, the first perception of the environment is acquired or an input is introduced. The execution zone is called also the *Von Neumann's Cycle* (VNC). The VNC is a loop with or without a stopping condition. Each execution of the VNC is a *cycle of the machine*. Through of the VN_{Alg} we can classified the machines, through of it VN_{Alg}, but to explain how it is done is out of the aims of this paper. The operator of computation $\vdash$ is used only in the VNC. The VNC is a sequence of operators $\vdash$.

Physical State Machine, psm. The psm is a matrix with rows and columns. Each column is an element of $ConfM \times \{0, 1\}$. A column c_i can be seen as a pair $c_i = (c_i^1, c_i^2)$ where $c_i^1 \in ConfM$ is a configuration of the FM and $c_i^2 \in \{0, 1\}$. When $c_i^2 = 1$ means that the configuration c_i^1 is active and when $c_i^2 = 0$ means that the configuration c_i^1 is not active.

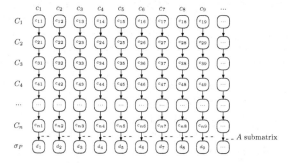

Fig. 5. Physical state of the Machine fM, psm_{fM}

The rows of the psm matrix are split in two kinds of entries. The firsts rows of the psm, labelled $C_1, C_2, ..., C_n$, are the components of the FM (elements of $CompM_B$) and the last row, denoted by $\sigma_P = (d_1, ..., d_{|ConfM|})$, is an element of the Cartesian product $\{0,1\}^{|ConfM|}$. σ_P represents, by signalization with the number one, the set of configurations that are active. We can also to define an invariant submatrix, A, of all configurations, $ConfM$, of a FM. We can study the behavior of a FM, in iteration or time, through of function $b(t) = psm(A, \sigma_P(t))$ where $\forall t \sigma_P(t) \in \{0,1\}$.

Finite Automaton, $\mathcal{A} = (Q, I, F, A, \Delta)$ with $\Delta \subseteq Q \times A \times Q$. This finite automaton, present in the CSFM definition, serve to describe and implement, automatically, the instructions of a FM. The set of states, Q, is the power set of the set of configurations, $ConfM$, $\mathcal{P}(ConfM)$. $Q = \mathcal{P}(ConfM) = \{c \in C_1 \times ... \times C_n : c \in ConfM\}$ and $A = InstM$. The operation NOP is defined in the following way. Suppose $c_P \subseteq ConfM$, from NOP you can obtain any $c'_P \subseteq ConfM$ where $c'_P \subseteq c_P$. Therefore, $(c_P, NOP, c'_P) \in \Delta$. Thus, we have a reason, through of the operator NOP, for the fact $c_P \vdash c'_P$. $I = \mathcal{P}(ConfM_i)$ and $F = \mathcal{P}(ConfM_f)$. For last the set of transitions, Δ, is $\Delta = \{(c_P, NOP, c'_P) : c'_P \subseteq c_P\} \cup \{(c_P, inst, c'_P): (inst \in InstM - \{NOP\}$ is a k-partial function}.

Computation Operator Algorithm (COA), $\vdash$. A computation of the FM is a finite sequence of $\vdash$'s. In that sequence the use of each one of the $\vdash$'s is called a *step of computation*. This step, in true, is a process of decision of the FM and is described in algorithmic way, called *Computational Operator Algorithm* (abbreviated COA). This algorithm shows by the existence of choices in the motive(s), by overwriting of abstract methods, which the FMs are highly versatile and because of it has high capacity to be a good model for engineering problems. Now we describe the COA Algorithm. The COA allows to give a step of computation $c_{P_{input}} \vdash c_{P_{output}}$.

COA Algorithm.

(state q_0) step 0. Take $c_{P_{input}}$. Translated the $c_{P_{input}}$ in a set of configurations c_{P_0} such that $c_{P_0} \in ConfM_i$ ($motive_1$). If $motive_{10}$ is true Then go to step 9 Else go to step 1.

(state q_1) step 1. Update psm, putting c_{P_0} in the column c_P, $c_P \leftarrow c_{P_0}$. Go to step 2

(state q_2) step 2. *counter* $\leftarrow 0$. Go to step 3

// *A step of computation* $\vdash$. *Here we are between steps 3 up 9. This is a cycle for the processing*

(state q_3) step 3. Read the c_P from the psm, $c_{P_i} \leftarrow c_P$. Go to step 4

(state q_4) step 4. Are you going to use an instruction? (motive$_2$) If yes Then go to instruction Else { $c_{P_r} = \emptyset$ and step 5}.

(state q_5) step 5. Are you going to use a NOP? (motive$_3$) If yes Then go to NOP Else { $c_{P_j} = \emptyset$ and go to step 6}

(state q_6) step 6. build $c_{P_i} = c_{P_r} \cup c_{P_j}$. Go to step 7

(state q_7) step 7. Update psm. Go to step 8

(state q_8) step 8. If (motive$_4$) is true Then { *counter* $\leftarrow$ *counter* $+1$ and go to step 3} Else {(motive$_9$)} and {$c_{P_{output}} \leftarrow c_{P_i}$ go to step 9}}

// *end the processing*

(state q_9) step 9. end

instruction:

(state $q_{4.i}$) i) Choose an instruction $inst \in InstM$. Suppose without generality that $inst$ is a k-partial function (motive$_5$). Go to step 4.ii)

(state $q_{4.ii}$) ii) Choose k configurations that are elements of c_{P_i}, $c_1, .., c_k \in c_{P_i}$, $\boldsymbol{c} = (c_1, .., c_k)$ (motive$_6$). Go to step 4.iii)

(state $q_{4.iii}$) iii) Apply $I(\boldsymbol{c}) = c_{P_r}$ (motive$_7$). End instruction.

NOP:

(state $q_{5.i}$) i) Choose a set of configuration, c_{P_j}, such that $c_{P_j} \subseteq c_{P_i}$ (motive$_8$). Go to step 5.ii)

(state $q_{5.ii}$) ii) End NOP

In the algorithm, above written, you have eleven methods called motives (labeling, motive$_0$ up to motive$_{10}$). The motives are abstract methods that when overwriting can be responsible to become the algorithm deterministic, since this algorithm is wildly indeterministic. The implementation of the motives should be consequence of the problem that you need solve. For a developer that wants to solve some problem using FM Technology he needs to create the psm table of the problem, to give the instructions of the Machine and overwrite the motives. The motive$_0$ is a motive placed in the setting zone of the VN_{Alg} and it is responsable to give to the FM the initial settings of the game such as: how many configurations, the machine has, in total (confgL), who is the first player (FM or HB), And what is the size of the game board. All this information is introduced in motive$_0$ method through of its arguments[7]. In the motive$_0$ is included also the initialization of the class that is necessary for the FM work.

---

[7] For example the string "4;5;512;FM". This means that the game board is a matrix of 4×5, row=4 and column=5, confgL=512, and the first player is the FM, firstplayer=FM.

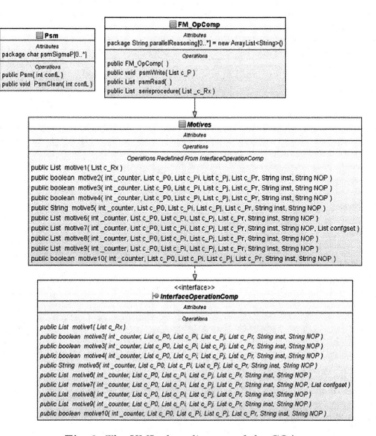

Fig. 6. The UML class diagram of the COA

Trusting Norms: A Conceptual Norms' Trust Framework for Norms Adoption in Open Normative Multi-agent Systems

Nurzeatul Hamimah Abdul Hamid[1], Mohd Sharifuddin Ahmad[1], Azhana Ahmad[1], Aida Mustapha[2], Moamin A. Mahmoud[1], and Mohd Zaliman Mohd Yusoff[1]

[1] Center for Agent Technology, College of Information Technology, Universiti Tenaga Nasional, Selangor, Malaysia
[2] Faculty of Computer Science and Information Technology, Universiti Tun Hussein Onn, Johor, Malaysia
nurzeatul@tmsk.uitm.edu.my,
{sharif,azhana,moamin,zaliman}@uniten.edu.my,
aidam@uthm.edu.my

Abstract. Norms regulate software agent coordination and behavior in multi-agent communities. Adopted norms have impacts on agent's goals, plans, and actions. Although norms generate new goals that do not originate from an agent's original target, they provide an orientation on how the agent should act in a society. Depending on situations, an agent needs a mechanism to detect correctly and adopt norms in a new society. Researchers have proposed mechanisms that enable agents to detect norms; however, their works entail agents that detect only one norm in an event. We argue that these approaches do not help agents' decision making in cases of more than one set of norms detected in the event. To solve this problem, we introduce the concept of norm's trust to help agents decide which detected norms are credible in a new environment. We propose a conceptual norm's trust framework by inferring trust using filtering factors of adoption ratio, authority, reputation, norm salience and adoption risk to establish a trust value for each detected norms in the event. This value is then used by an agent in deciding to emulate, adopt or ignore the detected norms.

Keywords: Normative agents, norm adoption, norms' trust, trust in normative multi-agent systems.

1 Introduction

A norm consists of a set of normal or expected behaviors of a society. Norms permit, prohibit and oblige behavior of agents according to their roles and responsibilities in a normative multi-agent system [1], [2]. Luck et al. [3] describe agents internalize norms as normative attitudes and then use them in its reasoning. Adopted norms have several impacts on an agent's goal generation and selection, plans and actions [4]. They generate normative goals that give the agent a clear orientation to behave which

© Springer International Publishing Switzerland 2015   149
S. Omatu et al. (eds.), *Distributed Computing & Artificial Intelligence, 12th Int. Conference*,
Advances in Intelligent Systems and Computing 373, DOI: 10.1007/978-3-319-19638-1_17

is not part of the agent's original goal. Norms aid an agent to prioritize its goals and act as a criterion to select what goal to consider first. In this case, the normative goal with a low priority has a high possibility of violation and vice versa. Consequently, it has an indirect impact on the agent's plan generation [4].

In ideal circumstances, norms are defined and programmed off-line in which agents are assumed to have sufficient knowledge about their environment [6]. However, in open multi-agent systems (OMAS), this assumption becomes a limitation and does not reflect the real-world environment. Heterogeneous characteristics in OMAS allow an agent to be part of many normative societies simultaneously, exhibiting self-interest with different goals [7], [15]. In such condition, an agent may encounter conflicting norms among groups. Contrary to closed MAS, OMAS implements indirect sanction, which progressively affects the agent's reputation and emotion [2]. Norm enforcement means any non-compliance behavior may negatively influence other agents from complying with such norms. Hence, the agent must be able to detect all norm variations used to execute a task in the society to avoid the adverse effect of failure to comply with the norm of the society. The credible detected norm is the one that is perceived as beneficial to the agent in achieving its goals [8].

Nonetheless, this is not a trivial task because the agent does not possess a global view of the environment including the norms that are currently being practiced by local agents. Moreover, according to Conte et al. [10], it is possible for an agent to detect versions of similar norms due to different interpretations of other agents' behaviors. Several works have proposed algorithms (i.e., Mahmoud et al. [8]; Savarimuthu et al. [11]; Andrighetto, Villatoro, and Conte [12]) to detect norms.

However, these algorithms suffer from two main limitations. Firstly, the detection process assumes only one set of norms enacted by local agents. Secondly, the verification of the detected norms is carried out only by asking the nearest local agent. By doing so, they assumed that all agents in the environment are trustworthy [9], [13], [14]. To overcome these limitations, we propose a new framework that incorporates a norms' trust concept. The advantage of adopting a norms' trust is to minimize the uncertainty (better prediction) associated with agents' decision making [19], [20]. Moreover, a stable state of norm (i.e., it is practiced by the majority and becoming more sustainable) in NMAS can be achieved [5]. In this framework, trust as an attribute of norm is introduced. Consequently, this paper focuses on the following objectives:

- To conceptualize a framework that enables an agent to decide the credible norms when more than one norm is detected in an event.
- To propose a norms' trust concept in supporting an agent's decision making process whenever the agent has to decide which norms to adopt, emulate or ignore.

The next section introduces the norms' trust concept and the definitions of trust in existing works. We discuss the concept of inferred trust and filter that influence the computation of norms' trust value. Section 3 proposes the norms' trust framework along with formalization and Section 4 concludes the paper.

2   Conceptualizing the Concept of Norm's Trust

We often rely on trust to decide whom to interact and cooperate. Researchers discuss the notion of trust across many fields (i.e. sociology, business, management, computer science) with a variety of meanings. Consequently, trust is an important mechanism for prediction and handling uncertainty in agent-based systems [19], [20].

In a new environment, whenever we face conditions that make us unsure of how to behave, we often observe what others do and imitate them. Indirectly, we gather information from available resources (i.e. by asking a policeman or simply ask a local people), and deliberate and decide whether to emulate some behaviors. On some occasions, depending on whether we emulate a correct or incorrect behavior, we learn a trusted behavior. Subsequently, over time, our trust towards the behavior may grow or decay. Once the trust is stable, we spread it by recommending others. This type of social control corresponds to non-normative behavior, as it affects an individual's reputation. When an agent performs a non-compliance behavior (i.e. violate a norm), other agents sanction it by generating a bad reputation. The reputation information spreads in the society causing other agents to refuse in cooperating with the agent in the future due to the poor reputation.

Josang et al. [17] refer trust as the degree of an agent's willingness to rely on something or someone in a situation. In this context, reliance indicates a positive possibility, whereby risk is the possible negative consequence, of an action. According to McKnight and Chervany [21], a trusting behavior discloses risk and then affects the trusting intentions. Specifically, Castelfranchi and Falcone [18] state that the risks that relate to agent's interaction are the risk of not achieving its goal, the risk of wasting efforts or investment and the risk of unexpected harms. In order to reduce these risks, an agent evaluates the trust only based on the combination of evidences and direct personal experiences [17].

2.1   Norm's Trust Value

We introduce the norm's trust concept based on an inferred trust of an agent that trusts another agent that practices a norm. Fig. 1 illustrates the trust infer process that applies to a particular norm. This process incorporates five filters; Authority, Image, Adoption Ratio, Norm's Salience and Adoption Risk. Based on Fig. 1, if agents B, C, and D perform a set of behaviors that is known as a norm, n_1, and an agent X can infer the trust value of the agents B, C, and D, then agent X can infer the norm's trust value of n_1. In this context, agent X first observes a set of behaviors that is performed by agent B, C, and D. Then agent X evaluates trustworthiness of the agents B, C, and D. If agent X trusts agents B, C, and D, then X can infer the norm's n_1 trust value based on the five filters.

2.2   Factors that Influence a Norm's Trust Value

The existing trust and reputation models for MAS use several information sources such as direct experience, witness information, sociological information and prejudice

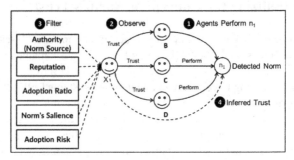

Fig. 1. Inferred norm's trust through filters

that influence a trust value. However, the contexts of trust in these models are used to evaluate an agent's trustworthiness [13], [14].

In this study, we analyze these sources together with norms' adoption motives. Based on this analysis, we identify five main factors that influence norms existence in a society:

1. *Adoption Ratio* - A norm's trust value can also be verified by observing others in a society. Adoption ratio refers to the ratio of the number of agents complying with a norm to the total number of agents in a society [22]. In this case, if a majority of a population performs a particular norm, then the adoption ratio is high. The norm is verified, if it is practiced by a significant number of the population, which exceeds the adoption ratio threshold, γ. This information can be obtained via two ways. First is via direct interaction (DI), with which is information gathered by an agent which is directly engaged with another agent in a society [13]. Second is via direct observation (DO), by observing other agents' interactions and behaviors in the society [13].

2. *Authority* - When a new agent joins a society, one of the resources for determining the trust value of a particular norm is by observing authorized bodies. Authorized bodies are trusted because they represent their societies and have the power to reward or sanction societies' members. Consequently, a norm that has been verified by an authorized body has a higher trust value [25]. Therborn [25] suggests that an individual is more likely to adopt a particular norm if he/she identifies that the source is credible (e.g., organization's authority, parents).

3. *Reputation* - Reputation is a type of social evaluation. Conte et al. [18] describe reputation as a meta-belief that acknowledges the existence of a positive evaluation of a person that diffuses within a society. In short, it is a belief about others' evaluation. Falcone et al. [19] define reputation as what is generally believed about a person. Interactions between people generate reputation values. Each member of a society sets reputation values for others based on the experience gained from interactions. According to Therborn [25] and Bicchieri [26], the sources of norms play an important role in influencing other individual to follow a norm. Thus, we

believe that the reputation of a local agent that practices a norm in a new environment influences the norms' trust value. The norm is verified, if it is observed to be enacted by a reputable agent.

4. *Norm Salience* - Norm salience provides information on the importance of a norm's phase based on a norm life cycle [13]. If the norm is salient and at a stable phase, the probability of an agent to adopt the norm is higher [13], [27]. At this stage, based on the norm life cycle, we generalize the salience level into high and low.

5. *Adoption Risk* – The adoption risk filter assesses the possibility of failure to achieve a goal if a potential norm is adopted [20].

3   A Proposed Norm's Trust Framework

This section introduces a norm's trust framework, in which an agent determines a trust value of a particular norm by utilizing the factors. Subsequent to determining the trust value, the agent can evaluate the trust level (low, medium, high), which influences its decision to adopt, emulate or ignore the detected norm. Fig. 2 illustrates the proposed framework that comprises of four main stages. Based on the agent's observations, Stage 1 deal with the norm detection module detects and classifies the set of actions performed by a group of local agents. The agent collects information from the environment and other local agents through communication and observation. At this stage, the detected norm is referred to as potential norms. Ultimately, the agent can reason and decide to comply with or even adopt a potential norm. In Stage 2, the agent analyzes the potential norms from Stage 1. A norm classification process assesses the adoption ratio of the potential norm. In Stage 3, we use norm's salience level and norms' trust filters to evaluate the norms' trust value. The norms' trust value determines the adoption decision and the willingness level in Stage 4.

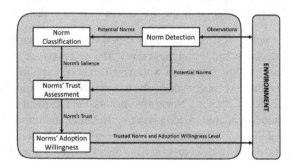

Fig. 2. The norms' trust framework architecture

In this work, we propose three cases of decision for a detected norm; adopt, emulate, or ignore. Decisions are influenced by a trust value which is determined by the identified factors. We consider the five filters (Authority, *A*; Reputation, *R*;

Adoption Ratio, *AR*; Norm Salience, *S* and Adoption Risk, *T*) to determine the norms' trust value, NT. NT_H is the highest value of a norm's trust (e.g., 1); NT_L is the lowest value of trust (e.g., 0); ρ is the norm trust threshold ($0 \leq \rho \leq 1$; $\rho \neq 0$) and ε is the adoption ratio threshold ($0 \leq \varepsilon \leq 1$; $\varepsilon \neq 0$). The five levels of norm trust are:

- **Level 1 - Full trust**, NT_F: An agent completely trusts a norm when all the five parameters (A, R, AR, S, T) produce a high value of norm's trust. There is no conflict between the values of the parameters and the agent positively verifies the norm with all factors. The agent, α, entirely trusts the norm, η, if and only if all the five parameters indicate high values of trust in the norm, η.
- **Level 2 - High trust**, NT_H: An agent trusts a norm when many of the parameters (A, R, AR, S, T) produce a high value of norm's trust. In this case, the agent positively verifies the norm with many factors exceed the thresholds and accepted level with minor conflicts/risks. The agent, α, highly trusts the norm, η, if and only if the parameters indicate high values of trust in the norm, η.
- **Level 3 - Medium trust**, NT_M: A norm is moderately trusted when the agent only observes a majority of agents practices the norm and it is not able to positively verify the norm or when there are conflicts between the results of the parameters. However, this decision is based on the Adoption ratio, AR, parameter. If the norm is observed to be practiced by a majority of agents or exceeds the AR threshold, ε, the norm's trust value is medium even though other parameters' values are negatives. This means that the agent, α, moderately trust the norm, η, if and only if a majority of agents is observed practicing the norm, η.
- **Level 4 - Low trust**, NT_L: A norm is poorly trusted when an agent observes only a minority of the agents practices the norm. The trust value does not exceed the AR threshold and it negatively verifies the norm parameters (A, R, AR, S, T).
- **Level 5 - Distrust**, NT_D: An agent distrusts a norm when all the five parameters negatively produce a very low value. The agent, α, distrusts the norm, η, if and only if all the five parameters indicate low values of trust in the norm, η.

The norms' trust value, NT, and the norm salience level are mapped to determine the norm adoption decision and willingness level. The decisions are as in Table 1.

Each decision has a willingness range with which a norm is applied. The willingness level to ignore, emulate or adopt depends on the NT threshold value. Table 2 summarizes the decisions' options. Normative agents have explicit knowledge about norms and autonomously capable of complying or violating and reasoning with norms. They also have capabilities for detecting and accepting norms. Specifically, we believe that trust is also an element that supports an agent's decision making process in multiple norms cases.

Table 1. Agents' decision and willingness level definitions

Decision	Willingness	
	Would	**Will**
Adopt - an agent internalizes a norm and adds it to its cognitive structure. The decision to adopt a norm entails all or most factors confirming the norm's trustworthiness.	An agent does not fully trust the norm. The agent *would* adopt a norm if its norm's trust value is between the NT threshold, ρ and 1. It has the choice to emulate it only without adopting. However, the agent, in such case, is motivated by the high value of trust to adopt it. $NT_H: \rho < NT < 1$	An agent has no choice but to adopt the norm because such norm could belong to conventional norms that are commonly practiced in a society. The agent *will* adopt a norm if its norm's trust value is equal to the highest possible value, 1. $NT_F: NT = 1$
Emulate - The decision to emulate a norm occurs when there is a conflict between the factors. E.g, a reputable agent, confirms that a queuing norm is not trustworthy. However, the agent observes that at some place the norm is practiced by the majority. In this case, the agent would only emulate the norm, but not adopt or ignore it. However, once it observes the norm is performed at the same place/context for the next occurrence, the norm's trust value increases.	An agent *would* emulate a norm when the NT value is medium or low. The medium value is the approximate threshold value, ρ and the low value is between the threshold value, ρ and 0. $NT_M: NT \cong \rho$ $NT_L: 0 < NT < \rho$	An agent *will* emulate a norm when the NT value is equal to 1; or between the threshold value, ρ and the highest possible value, 1. $NT_H: \rho < NT < 1$
Ignore - The decision to ignore occurs when there is a common negative agreement among the factors in confirming the norm's trust value. In this case, the agent ignores the detected norm.	An agent *would* ignore a norm when the NT value is between the threshold value, ρ and 0: $NT_L: 0 < NT < \rho$	An agent *will* ignore a norm if its norm's trust value is equal to the lowest possible value, 0. $NT_D: NT = 0$

Table 2. The Summary of Norm's Trust – Decision

Condition	Norm's Trust (NT) Level	Decision		
		Ignore	**Emulate**	**Adopt**
$NT = 1$	Full		✓ *(will)*	✓ *(will)*
$NT > \rho$	High		✓ *(will)*	✓ *(would)*
$NT \cong \rho$	Medium		✓ *(would)*	
$NT < \rho$	Low	✓ *(would)*	✓ *(would)*	
$NT = 0$	Distrust	✓ *(will)*		

4   Conclusion and Further Work

In this paper, we present our work-in-progress that establishes the concept of norm's trust to overcome the problems of multiple norms detection. By having a mechanism to compute the norms' trust value, it facilitates a normative agent decision-making process in the norm adoption stage. We identify five parameters that influence a norm's trust value; authority, reputation, adoption ratio, norm salience and adoption risk. Norm's trust value is used to facilitate a normative agent's decision making process to adopt, emulate or ignore the detected norms. In our future work, we shall extend this work by formalizing the norms' trust concept and develop a method to calculate the value of each factor. Subsequently, we shall validate the framework using agent-based simulation.

Acknowledgments. This project is sponsored by the Malaysian Ministry of Education (MoE) under the Fundamental Research Grant Scheme (FRGS) No. FRGS/1/2014 /ICT01/UNITEN/02/.

References

1. Criado, N., Argente, E., Botti, V.: Open issues for normative multi-agent systems. AI Communications 24(3), 233–264 (2011)
2. Hollander, C.D., Wu, A.S.: The Current State of Normative Agent-Based Systems. Journal of Artificial Societies and Social Simulation 14(2), 6 (2011)
3. Luck, M., Mahmoud, S., Meneguzzi, F., Kollingbaum, M., Norman, T.J., Criado, N., Fagundes, M.S.: Normative Agents. In: Ossowski, S. (ed.) Agreement Technologies, vol. 8, pp. 209–220. Springer Netherlands (2013)
4. Castelfranchi, C., Dignum, F., Jonker, C.M., Treur, J.: Deliberative Normative Agents: Principles and Architecture. In: Jennings, N.R., Lespérance, Y. (eds.) ATAL 1999. LNCS (LNAI), vol. 1757, pp. 364–378. Springer, Heidelberg (2000)
5. Andrighetto, G., Castelfranchi, C., Mayor, E., McBreen, J., Lopez-Sanchez, M., Parsons, S. (Social) Norms Dynamics. In: Andrighetto, G., Governatori, G., Noriega, P., van der Torre, L.W.N. (eds.) Normative Multi-Agent Systems, Dagstuhl Follow-Ups, Schloss Dagstuhl, vol. 4, pp. 135–170 (2013)
6. Criado, N., Argente, E., Noriega, P., Botti, V.: Reasoning about norms under uncertainty in dynamic environments. International Journal of Approximate Reasoning (2014) (in press)
7. McDonald, R.I., Fielding, K.S., Louis, W.R.: Energizing and De-Motivating Effects of Norm-Conflict. Personality and Social Psychology Bulletin 39(1), 57–72 (2013)
8. Mahmoud, M.A., Mustapha, A., Ahmad, M.S., Ahmad, A., Yusoff, M.Z.M., Hamid, N.H.A.: Potential Norms Detection in Social Agent Societies. In: Omatu, S., Neves, J., Rodriguez, J.M.C., Paz Santana, J.F., Gonzalez, S.R. (eds.) Distrib. Computing & Artificial Intelligence. AISC, vol. 217, pp. 419–428. Springer, Heidelberg (2013)
9. Huynh, T., Jennings, N., Shadbolt, N.: An integrated trust and reputation model for open multi-agent systems. Autonomous Agents and Multi-Agent Systems 13(2), 119–154 (2006)

10. Conte, R., Castelfranchi, C., Dignum, F.P.M.: Autonomous Norm Acceptance. In: Papadimitriou, C., Singh, M.P., Müller, J.P. (eds.) ATAL 1998. LNCS (LNAI), vol. 1555, pp. 99–112. Springer, Heidelberg (1999)
11. Savarimuthu, B.T.R., Cranfiled, S., Purvis, M.A., Purvis, M.K.: Identifying prohibition norms in agent societies. Artificial Intelligence and Law 21(1), 1–46 (2013)
12. Andrighetto, G., Villatoro, D., Conte, R.: Norm internalization in artificial societies. AI Communications 23(4), 325–339 (2010)
13. Pinyol, I., Sabater-Mir, J.: Computational trust and reputation models for open multi-agent systems: a review. Artificial Intelligence Review 40(1), 1–25 (2013)
14. Jung, Y., Kim, M., Masoumzadeh, A., Joshi, J.B.D.: A survey of security issue in multi-agent systems. Artificial Intelligence Review 37(3), 239–260 (2012)
15. Reuben, E., Riedl, A.: Enforcement of contribution norms in public good games with heterogeneous populations. Games and Economic Behavior 77(1), 122–137 (2013)
16. Conte, R., Paolucci, M.: Reputation in artificial societies: Social beliefs for social order, vol. 6. Springer (2002)
17. Jøsang, A., Ismail, R., Boyd, C.: A survey of trust and reputation systems for online service provision. Decision Support Systems 43(2), 618–644 (2007)
18. Castelfranchi, C.: Falcone. R.: Trust theory: A socio-cognitive and computational model. Wiley Publishing (2010)
19. Falcone, R., Castelfranchi, C., Cardoso, H.L., Jones, A., Oliveira, E.: Norms and Trust. In: Ossowski, S. (ed.) Agreement Technologies, vol. 8, pp. 221–231. Springer Netherlands (2013)
20. Ramchurn, S.D., Huynh, D., Jennings, N.R.: Trust in multi-agent systems. The Knowledge Engineering Review 19(01), 1–25 (2004)
21. McKnight, D.H., Chervany, N.L.: The Meanning of Trust, in Technical Report MISRC Working Paper Seires 96-041996, Management Information Systems Research Center, University of Minnesota
22. Wooldridge, M.: An Introduction Multi-agent Sytems. John Wiley & Sons Ltd. (2009)
23. Therborn, G.: Back to Norms! on the Scope and Dynamics of Norms and Normative Action. Current Sociology 50(6), 863–880 (2002)
24. Bicchieri, C.: The Grammar of Society: The Nature and Dynamic of Social Norms. Cambridge University Press, New York (2006)
25. Savarimuthu, B.T.R., Padget, J., Purvis, M.A.: Social Norm Recommendation for Virtual Agent Societies. In: Boella, G., Elkind, E., Savarimuthu, B.T.R., Dignum, F., Purvis, M.K. (eds.) PRIMA 2013. LNCS, vol. 8291, pp. 308–323. Springer, Heidelberg (2013)

Do Human-Agent Conversations Resemble Human-Human Conversations?

David Griol and José Manuel Molina

Computer Science Department
Carlos III University of Madrid
Avda. de la Universidad, 30, 28911 - Leganés (Spain)
{david.griol,josemanuel.molina}@uc3m.es

Abstract. In this paper, we are interested in the problem of understanding human conversation structure in the context of human-agent and human-human interaction. We present a statistical methodology for detecting the structure of spoken dialogs based on a generative model learned using decision trees. To evaluate our approach we have used a dialog corpus collected from real users engaged in a problem solving task. The results of the evaluation show that automatic segmentation of spoken dialogs is very effective not only with models built using separately human-agent dialogs or human-human dialogs, but it is also possible to infer the task-related structure of human-human dialogs with a model learned using only human-agent dialogs.

Keywords: Domain Knowledge Acquisition, Dialog Structure Annotation, Conversational Agents, Spoken Interaction, Spoken Dialog Systems.

1 Introduction

Research on data-driven approaches to dialog structure modeling is relatively new and focuses mainly on recognizing a structure of a dialog as it progresses. Dialog segmentation can be defined as the process of dividing up a dialog by one of several related notions (speaker's intention, topic flow, coherence structure, cohesive devices, etc.), identifying boundaries where the discourse changes taken into account such as specific criteria. This detection is usually based on combining different kinds of features, such as semantic similarities, inter-sentence similarities, entity repetition, word frequency, linguistic features, and prosodic and acoustic characteristics.

Different studies have been carried out for identifying discourse segments in spoken and written documents. For instance, TextTiling [1] is a two step algorithm for the segmentation of texts into discourse units that are meant to reflect the topic flow on the text. Yamron [2] presented an approach to segmentation that models an unbroken text stream as an unlabeled sequence of topics using Hidden Markov Models. Ponte presented in [3] an approach based on information retrieval methods that map a query text into semantically related words and phrases. Passoneau and Litman [4] developed an algorithm for identifying topic boundaries that uses decision trees to combine multiple linguistic features extracted from corpora of spoken text.

There is also a wide range of natural language processing applications for which discourse segmentation assists in. For instance, Angheluta, Busser and Moens adapted a three-step segmentation algorithm for automatic text summarization [5]. Walker applies this kind of techniques for anaphora resolution [6]. Different studies show the benefits of using discourse segmentation for question answering tasks [7].

In this paper, we present a machine learning approach for the automatic segmentation of spoken dialogs. The objective is to detect sequences of turns that accomplish a specific objective (tasks and subtasks) inside the dialog flow. These parts can be necessary for obtaining the final goal of the dialog, and also general parts not strictly related to the domain of the dialog system (greetings, error recovery, etc.). The detection of the dialog structure is useful to develop dynamic and adaptable conversational agents. Modeling subdialog structures is also useful to extend these agents to deal with more complex tasks.

In the literature, there are different studies for comparing human-agent (HA) and human-human (HH) interactions. Most of them compare specific features of both kind of conversations [8]. In our work, we try to infer the dialog structure of HH corpora by means of an active learning approach based on training an initial dialog model using the HA corpus and then use this model to learn the structure of HH conversations. To achieve this goal, two problems need to be addressed: i) creating a dialog representation that is suitable for representing the required domain-specific information, and ii) developing a machine learning approach that uses this representation to capture information from a corpus of in-domain conversations.

This field presents as a main challenge the need of detecting the dialog segment using different information sources that are provided by both user and system entities during the course of the dialog (semantic information, confidence scores, task-dependent and independent information, etc.). Our methodology is based on the use of decision trees classification to integrate all these different characteristics. This technique has been applied for the development of a discourse segmentation module for a spoken conversational agent in a customer support service/help-desk domain, using HA and HH interactions acquired for this task.

2   Our Methodology for Task/Subtask Prediction

We consider a task-oriented dialog to be the result of incremental creation of a shared plan by the participants [9]. This shared plan consists of several tasks and subtasks. The goal of subtask segmentation is to predict if the current utterance in the dialog is part of the current subtask or it starts a new subtask. We model this prediction problem as a classification task as the following equation shows:

$$\hat{S}_i = \underset{S_i \in \mathcal{S}}{\mathrm{argmax}} \, P(S_i | U_1 \cdots U_{i-1}, S_1 \cdots S_{i-1})$$

where set S contains all the possible kinds of tasks/subtasks defined for the dialog segmentation and U_n is the semantic representation of the user utterance at time n in terms of the list of features provided by the Spoken Language Understanding (SLU) module of the conversational agent. The prediction of the current task, that is a local process, takes into account the previous history of the dialog, that is to say, the sequence of user turns and dialog segments preceding time i.

The lexical, syntactic and semantic information associated to the speaker u's ith turn (U_i) is usually represented by means of different information sources:

- the words uttered;
- dialog acts, which represent the meaning of an utterance at the level of the speaker's intention in producing that utterance (e.g., *Acceptance, Not-Understood, Tag-Question,* or *Apology*).
- part of speech tags, also called word classes or lexical categories. Common linguistic categories include noun, adjective, and verb, among others;
- predicate-argument structures, used by SLU modules in various contexts to represent relations within a sentence structure. They are usually represented as triples (subject-verb-object).
- named entities: sequences of words that refer to a unique identifier. This identifier may be a proper name (e.g., organization, person or location names), a time identifier (e.g., dates, time expressions or durations), or quantities and numerical expressions (e.g., monetary values, percentages or phone numbers).

As a practical implementation of this equation, we propose the use of two modules. The first module deals with the detection of the specific problem described by the user. This detection is based on the specific semantic information regarded to the task that is provided by the SLU module, that is to say, the attribute-value and predicate features for the reference annotations. This module also updates a register that contains the complete list of features provided by the SLU module through the dialog history since the current moment.

Until a specific problem is detected, a generic model learned with all the dialogs in the training partition is used for the selection of the segment of the dialog. Once the problem has been detected, a specific model learned using only the dialogs that deals with such a problem is used for the segment detection. The C4.5 decision tree learning algorithm has been used for the learning of these models, using the Weka machine learning software for classifying the complete list of features contained in the history register. Using these models, the current segment of the dialog is selected by taking into account the previous dialog segment detected for the module and the complete list of features provided by the SLU module. Figure 1 shows a graphical scheme of the proposed methodology.

3   Case Application: The *Facilisimo* Spoken Dialog System

We have applied our proposal to the problem solving domain of the *Facilisimo* spoken dialog system, which acts as a customer support service to help solving simple and routine software/hardware repairing problems, both at the domestic and professional levels. The system has been developed using an hybrid dialog management approach that combines the VoiceXML standard with user modeling and statistical estimation of optimized dialog strategies [10].

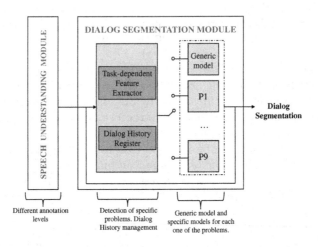

Fig. 1. Graphical scheme of the proposed architecture for task and subtask detection

The definition of the system's functionalities and dialog strategy was carried out by means of the analysis of 150 human-human (HH) conversations provided by real assistants attending the calls of users with a software/hardware problem at the City Council of Leganés (Madrid, Spain). The labeling defined for this corpus contains different types of information, that have been annotated using a multilevel approach similar to the one proposed in the Luna Project [11]. The first levels include segmentation of the corpus in dialog turns, transcription of the speech signal, and syntactic preprocessing with POS-tagging and shallow parsing. The next level consists of the annotation of main information using attribute-value pairs. The other levels of the annotation show contextual aspects of the semantic interpretation. These levels include the predicate structure, the relations between referring expressions, and the annotation of dialog acts.

The attribute-value annotation uses a predefined domain ontology to specify concepts and their relations. The attributes defined for the task include *Concept, Computer-Hardware, Action, Person-Name, Location, Code, TelephoneNumber, Problem*, etc.

Dialog act (DA) annotation was performed manually by one annotator on speech transcriptions previously segmented into turns. The DAs defined to label the corpus are the following: i) Core DAs: *Action-request, Yes-answer, No-answer, Answer, Offer, ReportOnAction, Inform*; ii) Conventional DAs: *Greet, Quit, Apology, Thank*; iii) Feedback-Turn management DAs: *ClarificationRequest, Ack, Filler*; iv) Non interpretable DAs: *Other*.

The original FrameNet[1] description of frames and frame elements was adopted for the predicate-argument structure annotation, introducing new frames and roles related to hardware/software only in case of gaps in the FrameNet ontology. Some of the frames included in this representation are *Telling, Greeting, Contacting, Statement, Recording, Communication, Being operational, Change operational state*, etc.

An example of the attribute-value, dialog-act and predicate structure annotations of a user utterance is shown below:

Hi, I have a problem with my printer.
Attributes-values: *Concept*:problem; *Hardware*:printer;
Dialog acts: *Answer*;
Predicate structure: (*Greeting*)(*Problem_description*) *Device Problem*

The basic structure of the dialogs is usually composed by the sequence of the following tasks: *Opening, Problem-statement, User-identification, Problem-clarification, Problem-resolution*, and *Closing*. This set of tasks contains a list of subtasks, such as *Problem-description, Problem-Request, Problem-Confirmation, Brand-Identification, Model-Identification, Help-Request, Message-Confirmation, Name-Identification, Resolution-Confirmation*, etc. The shared plan is represented as a data register that encapsulates the task structure, dialog act structure, attribute-values and predicate-argument structure of utterances.

During the *Problem-statement* task, the caller explains the problem the reasons why he/she calls the help-desk. In the *User-identification* task, the operator asks for additional information regarding the identity of the caller. Once the caller has described the problem, the operator can ask for additional information to clarify it during the *Problem-clarification* task.

During the *Problem-resolution* task, the operator asks the user to perform specific tests. We have defined nine different subtasks inside this generic segment, given that our goal is to detect not only that the dialog is in this segment, but also what are the specific problem that has to be resolved: *Printer* (P4), *Network connection* (P5), *PC going slow* (P6), *Monitor* (P7), *Keyboard* (P8), *Mouse* (P9), *CD-DVD player* (P10), *Power supply* (P11), and *Virus* (P12). The operator assigns a ticket number for the current call if the problem has not been solved after this task. The user must take note of this number and inform about this to the operator. The dialog ends at the *Closing* phase, in which the operator also tries to give a useful advice related to the described problem.

A total of 150 dialogs was acquired by 25 recruited users using the developed system (Human-Agent dialogs, HA). Ten different dialog scenarios inspired from the services provided in the human-human dialogs were designed for the acquisition. The complete set of HH and HA dialogs were manually labeled including

---

[1] https://framenet.icsi.berkeley.edu/fndrupal/

```
S1: Welcome to the Facilisimo system. How may I help you?
U1: Hello, my printer does not print correctly.
S2: Tell me your name and surname.
U2: My name is Ricardo Santamaria.
S3: Can you tell me your phone extension?
U3: Nine four sixty-three.
S4: Can you tell me the brand and model?
U4: HP Officejet Pro 8610.
S5: Is the printer on now?
U5: It is on.
S6: Can you tell me the message on the printer display?
U6: Paper jam is written.
S7: Can you see the active printer from the control panel of your PC?
U7: It is active.
S8: Please try the following steps. Say next after each step. Turn the printer off, and then on again.
U8: Next.
S9: Print a Printer Status Report.
U9: Next.
S10: Has this process solved your problem?
U10: No.
S11: Remove any loose paper from the main and optional trays.
U11: Next.
S12: Clear any jammed paper from the rear of the printer.
U12: Next.
S13: Has this process solved your problem?
U13: Yes.
S14: Anything else?
U14: No, thank you.
S15: Most of these steps will also work for troubleshooting an office copier. Have a nice day!
```

Fig. 2. Example of a dialog acquired with recruited users and the *Facilisimo* dialog system (translation from Spanish to English)

this task/subtask information. This information was incorporated for each user and system turn in the dialogs. Figure 2 shows the transcription of a dialog included in the HA corpus.

4   Experiments and Results

The developed methodology for the task detection has been evaluated by means of the HA and HH dialogs of the described corpora. As HH dialogs are spontaneous, they present several differences with regard to the HM dialogs. The main one is the great difference in the average number of turns (16.18 turns in the HM corpus and 25.71 for the HH dialogs). This is because HH dialogs present other minor topics (like small talks about other persons, previous problems, holidays, etc), a high frequency of interruptions, cut-off phrases, and overlapped contributions. This makes that the 18.24% of the utterances of the HH corpus have been labeled as *Out of the Task*.

Analyzing the annotation available for the DA level, we measured that in average an HH dialog is composed of 37.9 ± 7.3 (Std. Dev.) DAs, whereas a HA dialog is composed of 21.9 ± 5.4. The difference between average lengths shows how HH spontaneous speech can be redundant, while HA dialogs are more limited to an exchange of essential information. The standard deviation of a conversation in terms of DAs is considerably higher in the HH corpus than in the HA ones. This can be explained by the fact that the HA dialogs follow a unique, previously defined task-solving strategy that does not allow digressions.

The evaluation of the statistical dialog segmentation technique was carried out *turn by turn* using a five-fold cross validation process. Each one of the two corpus was randomly split into five subsets. Each trial used a different subset taken from the five subsets as the test set, and the remaining 80% of the dialogs was used as the training set. Table 1 shows the results of the application of this methodology for the HA and HH corpora. The results show how the prediction is improved once the different SLU features are incorporated to the model. As can be seen, the proposed methodology successfully adapts to the requirements of the HA dialogs, since a 0.98 F-measure is obtained, measuring the dialog segments provided by the developed module that are equal to the segment annotated in the corpus for the HA dialogs. This value is reduced to 0.79 for the HH dialogs, since the *Out of the Task* class is usually confused with the rest of dialog segments related to the task. Therefore, the methodology adapts to the very different nature that has been described for both kind of dialogs.

Finally, we learned a model with the total 100 HA dialogs and evaluated it using the total 100 HH dialogs. This experimentation was designed to evaluate if a model learned with HA dialogs can detect the task-related structure of spontaneous HH conversations. The main challenge of this experiment is that only a maximum of 81.76% can be achieved due to the 18.24% *Out of the task* that is only present in the HH corpus. As can be observed, the model successfully adapts to detect the task-related parts in the HH dialogs, achieving a 0.67 F-measure.

Table 1. Average results of the evaluation of the proposed dialog segmentation technique for the *Facilisimo* system

	Precision	Recall	F-measure
HA corpus for learning and evaluating			
Attribute-Values	0.89	0.87	0.88
DAs + Attribute-Values	0.94	0.92	0.93
Complete set	0.97	0.98	0.98
HH corpus for learning and evaluating			
Attribute-Values	0.72	0.60	0.66
DAs + Attribute-Values	0.86	0.71	0.78
Complete set	0.87	0.72	0.79
HA corpus for learning - HH corpus for evaluating			
Attribute-Values	0.62	0.55	0.58
DAs + Attribute-Values	0.69	0.61	0.65
Complete set	0.73	0.63	0.67

5   Conclusions

We have presented in this paper a statistical approach for automatically dialog segmentation of spoken dialogs. This approach uses feature selection to collect a set of informative features into a model that includes both the information provided by the user and the system prompts. This model can be used to predict where boundaries occur in the dialog, using this information in the selection of the next system prompt.

The results of the evaluation of this methodology in a practical problem solving task show that the statistical approach successfully adapts to the requirements of the task, not only separately for human-agent and human-human dialogs acquired for this project, but also it is possible to successfully detect the task-related information that is present in spontaneous human-human dialogs by learning a model only with human-agent dialogs.

As a future work, we want to perform a more detailed analysis of the situations that have been labeled as *Out of the Task*, studying if the task detector module is able to differentiate them. The experiments reported in this paper have been performed on transcribed speech. We want also to assess the performance of dialog structure prediction on recognized speech.

Acknowledgements. This work was supported in part by Projects MINECO TEC2012-37832-C02-01, CICYT TEC2011-28626-C02-02, CAM CONTEXTS (S2009/TIC-1485).

References

1. Hearst, M.: Multi-paragraph segmentation of expository text. In: Proc. ACL, pp. 9–16 (1994)
2. Yamron, J.: Topic detection and tracking segmentation task. In: Proc. Broadcast News Transcription and Understanding Workshop (1998)
3. Ponte, J., Croft, W.: Text segmentation by topic. In: Peters, C., Thanos, C. (eds.) ECDL 1997. LNCS, vol. 1324, pp. 113–125. Springer, Heidelberg (1997)
4. Passoneau, R., Litman, D.: Discourse segmentation by human and automated means. Computational Linguistics 23, 103–139 (1997)
5. Angheluta, R., Busser, R.D., Moens, M.: The use of topic segmentation for automatic summarization. In: Proc. ACL Workshop on Automatic Summarization, pp. 66–70 (2002)
6. Walker, M.A.: Centering, anaphora resolution, and discourse structure, pp. 401–435. Oxford University Press (1998)
7. Chai, J., Jin, R.: Discourse structure for context question answering. In: Proc. HLT-NAACL Workshop on Pragmatics of Question Answering, pp. 23–30 (2004)
8. Doran, C., Aberdeen, J., Damianos, L., Hirschman, L.: Comparing several aspects of human-computer and human-human dialogues. In: Proc. SigDial (2001)
9. Bangalore, S., Fabbrizio, G.D., Stent, A.: Learning the Structure of Task-driven Human-Human Dialogs. IEEE Trans. Audio Speech Lang. Processing 16(7), 1249–1259 (2008)
10. Griol, D., Molina, J., Callejas, Z.: Bringing together commercial and academic perspectives for the development of intelligent AmI interfaces. JAISE 4(3), 183–207 (2012)
11. Stepanov, E., Riccardi, G., Bayer, A.: The Development of the Multilingual LUNA Corpus for Spoken Language System Porting. In: Proc. LREC, pp. 2675–2678 (2014)

Artificial and Natural Intelligence Integration

Juan C. Alvarado-Pérez[1,2] and Diego H. Peluffo-Ordóñez[2]

[1] Universidad de Salamanca, Spain
jcalvarado@usal.e
[2] Universidad Cooperativa de Colombia, Pasto
diego.peluffo@campusucc.edu.co

Abstract. The large amount of data generated by different activities -academic, scientific, business and industrial activities, among others- contains meaningful information that allows developing processes and techniques, which have scientific validity to optimally explore such information. Doing so, we get new knowledge to properly make decisions. Nowadays a new and innovative field is rapidly growing in importance that is Artificial Intelligence, which involves computer processing devices of modern machines and human reasoning. By synergistically combining them –in other words, performing an integration of natural and artificial intelligence-, it is possible to discover knowledge in a more effective way in order to find hidden trends and patterns belonging to the predictive model database. As well, allowing for new observations and considerations from beforehand known data by using data analysis methods as well as the knowledge and skills (of holistic, flexible and parallel type) from human reasoning. This work briefly reviews main basics and recent works on artificial and natural intelligence integration in order to introduce users and researchers on this field integration approaches. As well, key aspects to conceptually compare them are provided.

Keywords: Data mining, visualization, machine learning.

1   Introduction

The human being is intuitive and deductive as well as has a convergent and divergent thought, having then the virtue of interpreting the world possible in a holistic, panoramic, parallel, flexible and contextual way, allowing for a quickly integrated sight of problems and taking out general concept for doing mind maps and establishing dynamic relationships among objects. All aforementioned is provided by imagination, creativity and spontaneity being characteristics that make a person able to give different meanings to the same term (i.e. the word love may be mean something different regarding the point of view). Also, such a virtue makes a human being able to interpret different and ambiguous feelings such as irony, moods and sarcasm [1], which cannot be interpreted by machine as feelings. This is due to the data interpretation by a machine that is deterministic. Nevertheless, there exists inherent subjectivism associated with the analyst experience, which may bias the outcomes when analyzing data. To deal with this drawback, objective processes become necessary for supporting the

© Springer International Publishing Switzerland 2015
S. Omatu et al. (eds.), *Distributed Computing & Artificial Intelligence, 12th Int. Conference*,
Advances in Intelligent Systems and Computing 373, DOI: 10.1007/978-3-319-19638-1_19

process of decision making [2]. A human being is rounded by different aroma, tastes, smelling, feelings, sounds and colors that are perceived and internalize through senses, who build a subjective concept from them. However, visual stimulus are the ones that give the most information to the brain so this process makes the sight the widest bandwidth sense, because it permits to perceive a lot of varied information from the environment [3]. Broadly speaking, it means that "a picture is worth a thousand words" so if representative, such picture will facilitate human to build a mind map for concepts or problems learning.

At present, modern technology is making that images tell histories from big data, taking them into a multimedia and multidimensional world where human being is able of manipulating and interacting with data to build a complex-reality representation in a graphic fashion, as well as identifying underlying patterns and/or models [4]. Our world perception has been encoded by different symbols such as: letters, numbers, diagrams, general data, etc., which should be intelligible for human beings. Thus, some visualization approaches are used that provide a real semiotic ability for human interpretation, so it uses different types of resources such as colors, forms, textures and geometrical representations based on psychological phenomena like color interpretation, preattentive process, context and reference interpretation [5]. There several tools for intelligent visualization [6], providing a semiotic relationship with explored data –by telling the data's story in a graphic, suitable and efficient way. This work presents a briefly review on artificial and natural intelligence integration.

The remaining of this paper is organized as follows: Section 2 starts by introduction what artificial and natural intelligence integration consists in. Then, section 3 outlines ways to integrate Information Visualization and Data Mining. Finally, section 4 draws the conclusions and final remarks.

2   Artificial and Natural Intelligence Integration

The modern computer and its emergent systems have implicitly generated a scenario where a human being and machine competition takes place, in which the latter has gradually increased its involvement into exclusively human beings activities. However, a machine might not be seen as a human being replacement since they both have special characteristics (as mentioned above) making them essentially different from each other. In this connection, it is worth to underline that under a proper integration of artificial –by machines- and natural –by human beings- intelligence may perform better and more robust processes. This means that an integrated system can be formed so that improved outcomes are reached. For instance, computers are deterministic, objective, tireless and untiring, as well as have a high processing ability. Meanwhile, human beings have flexible, parallel and dynamic thought within a holistic context viewpoint. In this sense, by joining the characteristics from both approaches, versatile and suitable systems able to generate solutions benefitting society can be accomplished. Even though the two studied approaches are essentially different, they may have similar properties to be exploited to improve the system performance. On one hand, computers can be flexible (just like human beings) by adding to them some autonomous learning techniques such as fuzzy logic [7]. On the other hand, human brain is divided into two hemispheres, being the right one in charge of creativity and feeling, meanwhile the left one provides the logic [8], which is closer to the way how machines process the information. Indeed, many learning and processing systems for computers emulate somehow the brain behavior as it is the case of neural networks [7].

By doing an adequate synergetic integration between human beings skills and computer abilities involving the best properties from each one, artificial intelligence along with natural intelligence may reach both accurate calculation ability as well as holistic and flexible learning processes. Then, aimed at providing a robust and proper data exploration, an integrated system may discover innovative and useful knowledge in a more efficient way in comparison with traditional approaches [9]. Within integration frameworks, Data Mining (DM) algorithms are able to exhaustively explore and analyze data by taking advantage of information visualization (IV) techniques that facilitates human interpretation. In fact, visualization processes are aimed at providing users with an efficient way to interpret data making the DM outcomes more intelligible and facilitating mind map building in a graphic way [10] –in this way the holistic and flexible human being thought is involved. In this sense, automatic learning and DM is on the computer side while the IV on the human interpretation side as explained in Fig. 1. Integrated systems work iteratively and interactively demanding often a high computational cost as well as a strongly collaborative work between analyst and machine [11].

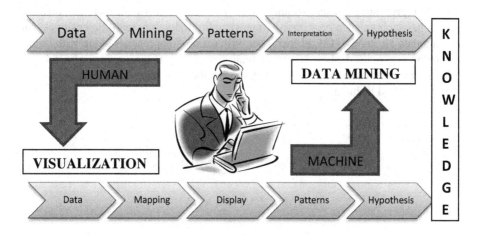

Fig. 1. Integrating human beings in machine, through knowledge discovery process

3   Ways to Integrate Information Visualization and Data Mining

DM and IV are fields that have been independently created and developed, being the former able to process data outputting if-then rules and patterns while the latter represents data in an easy-to-interpret fashion without using additional algorithm support. By construction, these methods were no intended to interact with each other.

Nonetheless, approaches to integrate DM and IV have been proposed [12]. Following are described approaches to integrate DM and IV, namely: Computer visualization, visually improved data mining.

3.1   Computer Visualization (V++)

The V++ objective is to make a data representation allowing for detecting and drawing patterns through computationally optimal views. This approach is mainly a visual technique that takes advantage of computer algorithms such as those based on DM. Visualization attempts to reveal and visualize the most relevant patterns, that is to say those ones being more informative regarding some criteria. Thus, intended to facilitate human interpretation, visualization allow for low-dimensional data representation (2 or 3 dimensions). Nonetheless, when dealing with big data, the original data space embedding space into a lower dimension space is not a trivial task. For instance, just look at a 2D conventional screen, which itself is inherently limited by its available rectangular space. Indeed, the most common (being feasible and real) representation relies on a 2D plotting whether n-dimensional data (n >3) or even one-dimensional data (normally an extra corresponding axis is used, i.e. the axis of ordinal numbers being the simplest one). On this regard, available 2D representation must be optimally utilized for visualization purposes, and then an intrinsic bi-dimensional space for properly representing of data should be determined. To this end, there are different approaches such as: data projection, dimensionality reduction, multidimensional scaling, and data clustering [13]. Properly set, visualization approaches along with human interaction may reach the best 2D representation according to the user's needs. In Fig. 2, the V++ process is shown. Notice that the symbol ++ means that visualization is improved by DM techniques.

Fig. 2. V++ Process

3.2   Visually Improved Data Mining (M++)

The core of this type of combination relies on the DM techniques. Meanwhile, VI is an additional stage to visualize and validate the DM outcomes. This is done through an interactive interface allowing for not only interpreting the extracted model by the DM, but also visually verifying, assessing and refining the model outputted by the DM techniques. In this way, a clear interaction between analyst and model takes place, which allows a deeper comprehension of the underlying data information [14]. In other words, DM is in charge of innovative and useful patterns discovery, while IV only generate a visual representation of them. Fig 3. shows graphically the stages of the M++ process. Symbol ++ denotes DM is supported by IV.

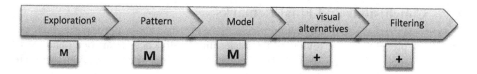

Fig. 3. Process in M++

3.3 Data Mining and Visualization Integration (VM)

VM techniques involve IV and DM, in which both fields are equally important. In fact, they combine in a synergic way the parallel thought from the human being with the processing capability of computers. Doing so, the best from the machine and human world converge to a meaningful data analysis and interpretation [15]. Two types of VM approaches are explained below.

Integration Black Box with Feedback (VM-IBBF): In this approach, algorithms are black boxes, that is, the analyst cannot observe the way the DM algorithm internally works. Nevertheless, user may modify parameters whose effect is automatically reflected in the IV process, yielding then an adaptive approach. In other words, VM-IBBF creates a feedback loop between IV and DM reaching more meaningful alternatives and models, according to the user's criterion.

Integration White Box (VM-IWB): This type of integration requires a strong joint effort between analyst and machine. Hence, every single stage of the DM algorithms can be observed so that model building process is easily accessed and modified. So, VM-IMB (Fig. 4.) provides better understanding and allows users customize the algorithms.

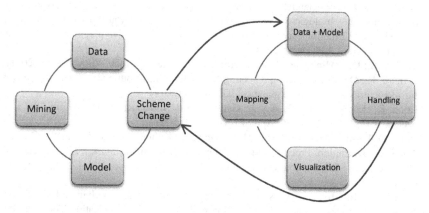

Fig. 4. Interactive and Iterative process in the Integration White Box: VM

4   Conclusions and Final Remarks

This work presents a brief review of the basics and approaches for artificial and natural intelligence integration. Within the context of this paper is concerned in, artificial intelligence is associated with DM (how to computer explore and analyze data). Correspondingly, natural intelligence refers to IV which allows for an intelligible human interpretation. The emerging field of IV aims to describe, explore and analyze data in order to discover underling knowledge from data. The processes for knowledge extraction and schematic representation of reality by visual models are transversal topics very useful in several fields involving data. Nonetheless, such processes performance is strongly dependent on the nature of data as well as the right use of the data processing techniques.

Given this, the synergic integration of artificial and natural intelligence represents a good alternative. Indeed, by properly mixing the holistic, flexible and parallel interpretation from human beings and the high computing storing capabilities of machines it is possible to discover relevant (even unexpected) new patterns within big data sets in an effective way. It is important to remark that due to the high demand of computational load, integration approaches are often implemented on parallel and distributed architectures such as computer clusters and grids, which in turn contribute to significantly improve the performance in terms of time processing and accuracy.

Acknowledgements. This work was supported by the project *"DESCUBRIMIENTO DE PATRONES DE DESEMPEÑO ACADÉMICO EN LAS COMPETENCIAS GENÉRICAS DE LAS PRUEBAS SABER PRO 2011-2"* from ICFES research program.

References

1. Diaz, N.V., Serrano-Garcia, I.: Pensabas que emocionarse era sencillo? Las emociones como fenómenos biológicos, cognoscitivos y sociales // Did you think it was easy to get excited about? Emotions such as biological, social and cognitive phenomena. Rev. Puertorriqueña Psicol. 13(1) (ene. 2014)
2. Tufféry, S.: Data Mining and Statistics for Decision Making. John Wiley & Sons (2011)
3. Pethuru, R.: Data Visualization: Creating Minds Eye. In: Handbook of Research on Cloud Infrastructures for Big Data Analytics. IGI Global (2014)
4. Cook, K., Earnshaw, R., Stasko, J.: Guest Editors' Introduction: Discovering the Unexpected. IEEE Comput. Graph. Appl. 27(5), 15–19 (2007)
5. Keim, D.A., Andrienko, G., Fekete, J.-D., Görg, C., Kohlhammer, J., Melançon, G.: Visual Analytics: Definition, Process, and Challenges. In: Kerren, A., Stasko, J.T., Fekete, J.-D., North, C. (eds.) Information Visualization. LNCS, vol. 4950, pp. 154–175. Springer, Heidelberg (2008)
6. Koh, L.C., Slingsby, A., Dykes, J., Kam, T.S.: Developing and Applying a User-Centered Model for the Design and Implementation of Information Visualization Tools. In: 2011 15th International Conference on Information Visualisation (IV), pp. 90–95 (2011)
7. Huang, M.-J., Tsou, Y.-L., Lee, S.-C.: Integrating fuzzy data mining and fuzzy artificial neural networks for discovering implicit knowledge. Knowl.-Based Syst. 19(6), 396–403 (2006)

8. Roselli, M.: Maduración cerebral y desarrollo cognoscitivo. Rev. Latinoam. Cienc. Soc. Niñez Juv. 1(1) (May 2011)
9. Kononenko, I., Kukar, M.: Machine Learning and Data Mining. Elsevier (2007)
10. Torres Ponjuán, D.: Aproximaciones a la visualización como disciplina científica. ACIMED 20(6), 161–174 (2009)
11. Alonso, F., Martínez, L., Pérez, A., Valente, J.P.: Cooperation between expert knowledge and data mining discovered knowledge: Lessons learned. Expert Syst. Appl. 39(8), 7524–7535 (2012)
12. Bertini, E., Lalanne, D.: Surveying the complementary role of automatic data analysis and visualization in knowledge discovery. In: Proceedings of the ACM SIGKDD Workshop on Visual Analytics and Knowledge Discovery: Integrating Automated Analysis with Interactive Exploration, pp. 12–20 (2009)
13. Peluffo-Ordóñez, D.H., Lee, J.A., Verleysen, M.: Short Review of Dimensionality Reduction Methods Based on Stochastic Neighbour Embedding. In: Villmann, T., Schleif, F.-M., Kaden, M., Lange, M. (eds.) Advances in Self-Organizing Maps and Learning. AISC, vol. 295, pp. 65–74. Springer, Heidelberg (2014)
14. Aguilar, D.A.G., Guerrero, C.S., Sanchez, R.T., Penalvo, F.G.: Visual Analytics to Support E-learning (ene. 2010)
15. Puolamäki, K., Bertone, A., Therón, R., Huisman, O., Johansson, J., Miksch, S., Papapetrou, P., Rinzivillo, S.: Mastering The Information Age – Solving Problems with Visual Analytics. In: Keim, D., Kohlhammer, J., Ellis, G., Mansmann, F. (eds.) Mastering the Information Age Solving Problems with Visual Analytics, Germany (2010)

CASOOK: Creative Animating Sketchbook

Miki Ueno, Kiyohito Fukuda, Akihiko Yasui,
Naoki Mori, and Keinosuke Matsumoto

Graduate School of Engineering
Osaka Prefecture University
1-1 Gakuencho, Nakaku, Sakai, Osaka 599-8531, Japan
ueno@ss.cs.osakafu-u.ac.jp

Abstract. Understanding drawn picture information is an important task because drawings are one of the most intrinsic representations of individuals, regardless of age, national origin, or culture. However, it is difficult to define an appropriate picture model for computers. To address this challenge, we propose the Creative Animating Sketchbook (CASOOK) application and two basic modules through which the application is constructed. We first build a drawing system as the primary module. It includes a usable interface, which enables users to obtain various features of pictures through drawing, thereby encouraging users to draw. We then extract this primary module with a support vector machine classifier utilizing a binary decision tree. Two-layer sketch recognitions are produced; these can predict complex object classes in free drawing, which comprises the second proposed module. By combining the two proposed modules, we construct the CASOOK system. Based on results from user and computer experiments, we confirm the effectiveness of the proposed modules.

Keywords: sketch recognition, drawing features, children's drawing, kansei engineering.

1 Introduction

Drawn picture information is an important topic in the artificial intelligence field. Various studies have analysed picture information by computer algorithms. However, three problems exist in these studies; they are outlined below.

- It is not sufficient to collect free drawings through various users, especially children and people who are not adept at drawing.
- Some researchers have analysed the features of existing pictures, such as colour histograms and picture composition, but they ignore the mode of drawing[1].
- An important feature of pictures is the object class. One study focuses on sketch recognition [2] and handwritten character recognition[3] to predict the object class. However, these approaches cannot be applied to free drawing because these latter pictures may be comprised of various classes of objects, such as animals, characters, symbols, and so on.

To solve these problems, we propose a system called *Creative Animating Sketchbook (CASOOK)*. CASOOK can be employed by various users; it recognizes what users have

© Springer International Publishing Switzerland 2015

S. Omatu et al. (eds.), *Distributed Computing & Artificial Intelligence, 12th Int. Conference,*
Advances in Intelligent Systems and Computing 373, DOI: 10.1007/978-3-319-19638-1_20

Fig. 1. Outline of CASOOK

depicted through drawing. In this paper, we present two key modules through which our system is constructed. The first module is the creation of a drawing system to apply to various users. The second module is two-layer sketch recognition through free drawing[4][5]. In an experimental evaluation, we apply the system to children in a private school.

We demonstrate the concept of CASOOK system in Section 2. We present the first module in Section 3 and the second module in Section 4. In Section 5, we describe user experiments, which confirm the effectiveness of our module. Finally, in Section 6, we provide our conclusions.

2   CASOOK

2.1   Concept

We propose a framework called Creative Animating Sketchbook (CASOOK) in which the two key modules are combined. CASOOK can communicate with users who intend to share drawings, through a usable interface, gaming functions, and two-layer sketch recognition.

CASOOK consists of two modules: an interaction module and an analysis module. The interaction module includes functions that encourage users to draw, and the analysis module contains methods to recognize user drawings. However, the latter module is not sufficient for recognizing drawings of people who are not adept at drawing. To solve this problem, CASOOK employs functions that interact with users.

2.2   User Interface

We explain the drawing system, which comprises the main component of CASOOK. We built CASOOK in Java. Fig. 1 depicts the main window of the drawing system.

Fig. 2. System object one   **Fig. 3.** System object two

Following are explanations of each part of the system, which correspond to the numbers in Fig. 1.

1. Buttons for choosing colour.
2. A check box to change the appearance of the colour mixing palette.
3. Buttons for choosing among four kinds of brushes.
4. A slider to change the size of the brush.
5. A canvas on which users freely draw.
6. A system object, which interacts with users' drawings.
7. Buttons to start or stop animation and clear the canvas.
8. A tab to change the panel of the playing animation and setting.

The system object and animation are related to the function presented in Section 3.

3   Interaction Module

Some applications [6][7][8][9] offer a gaming element to alleviate user boredom during drawing. However, limitations exist for the class of objects drawn by users to prepare the correct answers for the game. Therefore, they are not suitable for free drawing. To solve this problem, CASOOK contains an interaction module that includes a gaming function for extracting features through drawing. This function uses free drawing and thereby encourages users to draw, even if they are not adept at it. The system regards user strokes in drawings as one user object if the distance between the strokes becomes shorter than the threshold. The gaming functions in CASOOK as are outlined below.

Stage. CASOOK includes 56 stages from which users can choose one before starting to draw. Each stage is comprised of one background image and five system objects. The chosen stage becomes the canvas.

System Object. System objects interact with user objects; therefore, users can obtain new ideas for drawings. Some kinds of animation are added to the user object around system objects when the distance between the system object and user object is shorter than a specific value. System objects interact with the user object and enable users to obtain new ideas for drawings. An animation is added to the drawing when the distance

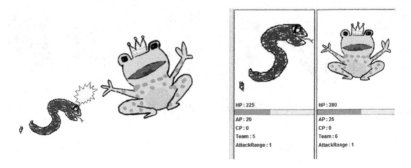

Fig. 4. Outline of battle mode

between user object and the system object becomes shorter than the threshold. There are three types of figures for system objects: a pole, circle, and others. Fig. 2 and Fig. 3 show trajectories of animations that are added to user objects around system objects.

Battle Mode. Fig. 4 shows an outline of battle mode. In this mode, a user object becomes an animated object, which has some parameters. Based on the colour and figure of the user object, the parameters of the hit point, ability point of attacking, ability point of healing, and number of teams are decided. When the animated object collides with other objects, the actions of those objects are determined as an attacking action or healing action, depending on whether the team numbers are the same. The action changes the parameter of the animated object. The object is removed from the canvas when the hit point of the object is 0.

Animation. Based on the recognition of a user object, some animation is added to the object. In a user experiment that we conducted, the type of animation used was decided based on the number of circles and lines in the drawing that were recognized by OpenCV [10]. After the experiment, we modified the animation function to assign a specific animation to the user object depending on what the user drew, specifically the class from the main classes of objects recognized by the analysis module.

4   Analysis Module

The analysis module is employed to recognize drawings. In this research, we constructed a method to recognize the class of object. Through user experiments, we found that almost all of the users' drawings could be classified into four classes: animals, faces, characters, and symbols; therefore, we defined these four classes as the main classes. We describe the dataset, features, and classifier as follows.

Dataset. 1000 drawings equally distributed among the four above-mentioned classes.
Features. An L-dimensional frequency histogram based on bag-of-features [2] is constructed and M-dimensional features specific to the concerned class are added. Thus, a drawings is mapped to an $(L + M)$-dimensional vector.

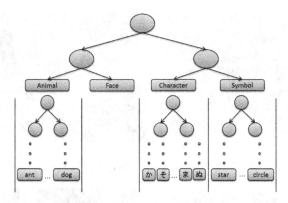

Fig. 5. Outline of two-layer recognition

Classifier. Drawings are recognized in two layers using a support vector machine based on a binary decision tree (SVM-BDT). Fig.5 shows the outline of the two-layer recognition method. The first layer is comprised of the main class recognition step for the upper part of Fig.5. After determining the main class, the second layer is applied to drawings and the precise class is obtained. The precise class is a more concrete class compared to the main class. For example, a precise class could be "cat," "dog," "frog," and so on, while its main class is "animal."

5   User Experiment

We conducted a user experiment in the classroom of a private school on April 18, 2014. We ran the hour-long experiment twice using different subject groups.

- Two users (two years old; one male and one female, sitting on the floor with parents)
- Five users (five years old; two males and three females, sharing a table with only children and each sitting on a chair)

5.1   User Experiment Flow

To perform the user experiment, we prepared four staff members, four tablets, a video camera, and a voice recorder. We observed users in the following flow.

1. Staff members introduced themselves and offered easy questions about users.
2. Staff members drew pictures on paper and asked users to guess what was drawn.
3. Users freely drew on the paper using coloured pencils.
4. Staff members drew pictures using CASOOK on tablets and informed users of the function of CASOOK.
5. Users freely drew by CASOOK on tablets.

Fig. 6. A two-year-old child drawing

Fig. 7. Five-year-old children drawing

Fig. 8. A drawing by five-year-old children

Fig. 9. Another drawing by five-year-old children

Fig. 10. Picture by user B

Fig. 11. Picture by user E

Fig. 12. Picture 1 by user F

Fig. 13. Picture 2 by user F

Fig. 14. Picture 1 by user G

Fig. 15. Picture 2 by user G

5.2   Results and Discussion

Fig.6 shows a drawing of a two-year-old child. Fig.7-fig.9 show drawings of five-year-old children. Fig.10 shows a picture drawn by the two-year-old child. Fig.11-fig.15 show pictures drawn by five-year-old children. Table 1 shows user profiles.

- The user's interest depended on the user's sex. Male children liked battle mode, whereas female children tended to be particular about drawing.

Table 1. User profiles

User	Age	User's sex	Mode of using CASOOK.
User A	2	Female	She likes to draw but did not use CASOOK because she was sleepy.
User B	2	Male	He drew scribbles but it was somewhat difficult for him to manipulate the tablet.
User C	5	Male	He likes battle mode.
User D	5	Female	She worried about the results of battle mode.
User E	5	Female	She was particular about drawings.
User F	5	Male	He wanted others to view his drawings.
User G	5	Female	She carefully selected various colours after completing the black line drawings.

Table 2. The number of the user's manipulation

User	Start	Stop	Clear canvas	Change colours	Erasers	Strokes
E	24	22	38	59	4	278
F	52	46	14	50	0	188
G	36	8	11	101	0	630

Table 3. Result of residual analysis

User	p-value
E	0.19
F	0.000032
G	0.0058

Users C and D are excluded from the table because they shared the same tablet.

- Sometimes what a user drew was not comprehensible to the others; however, those users did not admit this to the user. In such cases, to recognize drawings, we will improve the analysis module to recognize them. To this end, we will utilize other features, such as the position of the user object on the stage.
- We found that it was difficult for two-year-old children to manipulate the software on a tablet. On the other hand, five-year-old children easily used the tablets and were very much interested in CASOOK.
- Users wanted to play various animations depending on the class of the object.
- Table 2 shows the numbers of user's manipulating the software. CASOOK includes a function of automatically maintaining a user log; therefore, we counted the number of manipulations and analysed the mode of the user's drawing. User E often employed erasers, while user F pressed the start button and enjoyed the function of the animation many times. User G changed colours and was particular about the mode of drawing. The χ^2 test was performed for the proportion of the numbers of two manipulations. Manipulation 1 involved pressing the start button; Manipulation 2 involved changing the colour. As a result, the χ^2 value was $17.43(>9.21; p = 0.01, f = 2)$. There was a marked difference in the significance level of 1%. Table 3 shows the results of the residual analysis. There was no significant difference in the values of user E at the significance level of 5%. However, there was a significant difference in the values of users F and G at the significance level of 1%. Especially, the user F proportion of the numbers of two manipulations differed from the rest.

6 Conclusion

In this paper, we proposed the CASOOK system, which has a usable interface and entertains users while they draw in order to obtain various features during the drawing process. User experiments were performed to demonstrate the effectiveness of applying the system to children users. The results of the analysis conveyed the differences in users' manipulations without impeding free drawing. In our forthcoming presentation, we will present the results of the other user and computer experiments that we conducted for recognizing users' drawings. Following is an outline of key future work. We intend to:

- Analyse users' manipulations in more detail.
- Improve our two-layer sketch recognition algorithm.
- Assign appropriate animations to the given objects depending on the precise class of the objects.

Acknowledgment. This work was supported by Grant-in-Aid for JSPS Fellows(Grant Number:25·10978). We would like to thank all members of Shogakukan Inc. and ShoPro.

References

1. M. Suzuki and J. Gyoba : *Analyses of Factor Structure of Affective Impressions Produced by Line Drawings, Colors, Words and Combined Stimuli of Those Properties*, The Institute of Electronics, Information and Communication Engineering, Technical Report of IEICE. HIP, Vol.103(166), pp.57-62(2003)
2. M. Eitz, J. Hays and M. Alexa : *How Do Humans Sketch Objects?*, ACM Trans. Graph. (Proc. SIGGRAPH), Vol. 31, No. 4, pp. 44:1-44:10 (2012)
3. G. Vamvakas, B. Gatos, I. Pratikakis, N. Stamatopoulos, et al.: Hybrid Off-line OCR for Isolated Handwritten Greek Characters, in *Proceedings of the Fourth Conference on IASTED International Conference: Signal Processing, Pattern Recognition, and Applications*, SPPR'07, pp. 197–202 (2007)
4. G. Madzarov, D. Gjorgjevikj, I. Chorbev, et al.: A Multi-class SVM Classifier Utilizing Binary Decision Tree., *Informatica (Slovenia)*, Vol. 33, No. 2, pp. 225–233 (2009)
5. P. Bertola, N. Mori and K. Matsumoto : *Sketch Recognition for Interactive Multi-Agent System*, in *Institute of Systems Control and Information Engineers* 334-4, (2014)
6. R.S. Angel, T.R. Langston, G. Everson, and C. Militello. *The Official Pictionary Dictionary: The Book of Quick Draw*. Perigee Books (1989)
7. Drawdle : https://play.google.com/store/apps/details?id=oss.AndroidDrawdleFrame
8. Crayon physics : http://www.crayonphysics.com/
9. Draw a stickman : http://www.drawastickman.com/
10. G. Bradski : The OpenCV Library, *Dr. Dobb's Journal of Software Tools* (2000)

Collective Strategies for Virtual Environments: Modelling Ways to Deal with the Uncertainty

L. Castillo[1,*], C. López[2], M. Bedia[3], and F. López[2]

[1] Engineering Systems Dpt, University of Caldas, Manizales, Colombia
[2] Plastic and Physical Expression, University of Zaragoza, 50015 Zaragoza, Spain
[3] Computer Science Dpt, University of Zaragoza, 50018 Zaragoza, Spain
luisfercastillo@gmail.com

Abstract. Videogames offer new challenges and excellent application domains for AI technology and research. This paper shows a work on ant algorithms optimizing of collective strategies in a videogame environment. In general, ant algorithms are those that taking inspiration from the observation of ant colonies foraging behavior have developed optimization meta-heuristics. In the first part of the paper, we relate the strengths of ant strategies with the well-known exploration-exploitation problem. In the second part, a particular model on how to optimize group strategies in collective tasks is analyzed in light of the ideas previously examined. Finally, we will show how these ideas can constitute an improved stage in the problem of designing non-player characters in future videogames.

Keywords: videogames, exploration-exploitation dilemma, ant optimization.

1 Introduction

The use of artificial intelligence in games is increasing in game-development to this end and it is completely assumed by the videogame research community that Artificial Intelligence is going to become the core issue in creating realistic experiences in next generation video. Our game worlds are associated around intelligent agents, and as such we can apply "agent-based methodologies" using the metaphor "a participant is an agent" [1]. The classical AI agent taxonomy used by us is: Reflex Agents, Model-Based Agents, Goal-Based Agents and Utility-Based Agents, the classical AI methodologies associated are Fuzzy Systems, Neural Networks, Evolutionary Computing and Swarm Intelligence. All of them are based around symbolic representations of knowledge to create decision trees and model knowledge-based systems [2]. However, there could be differences between the type of intelligence that games will need and the real interests of the academic AI research. For example, AI research mainly focuses on the optimum and best solutions, whereas games require more practical solutions. A more advanced intelligence modeling should also adapt to the player's strategy allowing people to consider alternate

---

* Corresponding author.

© Springer International Publishing Switzerland 2015
S. Omatu et al. (eds.), *Distributed Computing & Artificial Intelligence, 12th Int. Conference*,
Advances in Intelligent Systems and Computing 373, DOI: 10.1007/978-3-319-19638-1_21

strategies to completing any goal in a game [3]. Artificial cognitive systems [4] – as opposed to "traditional" computer-based information processing systems – are defined by a number of general (interrelated) features and traits. Ideally, (i) they should be capable of interpreting (or "making sense of") whatever they are poised to sense in the environment they are operating; (ii) targeted learning (supervised or unsupervised, through interactions with their environment) to modify the way they operate and/or to improve their performance according to given criteria; pursuing goals and "what if…" deliberation; (iii) communicating and interacting between them [5], (iv) behaving "sensibly" and "robustly" under conditions of uncertainty e.g., through creative exploitation of novel situations in terms of previously learned regularities, through generalization and analogical reasoning [6]. The last point is the one in which we are interested.

In particular, in this paper, we are interested in modeling cooperative strategies for non-playable characters that learn from the rest of the members of the group and behave accordingly.

2   Ant Optimizing Strategies

Ant algorithms are distributing approaches able to deal with problems like the traveling salesman problem (TSP) and the quadratic assignment problem (QAP). There is currently a lot of ongoing activity in the scientific community to extend/apply ant-based algorithms to many different discrete optimization problems. Ant algorithms were inspired by the observation of real ant colonies. Ants are social insects whose behavior is directed more to the survival of the colony as a whole than to that of a single individual component of the colony. The pheromone trail allows the ants to find their way back to the food source (or to the nest). Also, it can be used by other ants to find the location of the food sources found by their nestmates. It has been shown experimentally that this pheromone trail following behavior can give rise, once employed by a colony of ants, to the emergence of shortest paths. Inspired on these, the ant colony optimization (ACO) meta-heuristics can be defined as a colony of artificial ants that cooperate in finding good solutions to difficult discrete optimization problems. There are different ways in which the behavior of a bot can be programmed. One of the most basic alternatives is to program it by means of a list of simple commands (for instance, by using if-then structures). Here, the programmers should know a priori a large list of possible situations to implement them. Other way is by Finite State Machines. This technique basically consists of a graph with (i) a set of states (for instance, walking, running, attacking, etc.) and (ii) their transitions. In particular, and related to our interests in this paper, there exist multiple experiences of using genetic algorithms to let virtual bots learn different tasks and perform different tasks. For instance, in [7] genetic algorithms were used to tune Counter Strike bot behavior and to calculate efficient paths between two endpoints in a landscape; in [8] genetic algorithms were chosen to evolve scripts to solve various tasks in Lemmings; in [9] they were used for implementing pathfinding techniques; in [10] to implement virtual entities and to define both their morphology and the neural circuitry involved, etc. Al, these initiatives have shown that genetic programming is a good tool for developing strategies in games. In this paper, we will use them to create a team of bots with capabilities to solve the problem of "capturing the flag", using techniques of long-distance communication, genetic algorithms tools and markovian chain notions. In the next part, we will show the proposal and evaluate the consequences and advantages of the model implemented.

3   Capturing the Flag: Other Instance of the E-E Dilemma

The trade-off between exploitation and exploration has been discussed for many years and in a big amount of scientific and technological fields. In particular, it is considered a classical discussion in technological issues where two competing tendencies exist. Pure exploration would correspond to trying all possible actions without any particular preference. The general lesson is that neither pure exploitation nor pure exploration is effective: an optimal trade-off must be found. It has been demonstrated numerically in a vast range of other cases.

The main objective of this paper is to show a series of experiments in the firstperson shooter videogame domain (Capture the Flag game) and it will be demonstrated how multi-agent systems solve problems of organizing and strategy. Our experiments will be a particular case of the "exploration vs exploitation dilemma", a paradox that appears in numerous situations where systems needs being adaptable and learnable at the same time and solutions of the dilemma are built evolving balances between parts. Our experiments will be programmed in the Unreal tournament 2004 (UT2004) environment. Several methods have been developed to solve problems from a multiagent-system approach but, in general, they systematically show a pair of incompatible aspects: (1) in order to maximize strategies, agents must solve tasks they know how to deal, but on the other hand, (2) they should try to discover new possibilities. If we focus on the first aspect, we would only exploit the information available with difficulties to be adaptable to new situations. For that reason, the ideal behavior will be built by a balance between two opposite strategies. Traditionally, the fact of finding a balance between these two factors is called the "exploration - exploitation dilemma" (EE). Some authors have remarked that different versions of this dilemma appear in different engineering domains [1]. In general, we could define both strategies as: (1) Exploration, including strategies as searching, experimentation, play, flexibility, innovation, etc., (2) Exploitation, where we can find refinement, choice, production, efficiency, selection, implementation and execution. In this paper, we show how to solve an EE dilemma in a videogame environment [11].

4   Model and Implementation

Our model consists of a set of agents that try to ``get the flag" and communicate to other partners (See Figure 1). Each bot can listen the transmissions from the others but only when he/she is not evolved in any other tasks. So, as a particular case of a EE dilemma, bots can do two different behaviors: moving around the map while they are looking for the flag, or waiting for other bots to capture the flag and being informed.

Fig. 1. Illustration of the environment in UT2004 where the experiments have been done

The aim of our team of bots is to find a flag and then normalize a path between them, through a trail that bots leave behind, for the rest of the team can also find them. At the time that the last bot finds one of two flags, give the experiment as complete. One of the challenges of working with ant algorithms is that they include a number of parameters that govern the behavior of our system:

1.   A bot crossing a road: This makes the value of the pheromone increases.
2.   Pheromone Evaporation: the amount of pheromone deposited in point is diluted over time, if that point is not visited by any ant.

Both behaviors are represented in the mathematical equations of subsection 3.1. and Figure 2.

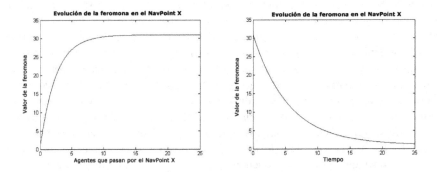

Fig. 2. Mathematical model of pheromone behavior: a) growing; b) decreasing

4.1   Artificial Pheromona's Model

One of the most important aspects of the system is to mediate pheromone level as it changes throughout the game. We will calculate the speed at which the pheromone increases or diluted, through two parameters included in the increment and decrement functions. For example, in the case of growthing:

$$I(t_i) = A(1 - e^{-t_i/\tau}) + 1 \tag{1}$$

where A is the upper limit of the intensity of the pheromone. For many bots to pass, the pheromone is still not growing; where τ is a parameter that sets the speed value inversely proportional to its growth. When points not pass any bot, the pheromone is losing strength:

$$I(t_d) = A(1 - e^{-t_d/\varepsilon}) + 1 \tag{2}$$

where A is the upper limit of the intensity of the pheromone and a parameter marking ☉ fast decay inversely proportional to its value. When applying these two functions simultaneously must take into account the value of the pheromone in the instant that applies each.

$$T_i = -\ln\left(\frac{1-I}{A} + 1\right) \cdot \tau \qquad\qquad T_d = -\ln\left(\frac{1-I}{A} + 1\right) \cdot \varepsilon \tag{3}$$

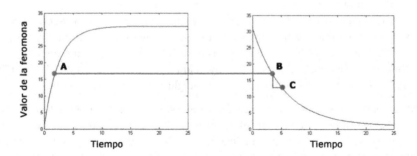

Fig. 3. Method of calculating the intensity value of the pheromone realtime

4.2   Genetic Evolution and Fitness

In a general situation we must find a balance between creating paths and the feedback from those already existing. This setting is what you tell us how good or bad it is a proposed solution.

$$F = (T_t - T_0) + \sum_{T_0}^{T_t} a(t) \tag{4}$$

We know what we want is that the route between the two flags and other possible routes strengthen roads are not strengthened. The fitness function remains as follows:

$$a(t) = \begin{cases} 0, p_h(t_i - 1) \le p_h(t_i) \\ 1, p_h(t_i - 1) > p_h(t_i) \end{cases} \tag{5}$$

Where Tt is the total time of the game. The parameter T0 is the time step in the simulation where a bot flag is first.

4.3   Results and Evaluation

How the model the group has been developed by using genetic algorithms, it is explained below in the next four stages.

- Representation: Each individual of the population consists of a bot identified by their coefficients $\{\tau,\varepsilon\}$
- Fitness function: The role of this function is to obtain a metric to guide the evolution. We have chosen a very general mathematical expression (4). and we have fixed in 35 s the maximum amount of time to find the flag.
- Population: All candidate configuration together are considered the population for the genetic algorithm. The population size is, thus, three bots (and their coefficients $\{\tau,\varepsilon\}$).
- Experimental settings: There are several game settings that are of importance and have been empirically fixed : (i) The experiment with the genetic algorithm was run 1600 times; (ii) The selection procedure is truncation selection on 50% of the population and mutation is done by adding Gaussian terms to the weights. (iii) Crossover is done by averaging the weight values of both parents.

After nearly 1600 generations, we obtain an ordered list of the best fitness. 1600 results the 80.14% solve the problem in less than 35 seconds limit set for the experiment. In order to have conclusions on the results, we calculate for each genotype quotient growth trend (Ctc) system. Higher values indicate that our system, building their capabilities increase and decrease of the pheromone tends to maintain high values at all points of sailing. Once calculated the ratios, we plot these ratios with respect to its assessment of fitness (Figure 4).

In Figure 4 we can observe three areas generate a range of ratios us a better or worse results: (i) Zone 1 [.05-.90]: In this range of ratios results are not desired, obtaining quite high fitness values; (ii) Zone 2 [0.90-1.20]: We are at the range where we have obtained the best results; (iii) Zone 3 [1.20-18: Here we can see some pretty scattered results, where we find situations of all kinds, although we found better solutions than in the first zone.

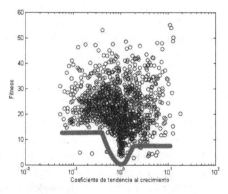

Fig. 4. Representation of the areas as success rate of genetic algorithm

In the table below we see the means of the values returned us the fitness function with respect to the area ranges in which we find.

Table 1. Representing success of genetic algorithm

Fitness	τ	⊙	T_0	T_t	Ajust.
2.656	3.77	7.17	1.63	3.28	2
3.422	1.39	5.24	1.09	3.52	4
3.485	7.42	7.72	2.17	4.66	4
3.516	5.81	7.50	2.16	4.67	4
3.688	0.93	5.12	2.17	4.86	5
3.766	1.22	4.83	1.03	3.30	4
3.766	3.61	9.55	2.05	4.31	4
3.953	4.02	10.14	3.41	5.86	5
3.953	9.33	3.40	2.97	5.42	5
4.125	9.43	9.94	6.36	8.99	5
4.203	6.17	5.69	1.13	3.83	6
4.266	9.20	9.15	1.67	4.44	6
4.312	1.27	5.80	3.67	6.48	6
4.453	1.09	4.31	1.91	4.86	7
4.531	3.70	0.92	3.28	6.31	7

In view of these results, we can see that the zone 2 has a tendency to growth ratios that are among the range 0.90-1.20, is where better results are obtained.

Table 2. Representation of the areas as success rate of genetic algorithm

Zone	mean	population
general	22.09	1595
1	24.01	707
2	15.66	235
3	22.11	653

Zone 1 consists of lower ratios, which makes the evaporation of the pheromone is faster than creating it. For good results, our ratios should be high. This implies that in our problem pheromones is desirable to maintain a high value and creating ways prevail with respect to the solution.

5   Conclusions

Although there have been attempts and successful experiences that have included or simulated intelligent systems into video games, we believe that the use of cognitive modeling techniques could provide video-games with an element of intelligence that existing games do not have. Our idea is to use recent ideas coming of natural cognitive systems together with recent advances in modeling of artificial cognitive systems, and discuss its relations in the artificial environments.

Our model brings forth the need to include the temporal dimension of agents and environment into current modeling frameworks. Adjustment speed, decay rates, deployment duration, patterns of intermittency, and so on crucially matter when it comes to real-world problem solving. Computational and representational approaches to cognition are prone to neglect such time-dependent phenomena and might often fail to account for natural behavior. Given strong cognitivist assumptions, one would be tempted to build models that first compute a near-maximum fitness solution and only then deliver an output command. We have shown, however, that the optimal solution to the "explore-exploit dilemma" uses non-maximal solutions by means of fast intermittent behavior in a manner that, in addition, requires only very simple control mechanisms. It could be further conjectured that, under certain constraints, intermittency and, perhaps more generally, recurrent agent-environment suboptimal interactions provide robust and simple solutions to many adaptive problems.

Acknowledgments. This research has been partially supported by the project TIN2011-24660, funded by the Spanish Ministry of science and Innovation.

References

1. Russell, S.J., Norvig, P.: Artificial Intelligence: A Modern Approach, 2nd edn. Prentice Hall (December 2002)
2. Kasap, Z., Magnenat-Thalman, N.: Intelligent virtual humans with autonomy and personality: State-of-the-art. In: Intelligent Decision Technologies. IOS Press (2007)
3. Stork, H.-G.: Report on the Workshop "Future Trends in Artificial Cognitive Systems". In: Frankfurt Conference Centre, December 9-10. ECVISION (2004)
4. Hoffman, R.R., Klein, G., Laughery, K.R.: The State of Cognitive Systems Engineering. IEEE Intelligent Systems 17(1), 73–75 (2002)
5. Woods, D.D., Hollnagel, E.: Joint cognitive systems: Patterns in cognitive systems engineering. Taylor & Francis (2006)
6. Du Boulay, B., Rebolledo-Mendez, G., et al.: Motivationally intelligent systems: Three questions. In: Second International Conference on Innovations in Learning for the Future, Future e-Learning 2008, Istanbul, pp. 1–10 (2008)

7. Cole, N., Louis, S.J., Miles, C.: Using a genetic algorithm to tune first-person shooter bots. In: Proceedings of the Congress on Evolutionary Computation, vol. 1, pp. 139–145. IEEE, Piscataway (2004)

8. Kendall, G., Spoerer, K.: Scripting the game of Lemmings with a genetic algorithm. In: Proceedings of the 2004 Congress on Evolutionary Computation, vol. 1, pp. 117–124. IEEE, Piscataway (2004)

9. Alliot, J.M., Durand, N.: A genetic algorithm to improve an Othello program. Artificial Evolution, 307–319 (1995)

10. Sims, K.: Evolving virtual creatures. In: Computer Graphics Proceedings. Annual Conference Series 1994. SIGGRAPH 1994 Conference Proceedings, pp. 15–22. ACM, New York (1994)

11. Berger-Tal, O., Nathan, J., Meron, E., Saltz, D.: The Exploration-Exploitation Dilemma: A Multidisciplinary Framework. PLoS ONE 9(4), e95693 (2014), doi:10.1371/journal.pone.0095693

Solar Intensity Characterization Using Data-Mining to Support Solar Forecasting

Tiago Pinto[1], Gabriel Santos[1], Luis Marques[1], Tiago M. Sousa[1], Isabel Praça[1], Zita Vale[1], and Samuel L. Abreu[2]

[1] GECAD – Knowledge Engineering and Decision-Support Research Center
Institute of Engineering – Polytechnic of Porto (ISEP/IPP), Portugal
{tmcfp,gajls,lumar,tmsbs,icp,zav}@isep.ipp.pt
[2] General Alternative Energies Group - IFSC – Instituto Federal de Santa Catarina, Brazil
abreu@ifsc.edu.br

Abstract. The increase of renewable based generation as alternative power source brings an added uncertainty to power systems. The intermittent nature of renewable resources, such as wind speed and solar intensity, requires the use of adequate forecast methodologies to support the management and integration of this type of energy resources. This paper proposes a clustering methodology to group historic data according to the data correlation and relevance for different contexts of use. Using the clustering process as a data filter only the most adequate data is used for the training process of forecasting methodologies. Artificial Neural Networks and Support Vector Machines are used to test and compare the quality of forecasts when using the proposed methodology to select the training data. Data from the Brazilian city of Florianópolis, Santa Catarina, has been used, including solar irradiance components and other meteorological variables, e.g. temperature, wind speed and humidity. Experimental findings show that using the proposed method to filter data used for training ANN and SVM achieved promising results, outperforming the approaches without clustering.

Keywords: Artificial Neural Network, Clustering, Data Mining, Machine Learning, Solar Forecasting, Support Vector Machine.

1 Introduction

The use of renewable energy sources is having a significant increase, encouraged by governmental policies and incentive programs whose concern is to avoid the exploitation of finite fossil fuel reserves and at the same time avoid environmental damages. In Europe a set of legislation was been defined having the well known "20-20-20" as targets [1]. The national targets will enable EU as a whole to reach a 20% renewable energy target for 2020: more than double the 2010 level of 9.8%. These targets reflect member states' different starting points and potential for increasing renewable generation, ranging from 10% in Malta to 49% in Sweden [1]. In this context, alternative sources of renewable and clean energy, such as tidal, wind and solar have become of great importance. However, the variable and intermittent nature of these resources

© Springer International Publishing Switzerland 2015
S. Omatu et al. (eds.), *Distributed Computing & Artificial Intelligence, 12th Int. Conference*,
Advances in Intelligent Systems and Computing 373, DOI: 10.1007/978-3-319-19638-1_22

poses a lot of challenges to several entities, e.g. utilities, system operators and market operators, especially when considering a significant market penetration rate [2].

Solar energy is clearly the most abundant resource available to modern societies. Usually summer months have smaller variability; however, even during some sunshine months sudden changes might occur. The variability of the solar resource is mostly due to cloud cover variability and atmosphere conditions.

This paper proposes the use of a clustering algorithm to filter the data that is most appropriate to be used in the training process of forecasting methodologies. The use of the specific historical data that most potentiates the optimization of the forecast is essential for the improvement of the forecasting quality of solar intensity, by excluding potential irrelevant or even prejudicial data, and using just the most relevant data in the training process. In order to test and validate the proposed approach, Artificial Neural Networks (ANN) [3] and Support Vector Machines (SVM) [4] are used and conclusions about the influence of the proposed data filtering method are taken.

2   Solar Forecasting

Despite its importance for the existence of life on earth, the sun is nowadays a source of clean energy that contributes to reduce the difficulty in fulfilling the energy demand. Photovoltaic (PV) and solar thermal are the main sources of electricity generation from solar irradiance. The management and operation of solar thermal energy plants with storage energy system need reliable predictions of solar irradiance with the same temporal resolution as the temporal capacity of the back-up system [5]. Advances in power semiconductor technology have allowed higher efficiencies in the conversion of solar energy into electrical energy trough photovoltaic cells and PV systems have reached the end-user and nowadays it is being used in several buildings to generate electricity.

The increase on the use of renewable energy sources (RES) affects the behavior of a considerable number of entities from the electricity sector and imposes economical and technical challenges. Forecasting renewable resources is important from the producers, retailers, aggregators, system operators and market operators' point of view. From the utility point of view, application of renewable sources can potentially reduce the demand for distribution and transmission facilities. Clearly, distributed generation located close to loads can reduce power flows in transmission and distribution circuits with two important effects: loss-reduction and the ability to potentially substitute network assets. Furthermore, the presence of generation close to demand can increase service quality of end customers [2]. From the power system operators' point of view, short-term forecasting is relevant for dispatching and regulatory purposes, to optimize the decision making by allowing corrections to unit commitment. Concerning market operators, the generation forecast is essential to plan transactions in the electricity market in order to assure the balance between supply and demand. From the economical point of view it is also important for electricity players to use this knowledge as competitive advantage in electricity trading.

Usually, to predict the generation of RES, two approaches are used: an approach based on physical models [6], using mathematical equations to describe physics and dynamics of the atmosphere that influence solar radiation; or an approach based on time series analysis by means of statistical models [7]. Physical models work well for medium- and long-term forecasting, while statistical models usually perform well for short-term forecasting. Several techniques have been applied to solar irradiation or solar power forecast, such as regression techniques, Auto Regressive Moving Averages, Auto Regressive Integrated Moving Averages, Artificial Neural Networks (ANN), Genetic Algorithms (GA) and Support Vector Machines (SVM).

References [6-8] provide good overviews on the state of the art in solar irradiance forecasting. In [9] a comparison on several forecasting techniques to predict solar power at a photovoltaic power plant in California is presented. In this work, ANN has proved to be a promising technique on this field, showing improved results while combined with GAs. The same conclusions about the use of ANNs were achieved by [10]. ANNs have also been successfully applied to the forecasting of other renewable sources based production types, such as the wind power, in [11]. The good results achieved by ANNs in the most varied fields provide an encouraging indication of ANNs' capability of coping with the approached problem. A weighted SVM methodology is proposed in [12], and [13] presents an alternative SVM method to predict solar power. SVM has proved to be a promising technique for solar forecasting.

3   Solar Intensity Characterization

The main objective of the proposed methodology is to filter the available data, in order to select only the most relevant data to be used by forecasting methodologies, such as ANN or SVM. The proposed way to achieve this is to apply a clustering methodology in order to separate the available data into different groups according to their correlation and similarity. Once the clustering process is completed, the achieved clusters of data are evaluated, so that an adequate balance between the number of used clusters and the associated data variability is achieved. This is necessary because the optimal result should be the use of the least number of possible clusters that allows the maximum data separation (least variability of data). When using an exaggerated number of data clusters, the available data to be used by the forecasting methodologies will decrease dramatically, which may cause the forecasting process to decrease its effectiveness (in the extreme case, the data variability in each cluster is null when the number of clusters is equal to the number of data observations; however, this results in using only one data observation as input for the forecasting process, which is obviously fatal for the forecasting process). On the other hand, if using an excessively low number of clusters, the separation of data will be very small, and consequently the variability of data in each cluster is high. For this reason an adequate balance between the number of clusters and the associated variability is vital, with the aim of achieving the smaller number of data sub-groups that results in the higher gain in terms of data variability reduction.

Clustering is the process of partitioning a certain database into clusters based on a concept of similarity or dissimilarity among data [14]. A good cluster partition must present objects with high similarity among them and low similarity among objects that belong to others clusters. Although there is a wide variety of clustering algorithms, there is no single algorithm that can, by itself, discover all sorts of cluster shapes and structures. For review on data clustering algorithms see [14].

One of the most popular and simple clustering algorithms, K-means, was first published in 1955. In spite of the fact that K-means was proposed over 50 years ago and thousands of clustering algorithms have been published since then, K-means is still widely used [15]. The K-Means clustering methodology considers a set of observations $(x_1, x_2, ..., x_n)$, where each observation is a d-dimensional real vector, and n is the number of considered observations. The clustering process aims at partitioning the n observations into k $(\leq n)$ clusters $C = \{C_1, C_2, ..., C_k\}$ so that the Within-Cluster Sum of Squares (WCSS) is minimized (1).

$$\min \sum_{i=1}^{k} \sum_{x \in C_i} ||x - \mu_i||^2 \tag{1}$$

where μ_i is the mean of points in C_i, i.e. the cluster *centroid*.

The dimension of the vector that characterizes each observation x_p, $p \in \{1, ..., n\}$ is equal to the sum of the individual dimensions of n vectors, where each of these n vectors contains the data that is referent to a different data variable, e.g. electricity market prices, wind speed, solar intensity, etc. With the objective of minimizing equation (1), the clustering process executes an iterative process between two steps: (i) the assignment step, where each observation x_p is assigned to the cluster $C^{(t)}$ whose mean value yields the minimum WCSS in iteration t, as presented in (2); and (ii) the update step, where the new means of each cluster are calculated, considering the newly assigned observations, determining the new *centroid* μ_i of each cluster, as in (3).

$$C_i^{(t)} = \{x_p : ||x_p - \mu_i^{(t)}||^2 \leq ||x_p - \mu_j^{(t)}||^2 \; \forall j, 1 \leq j \leq k\} \tag{2}$$

$$\mu_i^{(t+1)} = \frac{1}{|C_i^{(t)}|} \sum_{x_j \in C_i^{(t)}} x_j \tag{3}$$

The execution of the algorithm stops when the convergence process is completed, i.e. when the assignments of observations to different clusters no longer change. By minimizing the WCSS objective, in equation (1), the K-Means clustering methodology assigns observations to the nearest cluster by distance. This means that each observation will be grouped in the same cluster as the ones that are more similar.

The correct choice of k is often ambiguous, with interpretations depending on the shape and scale of the distribution of points in a data set and the desired clustering resolution of the user. In addition, increasing k without penalty will always reduce the amount of error in the resulting clustering, to the extreme case of zero error if each data point is considered its own cluster (*i.e.*, when k equals the number of data points). Intuitively then, the optimal choice of k will strike a balance between maximum compression of the data using a single cluster, and maximum accuracy by assigning each data point to its own cluster. If an appropriate value of k is not apparent from prior knowledge of the properties of the data set, it must be chosen somehow.

The Clustering Dispersion Indicator (CDI) [16] is a method to determine the quality of the division of objects into the different clusters. With this objective, these methods perform a dispersion analysis of the elements. In other words, they examine the similarities between the objects and check which groups are more compatible. CDI depends on the distance between the observations grouped in the same cluster and (inversely) on the distance from each observation to all other clusters' centroids, as described in (4), where R is the mean point of all observations, and n is the number of observations that have been assigned to each cluster.

$$CDI = \frac{\sqrt{\frac{1}{K}\sum_{k=1}^{K}\left[\frac{1}{2.n^{(k)}}\sum_{n=1}^{n^{(k)}}d^2(x^{(m)},\mu^{(k)})\right]}}{\sqrt{\frac{1}{2K}\sum_{k=1}^{K}d^2(x^{(k)},R)}} \qquad (4)$$

The distance between two observations x_i and x_j is calculated as defined in (5).

$$d(xi,xj) = \sqrt{\frac{1}{H}\times\sum_{h=1}^{H}(xi(h)-xj(h))^2} \qquad (5)$$

where H represents the size of the vector containing all elements of each observation.

When comparing the variance values, provided by CDI, that each different number of clusters originates, the first clusters will add much information (explain a lot of variance), but at some point the marginal gain will drop. This means that from this point the gain (in variance reduction) of adding an extra cluster is not as significant as before. The number of clusters is chosen at this point; *i.e.* the point where the variance from one number of clusters to the next stops decreasing sharply and starts stabilizing.

4 Experimental Findings

This case study has the goal of demonstrating the application of the proposed methodology to the problem of solar intensity forecasting. Results are shown for the forecasting with and without the use of the proposed methodology, for two distinct forecasting methodologies: ANN and SVM. The used ANN has been presented in [3], and it is characterized as a feed-forward ANN, with a variable number of hidden nodes in the intermediate layer. The SVM has been presented in [4]. The training of the ANN and SVM is updated in all iterations so that the most recent available data is always considered as input for the training. The used data refers to Florianópolis, Santa Catarina, Brazil. Data corresponds to the period from 1990 to 1999, including the values of Global, Direct, Diffuse and Extra-terrestrial Irradiance, in W/m^2; temperature in °C; humidity in %; and wind speed in m/s. More details are available in [17].

The clustering process has been performed considering different grouping approaches. The different approaches depend on the considered perspective when analysing the time series. For performance comparison purposes, the series has been analyzed separately for each season and for each month of the year, with the aim of extracting tendencies that are dependent on the time of the year. Additionally, years as a whole have been considered, as well as an analysis per period of the day where hours with different characteristics are grouped together. Fig. 1 shows the clustering results, for different grouping approaches, considering different number of clusters.

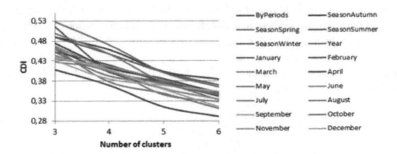

Fig. 1. CDI clustering evaluation for different clusters and distinct grouping approaches

From Figure 1 it becomes conclusive that the division by periods of the day is the best one, as the CDI values are clearly below all other approaches for all numbers of clusters. The splitting by periods of the day means that data is divided by hours of the day. In the current case, of solar intensity forecasting, the data profile for each day is similar throughout the time: null values during night-time; an increasing tendency during the morning; the peak values are found during the midday hours; and finally a decreasing tendency until nightfall. This grouping process assures that hours with similar profiles are all placed in the same group. This way, when forecasting a solar irradiance value for a certain hour of the day, only the data concerning the most similar hours are used, refraining from using irrelevant (or even misleading) data from the training process. Additionally, from Fig. 1 it is also visible that the *ByPeriods* approach CDI values present an stabilization of the variance decrease (an elbow in the plot) when the number of clusters is equal to five. The decrease in clusters variance until this point is more accentuated than the variance from this point on. This, by the "elbow criterion", explained in section 3, indicates that five is the ideal number of clusters for this case, into account the balance between the variance of the clusters and the number of clusters. Fig. 2 presents the clustering results of the *ByPeriods* approach using 5 clusters, for 12 sample days (1st day of each month of the year 1999).

From Fig. 2 it is visible that hours with similar solar intensity values are grouped in the same cluster. This is true even for days with very different values of solar intensity, such as days during the winter and during the summer. Since the solar tendency is similar during all year, all days are successfully classified in clusters that join nightly hours (null values), peak intensity hours, and hours with intermediate solar intensity. Using the separation of hours provided by the clustering process, the forecasting process is adapted, so that only the hours from previous days that are classified in the same cluster are considered as training data by the ANN and the SVM approaches. The considered ANN uses the topology defined in [3]. The SVM uses two different kernel functions: Radial Basis Function (RBF) and exponential Radial Basis Function (eRBF), with the following parameters, as detailed in [4]:

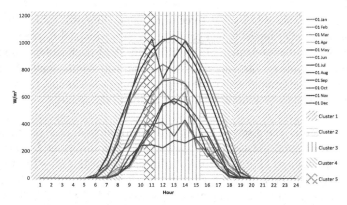

Fig. 2. Clustering results of the *ByPeriods* approach, for five clusters

- Kernel RBF: σ (angle) = 6, ε-insensitive = 0, limit = ∞, offset = 1;
- Kernel eRBF: σ (angle) = 18, ε-insensitive = 0, limit = ∞, offset = 1;

Table 1 shows the average Symmetric Mean Absolute Percentage Error (SMAPE) and the Standard Deviation (SD) achieved by the forecasts of ANN and SVM, with and without the use of the proposed clustering based data filtering, for 1464 executions referent to the 24 hours of each day of November and December, 1999.

Table 1. SMAPE and SD of ANN and SVM with and without the proposed clustering process

Methodology	Kernel	Without Clustering		With Clustering	
		SMAPE	SD	SMAPE	SD
ANN	-	34,02	13,26	29,71	12,83
SVM	RBF	23,36	9,13	21,84	9,18
	eRBF	21,48	7.46	19,23	7,42

From Table 1 it is visible that the forecasting error decreases when using the proposed methodology to filter training data of the ANN and SVM. By using only the most correlated and relevant data, the training process is refined, avoiding the use of reduntant data that damages the training, and consequently the forecasting quality.

5   Conclusions

This paper presented a new data mining based approach to filter the most relevant data to be used in the training process of ANN and SVM based solar intensity forecasting methodologies. The proposed methodology uses a clustering algorithm to group the historic data into different clusters, according to data similarity and solar intensity tendencies. This way the disparity of training values is reduced and the forecast process is facilitated. Results show that the application of the proposed method enables ANN and SVM approaches to improve the forecasting results quality. The best results quality is achieved when grouping data according to periods of each day.

As future work, the use of other forecasting methodologies, namely fuzzy inference systems and other methods that combine classification and forecasting, such as M5P. Additionally, the refinement of the ANN and SVM methodologies may lead to an improvement of the results, especially taking into account the methods referred in the related work. The comparison of the proposed approach with these previous works will provide the means for a more suitable validation of the results.

Acknowledgements. The research leading to these results has received funding from the People Programme (Marie Curie Actions) of the European Union's Seventh Framework Programme FP7/2007-2013/ under project ELECON - Electricity Consumption Analysis to Promote Energy Efficiency Considering Demand Response and Non-technical Losses, REA grant agreement No 318912 (PIRSES-GA-2012-318912). This work is also supported by FEDER Funds through COMPETE program and by National Funds through FCT under the projects FCOMP-01-0124-FEDER: PEst-OE/EEI/UI0760/2015 and PTDC/EEA-EEL/122988/2010.

References

1. European Commission, The 2020 climate and energy package,
   `http://ec.europa.eu/clima/policies/package/index.en.htm`
   (last accessed December 2014)
2. Silva, M., et al.: An integrated approach for distributed energy resource short term scheduling in smart grids considering realistic power system simulation. Energy Conversion and Management 64, 273–288 (2012)
3. Pinto, T., et al.: Dynamic Artificial Neural Network for Electricity Market Prices Forecast. In: IEEE 16th Int. Conference on Intelligent Engineering Systems, Portugal (June 2012)
4. Pereira, R., Sousa, T.M., Pinto, T., Praça, I., Vale, Z., Morais, H.: Strategic Bidding for Electricity Markets Negotiation Using Support Vector Machines. In: Bajo Perez, J., et al. (eds.) Trends in Practical Applications of Heterogeneous Multi-agent Systems. The PAAMS Collection. AISC, vol. 293, pp. 9–18. Springer, Heidelberg (2014)
5. Kopp, G., Lean, J.L.: A new, lower value of total solar irradiance: Evidence and climate significance. Geophysical Research Letters 38, L01706 (2011)
6. Inman, R., et al.: Solar forecasting methods for renewable energy integration. Progress in Energy and Combustion Science 39, 535–576 (2013)
7. Pelland, S., et al.: Photovoltaic and Solar Forecasting: State of the Art. International Energy Agency Photovoltaic Power Systems Programme (2013)
8. Diagne, M., et al.: Review of solar irradiance forecasting methods and a proposition for small-scale insular grids. Ren. and Sust. Energy Reviews 27, 65–76 (2013)
9. Pedro, H., Coimbra, C.: Assessment of forecasting techniques for solar power production with no exogenous inputs. Solar Energy 86, 2017–2028 (2012)
10. Ioakimidis, C.S., et al.: Solar Production Forecasting Based on Irradiance Forecasting Using Artificial Neural Networks. In: IEEE Industrial Electronics Society Conference (2013)
11. Quan, H., et al.: Short-Term Load and Wind Power Forecasting Using Neural Network-Based Prediction Intervals. IEEE Trans. NN and Learning Systems 25(2) (February 2014)
12. Xu, R., et al.: Short-term Photovoltaic Power Forecasting with Weighted Support Vector Machine. In: IEEE International Conference on Automation and Logistics (August 2012)

13. Zeng, J., Qiao, W.: Short-term solar power prediction using a support vector machine. Renewable Energy 52, 118–127 (2013)
14. Jain, A.K., Murty, M.N., Flynn, P.J.: Data Clustering: A Review. ACM Computing Surveys 31(3), 264–323 (1999)
15. Jain, A.K.: Data Clustering: 50 years beyond K-Means. Pattern Recognition Letters 31(8), 651–666 (2010)
16. Ilie, C.: Support Vector Clustering of Electrical Load Pattern Data. IEEE Transactions on Power Systems 24(3), 1619–1628 (2009)
17. Abreu, S.L., et al.: Qualificação e Recuperação de Dados de Radiação Solar Medidos em Florianópolis – SC. In: 8th Brazilian Congress of Thermal Engineering and Sciences, Porto Alegre (2000)

Automated Multi-agent Negotiation Framework for the Construction Domain

Moamin A. Mahmoud[1], Mohd Sharifuddin Ahmad[1],
Mohd Zaliman M. Yusoff[1], and Arazi Idrus[2]

[1] Centre for Agent Technology, College of Information Technology,
Universiti Tenaga Nasional, Kajang, Selangor, Malaysia
{moamin,sharif,zaliman}@uniten.edu.my
[2] Department of Civil Engineering, Universiti Pertahanan Nasional Malaysia
Kem Sungai Besi, 57000 Kuala Lumpur, Malaysia
arazi@upnm.edu.my

Abstract. In this paper, we propose an automated multi-agent negotiation framework for decision making in the construction domain. It enables software agents to conduct negotiations and autonomously make decisions. The proposed framework consists of two types of components, internal and external. Internal components are integrated into the agent architecture while the external components are blended within the environment to facilitate the negotiation process. The internal components are negotiation algorithm, negotiation style, negotiation protocol, and solution generators. The external components are the negotiation base and the conflict resolution algorithm. We also discuss the decision making process flow in such system. There are three main processes in decision making for specific projects, which are propose solutions, negotiate solutions and handling conflict outcomes (conflict resolution). We finally present the proposed architecture that enables software agents to conduct automated negotiation in the construction domain.

Keywords: Intelligent Software Agent, Multi-agent Systems, Agent and Negotiation, Automated Negotiation, Value Management, Construction Domain.

1 Introduction

In the construction domain, deciding on a new project is dependent upon a company's strategy. If the strategy is based on a decision made by a stakeholder, then it takes a very short time to decide. However, such decision has no value in terms of value management, because the decision-making process does not include other experienced stakeholders that hold different backgrounds.

Figure 1 considers a value management approach that emphasizes on involving various stakeholders in the decision-making process to arrive at a single valued solution. In other words, the various stakeholders with different backgrounds that have a stake in the project must contribute to the decision. In fact, these stakeholders often belong to different departments and possess different perspectives about the solutions according to their background and positions they hold.

© Springer International Publishing Switzerland 2015
S. Omatu et al. (eds.), *Distributed Computing & Artificial Intelligence, 12th Int. Conference*,
Advances in Intelligent Systems and Computing 373, DOI: 10.1007/978-3-319-19638-1_23

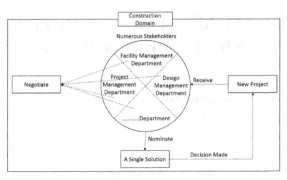

Fig. 1. A Decision Making based on Value Management in Construction Domain

For example, a project manager usually cares more about the cost of a project than the function while a design manager is more concerned about the function than the cost. Thus, for any decision to be made regarding a new project, stakeholders must propose a single optimal solution. However, a problem may arise when stakeholders propose more than one solution. In such situation, stakeholders need to negotiate on the proposed solutions and agree on a single solution. But the negotiation may not be easy and smooth because when stakeholders possess different backgrounds, often their views about the optimal solution for a particular project are different. Such differences cause conflicts in arriving at a decision. In addition, stakeholders may work at different branches throughout the country or other parts of the world which make a meeting for decision more difficult and costly. While applying Value Management on decision making in the construction domain is useful, it faces communication difficulties between stockholders and conflicting issues that require negotiation.

In this paper, we attempt to overcome these difficulties by proposing a Value-based Automated Negotiation Model utilizing the multi-agent system's approach. It enables software agents to conduct negotiations and autonomously arrive at a decision.

While this work is inspired by the work of Utomo [1], his study is only in conceptual level and lacks an intelligent agent architecture that aids an agent to interact with other agents and respond to its environment and eventually influences its autonomy level in decision making. Automated Negotiation as a very complicated system could not be efficiently used if agents have trivial architecture. Moreover, he does not incorporate the different negotiation styles to the agent architecture which could help the agents in mimicking humans' styles in negotiation. Briefly, the major development that we intend to do is to develop concrete agent architecture such as the Belief-Desir-Intention architecture and explore the potential components that an agent could employ to conduct useful and efficient negotiations. Consequently, we consult the various resources that are presented by Utomo [1] to come up with our framework.

The next section dwells upon the related work on automated negotiation. Section 3 presents the proposed framework. In Section 4, we discuss the decision making process flow. Section 5 presents the agent architecture and Section 6 concludes the paper.

2   Related Work

In this section, we discuss two prominent topics to this research which are, value management, and applications of negotiation in multi-agent systems.

Value Management (VM) is defined as "a structured, organized team approach to identify the functions of a project, product, or service that will recognize techniques and provide the necessary functions to meet the required performance at the lowest overall cost" [17]. VM works on identifying and eliminating unnecessary cost [18] but without affecting a quality parameter [19]. VM is based on data collection method from reliable resources and functional requirements to fulfill the needs, wants and desires of the customers [1]. According to Kelly and Male [20], VM is a multidisciplinary, team-oriented approach to problem solving.

The application of VM in decision making has been reported by many researchers [1, 21, 22]. One of the techniques that is relevant to VM is weighting and scoring in which a decision needs to be made in selecting an option from a number of competing options, and the best option is not immediately identifiable [1, 23, 24].

Intelligent software agents have been widely used in distributed artificial intelligence and due to their autonomous, self-interested and rational abilities, agents are well-suited for automated negotiation on behalf of humans [2]. According to Kexing [2], automated negotiation is a system that applies artificial intelligence and information and communication technology to negotiation strategies, utilizing agent and decision theories.

Numerous research have discussed the negotiation on multi-agents systems in various domains [3, 4, 5, 6, 7]. Few of them study the issues of conflict resolution and negotiation in construction domain [1, 8, 9]. Anumba et al. [9] presented two main negotiation theories; mechanical and behavior theories. The mechanical theory is inspired by game theories which are mathematical models relied on rational behavior assumption, while the behavior theory studies human behavior in negotiation.

Coutinho et al. [10] proposed a negotiation framework to serve collaboration in enterprise networks to improve the sustainability of interoperability within enterprise information systems. Utomo [1] presented a conceptual model of automated negotiation that consists of methodology of negotiation and agent based negotiation. Dzeng and Lin [11] presented an agent-based system to support negotiation between construction and suppliers via the Internet. Anumba et al. [12] proposed a collaborative design of light industrial buildings based on multi-agent systems to automate the interaction and negotiation between the design members. Ren et al. [4] developed a multi-agent system representing participants, who negotiate with each other to resolve construction claims.

3   A Conceptual Automated Negotiation Framework

From our initial investigation of the literature [1, 2, 3, 4, 8, 9], we observe that agents need to be integrated with six main components to conduct negotiations, which are classified, in this research, into internal and external components. The internal components are negotiation algorithm, negotiation style, negotiation protocol, and solution generator. The components are integrated with a BDI agent architecture (as discussed in Section 5). The external components are the negotiation base and the conflict resolution algorithm.

As shown in Figure 2, the negotiation algorithm presents a formal and intelligent procedure that maintains negotiations with other agents. Each agent is endowed with a negotiation style that represents the agent's approach to-negotiation. Each agent possesses one style, either competing style or collaborating style.

For agents to conduct negotiations systematically, they must have a negotiation protocol, which controls the negotiations process between agents. For example, agents could possibly negotiate individually or form groups (coalitions) before negotiating. They could also negotiate directly by sending messages to each other or deploy another method to share their inputs. Finally, agents must be able to generate solutions that conform to their interests and reap the benefits of negotiation.

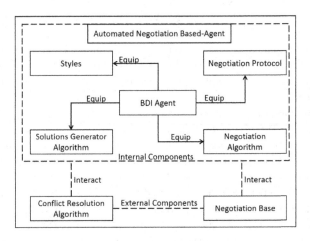

Fig. 2. Automated Negotiation Model based on Multi-agent System (AN-MAS)

As shown in Figure 2, the four components are called internal components.

The Negotiation Base represents the negotiation hub that contains suggested solutions of negotiations used by agents in sharing their solutions and form coalitions. The base reduces direct interactions between agents that would increase the network load.

The conflict resolution algorithm handles negotiations outcomes. If agents have not agreed on a single solution, the conflict resolution algorithm works on solving that conflict. Consequently, the proposed framework with the internal and external components manifests the Automated Negotiation Model based on the Multi-agent System (AN-MAS).

4   Decision Making Process Flow

A decision made by agents goes through several processes. These processes work by gradually reducing candidate solutions of a project until a single solution is reached. Consequently, in this work, the process of nominating a single solution from a set of solutions is called decision making.

There are three main processes in decision making for a specific project, which are propose solutions, negotiate solutions and handling conflict outcomes (conflict resolution).

- Propose solutions: In this process, each agent proposes solutions and ranks them from 1st to n^{th} solution where n is any natural number.
- Negotiate solutions: When ranked solutions are ready, agents negotiate by submitting their ranked solutions to each other. Since each agent's target is to maximize its utility by selecting a solution that has a better order, each agent prepares a plan. Using these plans, agents form coalitions among them based on similar plans. These coalitions continuously compare plans with each other until a single or more solutions converge after exhausting all attempts.
- Resolve conflict: If agent coalitions agree upon a single solution, then this process is forfeited, but if there are two or more conflicting solutions, then the conflicts need to be resolved. This process resolves conflicts based on each coalition's strength and its solutions' risks. From these two parameters, this process drops solutions until a single solution is reached.

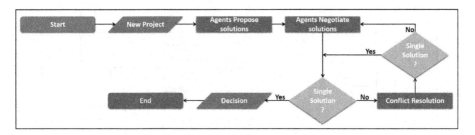

Fig. 3. Decision Making Flowchart

Figure 3 shows the decision making flowchart as described above. The process starts when agents receive a new project. The agents first propose solutions in ranked order. They then negotiate these solutions. If they agree upon a single solution, then the decision is made, otherwise, the conflict resolution process takes over to drop the weak and risky solutions. If the outcome of the conflict resolution process is a single solution then the decision is made. Otherwise, the agents negotiate the outcome of the conflict resolution process. Ultimately, one coalition's solution is accepted.

5   The Belief-Desire-Intention (BDI) Agent Architecture

This section presents an architecture that enables software agents to mimic human behaviors and styles in building an automated negotiation system in the construction domain. In this work, we develop BDI agents that are widely used by researchers to build intelligent agents.

The BDI agent consists of three main components that are affected by the environment; Belief, Desire, and Intention. Agents usually perform tasks within an environment and they exploit the environment to update their goals. The agents' beliefs are influenced by the environmental changes. The belief in turn updates their desires and the intentions.

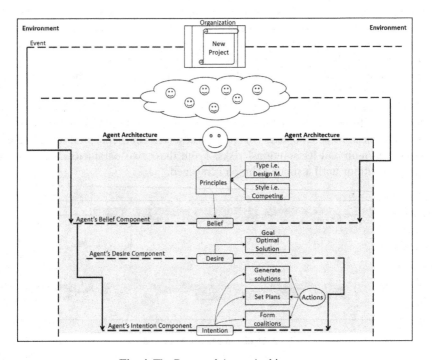

Fig. 4. The Proposed Agent Architecture

As shown in Figure 4, the proposed agent architecture consists of the outer area that represents the environment and the inner area that represents the decision making process. The environment constitutes the variables of new project information and agent's activities information, e.g. interactions, decisions, negotiations, coalitions. The belief component within the architecture is influenced by the environment and agent attributes, which include the agent type, e.g. Design Manager, and the agent style e.g. competing. The desire component represents an agent's goal.

In the construction domain, each agent attempts to ensure its first rank solution wins. If it is not possible, it works on the second rank and so on. This scenario represents its desire or goal. From the agent's belief and desire, it performs actions which represent the intention components. The intention component represents a bridge between the belief and the desire, in other words, it represents the practical steps to achieve the desire according to its belief about the environment and the attributes.

6   Conclusion and Future Work

To create a multi-agent automated negotiation model, agents need to be integrated with several components. In this paper, we identify four internal components (negotiation algorithm, negotiation style, negotiation protocol, and solution generators) integrated with the agent design and two external components (the negotiation base and the conflict resolution algorithm) within the environment. These components constitute the proposed framework.

We also discuss the decision making process flow in such system, consisting of three main processes which are propose solutions, negotiate solutions and handling conflict outcomes (conflict resolution). We finally reveal our proposed agent's architecture to conduct automated negotiation in the construction domain.

Since this work is in its theoretical stage, it only presents the conceptual underpinnings of pertinent issues in negotiation and does not present the experimental results. Such outcome will be presented in our future work.

In addition, for our future work, we shall study and propose mechanisms for the three methods needed by the decision making process which are Agent Proposes Solutions, Agent Negotiate Solutions and Conflict Resolution.

References

1. Utomo, C.: Development of a negotiation Support Model for Value Management in Construction. PhD Thesis, University Teknologi PETRONAS (December 2009)
2. Kexing, L.: A survey of agent based automated negotiation. In: 2011 International Conference on Network Computing and Information Security (NCIS), vol. 2, pp. 24–27. IEEE (2011)
3. Beer, M., d'Inverno, M., Jennings, R.N., Luck, M., Preist, C., Schroeder, M.: Negotiation in multi-agent systems. Knowledge Engineering Review 14(3), 285–289 (1999)
4. Ren, Z., Anumba, C.J.: Multi-agent systems in construction—state of the art and prospects. Automation in Construction 13, 421–434 (2004)
5. Wang, M., Wang, H., Vogel, D., Kumar, K., Chiu, D.K.W.: Agent-based negotiation and decision making for dynamic supply chain formation. Engineering Applications of Artificial Intelligence 22(7), 1046–1055 (2009)
6. Utomo, C., Idrus, A.: A Concept toward Negotiation Support for Value Management on Sustainable Construction. Journal of Sustainable Development 4(6) (2011)

7. Sanchez-Anguix, V., Julian, V., Botti, V., García-Fornes, A.: Tasks for agent-based negotiation teams: Analysis, review, and challenges. Engineering Applications of Artificial Intelligence 26(10), 2480–2494 (2013)
8. Anderson, R.M., Hobbs, B.F., Bell, M.L.: Multi-objective decision-making in negotiation and conflict resolution. In: Encyclopedia of Life Support Systems. Eolss Publishers, New York (2002)
9. Anumba, C.J., Ugwu, O.O., Ren, Z.: Agents And Multi-agent Systems in Construction. Spon Press (2005)
10. Coutinho, C., Cretant, A., Ferreira da Silva, C., Ghodous, P., Jardim-Goncalves, R.: Service-based negotiation for advanced collaboration in enterprise networks. Journal of Intelligent Manufacturing (2014), doi:10.1007/s10845-013-0857-4
11. Dzeng, R.J., Lin, Y.C.: Intelligent agents for supporting construction procurement negotiation. Expert Systems with Applications 27(1), 107–119 (2004)
12. Anumba, C.J., Ren, Z., Thorpe, A., Ugwu, O.O., Newnham, L.: Negotiation within a multi-agent system for the collaborative design of light industrial buildings. Adv. Eng. Software 34(7), 389–401 (2003)
13. Johnson, H., Johnson, P.: Task Knowledge Structures: Psychological basis and integration into system design. Acta Psychologica 78, 3–26 (1991)
14. Friend, M., Cook, L.: The new Mainstreeem. Instructor, 30–36
15. Matthews, J.: Implications for collaborative educators preparation and development: A sample instructional approach. In: Pounder, D.G. (ed.) Restructuring School of Collaboration: Promise and Pitfalls. State University of NewYork Press (1998)
16. Holley, W.H., Jennings, K.M.: The labour relations process, 8th edn., ch. 6. Thomsan/South-Westeren, Mason (2005)
17. SAVE International, value methodology standards (2001)
18. Kelly, J., Male, S.: Value Management. In: Kelly, J., Morledge, R., Wilkinson, S. (eds.) Best Value in Construction, ch. 5, pp. 77–99. Blackwell, Oxford (2002)
19. Mukhopadhyaya, A.K.: Value Engineering Concept, Techniques and Applications. Response Books, New Delhi (2003)
20. Kelly, J., Male, S.: Value Management in Decision and Construction, The Economic Management of Projects. Spon Press, London
21. Jaapar, A., Endut, I.R., Bari, N.A.A., Takim, R.: The impact of value management implementation in Malaysia. Journal of Sustainable Development 2(2) (2009)
22. Shen, Q., Chung, J.K.H., Li, H., Shen, L.: A Group Support System for improving value management studies in construction. Automation in Construction 13, 209–224 (2004)
23. Cariaga, I., El-Diraby, T., Osman, H.: Integrating Value Analysis and Quality Function Deployment for Evaluating Design Alternatives. Construction Engineering and Management 133(10), 761–770 (2007)
24. Qing, Y., Wanhua, Q.: Value Engineering Analysis and Evaluation. For the Second Beijing Capital Airport. Value World, Spring, SAVE International (2007)

A Comparative Study of the Dynamic Matrix Controller Tuning by Evolutionary Computation

Gustavo Maia de Almeida,[1] Marco Antonio de S.L. Cuadro[1],
Rogério Passos Pereira Amaral[1], and José Leandro F. Salles[2]

[1] Instituto Federal do Espírito Santo, Coordenadoria de Automação Industrial,
Campus Serra, Serra, ES, Brasil
[2] Universidade Federal do Espírito Santo, Departamento de Engenharia Elétrica,
Vitória, ES, Brasil

Abstract. The Dynamic Matrix Control (DMC) Algorithm is a control method widely applied to industrial processes. Evolutionary Computation (EP) is a vibrant area of investigation, with some of the least widely known approaches being Genetic Algorithm (GA), Ant Colony Optimization (ACO) and Particle Swarm Optimization (PSO) all of which can be used in optimisation problem. This work make a comparative study of the effectiveness of the three methods to optimize the tuning parameters of the Dynamic Matrix Controller for SISO (single-input single-output) and MIMO (multi-input multi-output) linear dynamical systems with constraints.

Keywords: Evolutionary Computation, Genetic Algorithm, Ant Colony Optimization, Particle Swarm Optimization, Dynamic Matrix Controller.

1 Introduction

Model Predictive Control (MPC) refers to a class of computer control algorithms that utilizes an explicit process model to predict the future response of the plant. A variety of processes, ranging from those with simple dynamics to those with long delay times, non-minimum phase zeros, or unstable dynamics, can all be controlled using MPC. MPC integrates optimal, stochastic, multivariable, constrained control with dead time processes to represent time domain control problems, see [1] and [2].

The MPC algorithms usually exhibit very good performance and robustness provided that the tuning parameters (prediction and control horizons and move suppression coefficient) have been properly selected. However, the selection of these parameters is challenging because they affect the close loop time response, being able to violate the constraints of the manipulated and controlled variables. In the past, systematic trial-and-error tuning procedures have been proposed, see [3] and [4]. Recently there are some works proposing alternative techniques of adjusting then automatically using genetic algorithms [5] and [6]. Another tuning strategy that can be implemented in a computer was proposed by [9],

© Springer International Publishing Switzerland 2015
S. Omatu et al. (eds.), *Distributed Computing & Artificial Intelligence, 12th Int. Conference,*
Advances in Intelligent Systems and Computing 373, DOI: 10.1007/978-3-319-19638-1_24

which developed an easy-to-use tuning guidelines for the Dynamic Matrix Control (DMC) algorithm, one of the most popular MPC algorithm in the industry. DMC uses the step response to model the process, and it is also applied in SISO and MIMO processes that can be approximated by first order plus dead time models.

In the [10] is made in a review of various techniques for tuning of the parameters of predictive controllers. The main goal of this work is make a comparative study of the effectiveness of the evolutionary computation [7] and [8] through of the GA [5], [6], ACO [11], [12], [13], [14] and PSO [11], [12], [13] to optimize the tuning parameters of the Dynamic Matrix Controller for SISO (single-input single-output) and MIMO (multi-input multi-output) linear dynamical systems with constraints.

2   Dynamic Matrix Controller (DMC)

In this section, we formulate the predictive control problem and we present its solution from the DMC algorithm. Let us consider a linear MIMO dynamic system with m inputs $(u_l, l = 1, 2, ..., m)$ and n outputs $(y_j, j = 1, 2, ..., n)$, that can be described by the expression:

$$y_j(k) = \sum_{l=1}^{m} \sum_{q=1}^{\infty} g_{jl}(q) \Delta u_1(k - q) \tag{1}$$

where g_{jl} is the step response of output j with respect to the input l, and $\Delta u_l(k = u_l(k - u_l(k - 1)))$, is the control signal variations. Let us denote by $\hat{y}_j(k + 1)$ the i step ahead prediction of the output j from the actual instant k; and $r_j(k + 1)$ is the i step ahead prediction of the set point with respect to the output j.

The basic idea of DMC is to calculate the future control signal along the control horizon h_c, i.e., determine $u_l(k + 1)$, for $l = 1, 2, ..., m$ and $i = 0...h_c - 1$, in such a way that it minimize the cost function defined by:

$$J = \sum_{j=1}^{n} \sum_{i=1}^{h_p} (\hat{y}_j(k + i) - r_j(k + 1))^2 + \sum_{l=1}^{m} \sum_{i=0}^{h_c-1} \lambda_j \Delta u_l^2(k + 1) \tag{2}$$

where λ_j is the move suppression parameters and h_p is the prediction horizon. In addition, the following constraints must be satisfied for $l = 1, ..., m$ and $j = 1, ..., n$:

$$\Delta u_l^{min} \leq \Delta u_l(k + i) \leq \Delta u_l^{max} \quad i = 0, ..., h_c - 1,$$
$$u_l^{min} \leq u_l(k + i) \leq u_l^{max} \quad i = 0, ..., h_c - 1 \tag{3}$$
$$y_j^{min} \leq \hat{y}_j(k + i) \leq y_j^{max} \quad i = 1, ..., h_p$$

Considering $\Delta u_l(k+1) = 0$, $i > 0$, $i > h_c$, $i \geq 1$, and assuming that exists an integer number N_S such that $\Delta u_l(k)$ for $k > N_S$, and $l = 1, ..., m$, the predicted output from the actual instant k is given by the following convolution:

$$\hat{y}_j(k+i) = \sum_{l=1}^{m} \sum_{q=\min\{1,i-h_c+1\}}^{N_s} g_{jl}(q)\Delta u_l(k+i-q) \tag{4}$$

Disconnecting in 4 the future control actions from the past control actions, and defining:

$$G_j(i) := [g_{j1}(i) \ g_{j2}(i) \ \cdots \ g_{jm}(i)]$$

$$f_j(k+i) := \sum_{l=1}^{m} \sum_{q=i+1}^{N_s} g_{jl}(q)\Delta u_p(k+i-q) \tag{5}$$

$$\Delta u(k+i) := [\Delta u_1(k+i) \ \Delta u_2(k+i) ... \Delta u_m(k+i)]^T$$

We can be represent the expression 4 by:

$$\hat{y}_j(k+i) = \sum_{q=\min\{1,i-h_c+1\}}^{N_s} G_j(q)\Delta u(k+i-q) + f_j(k+i) \tag{6}$$

where the first term in expression 6 is the forced response, that depends on the present and future control actions, and the second is the free response, that depends on the past control actions. In the sequel we formulate the control problem as a standard quadratic programming problem with constraint.

Using the definition above, we can represent the m predicted outputs along the prediction horizon h_p by the expression:

$$\hat{\bar{Y}} = \bar{G}\Delta\bar{U} + \bar{F} \tag{7}$$

where $\bar{G}$ has dimension $nh_p \times mh_c$.

From 7, we can represent the cost function 2 by the following expression:

$$\frac{1}{2}\Delta\bar{u}^T\bar{\mathcal{H}}\Delta\bar{u} + \bar{\mathcal{B}}^T\Delta\bar{u} + \bar{\mathcal{F}} \tag{8}$$

where: $\bar{\mathcal{H}} = 2(\bar{G}^T\bar{G} + \bar{\Lambda})$, $\bar{\mathcal{B}}^T = 2(\bar{f} - \bar{r})^T\bar{H}$ e $\bar{\mathcal{F}} = (\bar{f} - \bar{r})^T(\bar{f} - \bar{r})$.

The solution of the Predictive DMC control problem is divided in two steps: first, we calculate the prediction vector $\Delta\bar{u}$ by minimizing the quadratic function 8 with $2mh_c \times nh_p$ constraints given in 3. Next, we apply the current signal $\Delta u(k)$ in the input of the processes and we discard the future signals $\Delta u(k+i)$, $i = 1, ..., h_c - 1$ (this is called receding horizon control technic). The time response of the output $y(k)$ (overshoot, rise time and steady state error) is affected by the sample time T, move suppression parameters λ_j, control horizon (h_c) and the prediction horizon (h_p). On the other hand, the constraints in the quadratic programming problem increase proportionally with the prediction and control horizons, increasing the complexity of the calculations. Thus, it is so important to consider these facts to make the best tuning of the DMC parameters.

3   Genetic Algorithm

The Genetic Algorithm according to Rao [15] is a powerful optimization searching technique based on the principles of natural genetics and natural selection.

In the GA, normally the design variables, corresponding to the genomes in the natural genetic, are represented as binary strings and they are concatenated to form an individual, corresponding in turn to a chromosome in natural genetics. Other representations can be used. The main elements of natural genetics used for the searching procedure are, reproduction, crossover and mutation.

In order to obtain the optimal solutions, the population of the GA must be evaluated. This evaluation consists of calculating the fitness value of each individual by minimizing or maximizing an objective function, J. The objective function is defined based on the optimization problem; the fitness function, f.

The objective of the GA is to search for the global optimal solutions in the optimization problems. The fitness value is an indicator of the best and the worst individuals in the population. Therefore, the optimal solutions are obtained by selecting the best individuals in each generation of the optimization process; a generation is a complete cycle in which the population is generated, the genetic operators are applied and finally the population members are evaluated. The iterative cycles finish when the termination criterion is finally satisfied. There are many methods for selecting the best individuals according to their fitness value. For instance, the most popular methods applied in the GA are the proportional selection, the roulette wheel selection, the ranked selection and the tournament selection.

4   Ant Colony Optimization

ACO is an evolutionary meta-heuristic algorithm based on a graph representation that has been applied successfully to solve various hard combinatorial optimization problems [16]. The main idea of ACO is to model the problem as the search for a minimum cost path in a graph. Artificial ants walk through this graph, looking for good paths. Each ant has a rather simple behavior so that it will typically only find rather poor-quality paths on its own. Better paths are found as the emergent result of the global cooperation among ants in the colony.

The behavior of artificial ants is inspired from real ants. They lay pheromone trails on the graph edges and choose their path with respect to probabilities that depend on pheromone trails and these pheromone trails progressively decrease by evaporation. Ants prefer to move to nodes, which are connected by short edges with a high amount of pheromone. In addition, artificial ants have some extra features that do not find their counterpart in real ants, in particular, they live in a discrete world and their moves consist of transitions from nodes to nodes. Also, they are usually associated with data structures that contain the memory of their previous action. In most cases, pheromone not during the walk, and the amount of pheromone deposited is usually a function of the quality of the path. Finally, the probability for an artificial ant to choose an edge often depends not only on pheromone, but also on some problem-specific local heuristics.

At each generation, each ant generates a complete tour by choosing the nodes according to a probabilistic state transition rule. Every ant selects the nodes in the order in which they will appear in the permutation. For the selection of a node, an ant uses a heuristic factor as well as a pheromone factor the heuristic factor, denoted by η_{ij} and the pheromone factor, denoted by τ_{ij} are indicators of how good it seems to have node j at node i of the permutation. The heuristic value is generated by some problem dependent heuristics whereas the pheromone factor stems from former ants that have found good solutions. The next node is chosen by an ant according to the following rule that has been called Pseudo-Random-Proportional Action Choice Rule. With probability q_0, where $0 \leq q_0 \leq 1$ is a parameter of the algorithm, the ant chooses a node j from the set S of nodes that have not been selected so far which maximizes:

$$[\tau_{ij}]^\alpha [\eta_{ij}]^\beta \tag{9}$$

Where $\alpha \geq 0$ and $\beta \geq 0$ are constants that determine the relative influence of the pheromone values and the heuristic values on the decision of the ant.

5   Particle Swarm Optimization

The PSO technique is derived from research on swarm such as bird flocking and fish schooling. In the PSO algorithm, instead of using evolutionary operators such as mutation and crossover to manipulate algorithms, for a d-variable optimization Problem, a flock of particles are put into the d-dimensional Search space with randomly chosen velocities and positions knowing their best values. So far (*pbest*) and the position in the d-dimension space [17]. The velocity of each particle, adjusted accordingly to its own flying experience and the other particles flying experience[17].

The modified velocity and position of each particle can be calculate using the current velocity and distance from $Pbest_{i,d}$ to $gbest_d$ as show in the following formulas,

$$V_{i,m}^{(t+1)} = W.V_{i,m}^{(t)} + c_1 * rand() * (Pbest_{i,m} - x_{i,m}^{(t)}) + c_2 * rand() * (gbest_m - x_{i,m}^{(t)}) \tag{10}$$

$$x_{i,m}^{(t+1)} = x_{i,m}^{(t)} + v_{i,m}^{(t+1)}, \quad i = 1, 2, ..., n \ \ and \ \ m = 1, 2, ...d$$

Where: n is a number of particles in the group, d is a dimension, t is a pointer of iterations (generations), $V_{i,m}^{(t)}$ is a velocity of particle i at iteration 4, W is a inertial weight factor, c_1 and c_2 are acceleration constant, $rand()$ is a random number between 0 and 1, $x_{i,m}^{(t)}$ is a current position of particle i at iterations, $Pbest_{i,m}$ is a best previous position of the ith particle and $gbest_m$ is a best particle among all the particle in the population.

6 Application of GA, ACO and PSO to Tuning of DMC

The objective of an optimal design of DMC controller for a given plant is to select the best parameters $h_c^*, h_p^*, \alpha^*, \delta^*$ and λ^* of the DMC controller such that the performance indexes on the load disturbance response and transient response is minimum. The design of the DMC algorithm from GA, ACO and PSO can be represented as a flowchart in Fig. 1 (a), (b) and (c) respectively.

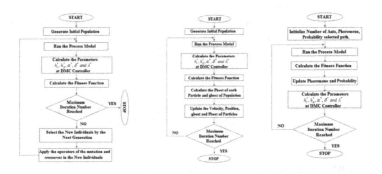

Fig. 1. The flowchart of the (a)GA,(b)ACO and (c)PSO

7 Numerical Results

In order to illustrate the differences between the DMC tuning methods presented in this work, it was simulated in Matlab the MIMO system described by the dynamic equation (11) with and without hard constraints in the control action $(-1 \leq u_i(k) \leq 1)$, in the control signal variation $(-0.1 \leq \Delta u_i(k) \leq 0.1$) and in the output response $(0 \leq y_i(k) \leq 1.00.)$

$$\begin{bmatrix} y_1(s) \\ y_2(s) \end{bmatrix} = \begin{bmatrix} \frac{12.8e^{-s}}{16.7s+1} & \frac{-18.9e^{-3s}}{21.0s+1} \\ \frac{6.6e^{-7s}}{10.9s+1} & \frac{-19.4e^{-3s}}{14.4s+1} \end{bmatrix} \begin{bmatrix} u_1(s) \\ u_2(s) \end{bmatrix} \tag{11}$$

The population size and the number of iterations performed by the GA, ACO and PSO programs are equals to 100, and their fitness function are defined by the ISE index.

When all constraints were taken into account, only ACO determined feasible solution, so that the constrained output did not exceed to 1. For this case, the tuning parameters determined by ACO were: $h_p = 7$, $h_c = 7$, $\alpha_1 = 0,76$, $\alpha_2 = 0,83$, $\delta_1 = 40$, $\delta_2 = 21$, $\lambda_1 = 69$ and $\lambda_2 = 7$. The Figure 2 shows the time responses of MIMO system (11) without constraints that was controlled by DMC with parameters tuned by GA, ACO and PSO.

The vector of tuning parameters $[h_c, h_p, \alpha_1, \alpha_2, \delta_1, \delta_2, \lambda_1, \lambda_2]$ determined by GA, ACO and PSO are respectively, $[1, 5, 0.91, 0.72, 31.19, 17.9, 33.89, 21.01]$, $[1, 3, 0.91, 0.72, 5.4, 2.2, 0.3, 9.7]$ and $[1, 3, 0.96, 0.21, 10.65, 8.41, 0.08, 16.76]$.

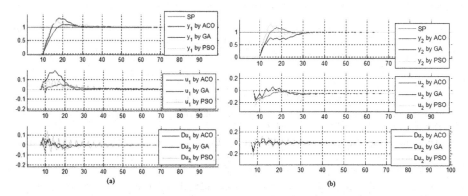

Fig. 2. Time responses:(a)y_1, u_1 and Δu_1 (b)y_2, u_2 and Δu_2

Table 1. Performance of the DMC controller tuned by evolutionary computation

Method	ISE	$Mp_1(\%)$	$Mp_2(\%)$	$t_{r1}(seg.)$	$t_{r2}(seg.)$	$t_{s1}(2\%)$	$t_{s2}(2\%)$
GA	15.62	16.38	0.01	6.2	15.99	46.4	38.5
ACO	14.54	9.78	17.83	5.87	3.8	28.9	32.1
PSO	13.44	26.19	30.85	3.33	2.7	52.5	41.2

Considering these tuning parameters, we can compare the performance of the MIMO process controlled by DMC in the Table 1, where Mp_i, t_{ri}, t_{si} are, respectively, the percentage overshoot of output i, the rise time of the output i and the settling time of the output i. We observe in Table 1 that the PSO method furnishes the ISE index lesser than ACO and GA, however its overshoot is higher than the others methods. We also observe in the time responses illustrated in the graphs of the Figure 2 that the DMC control problem tuned by GA and PSO give non-feasible solutions, since the output exceed the constraints (3).

8   Conclusion

Overall the results indicate that all three tuning methods can be used for tuning of the DMC controller, but the ACO method showed a great advantage over the others because it can control the system by respecting all hard constraints (3), in addition it presents the best settling time. In other hand, the PSO method furnishes the lowest ISE index and rise time but its performance presents the highest overshoot. The lowest overshoot in the outputs 1 and 2 were obtained from GA and ACO, respectively. The automatic tuning of DMC parameters performed by the evolutionary algorithms presented in this paper can be implemented on-line in a real plant and are more suitable for use in nonlinear processes which parameters can be updated automatically whenever there is a change in the operating point.

References

1. Camacho, E.F., Bordons, C.: Model Predictive Control, 2nd edn. Springer (2000)
2. Maciejowski, J.: Predictive Control with Constraints. Prentice Hall, Englewood Cliffs (2002)
3. Murath, P.R., Mellichamp, D.A., Seborg, D.E.: Predictive Controller Design for Single Input / Single Output System. Ind. Eng. Chem. Res. 27 (1998)
4. Rawlings, J., Muske, K.: The Stability of Constraints Recedings Horizon Control. IEEE Trans. Autom. Control 38 (1993)
5. Almeida, G.M., Salles, J.L.F., Filho, J.D.: Using Genetic Algorithm for tuning the parameters of Generalized Predictive Control. In: VII Conferéncia Internacional de Aplicaçöes Industriais INDUSCON (2006)
6. Almeida, G.M.: Controle Preditivo Sintonizado via Algoritmo Genético Aplicado em Processos Siderúrgicos. Tese de Doutorado, Universidade Federal do Espírito Santo (2011)
7. Coelho, C.A., Lamont, G., van Veldhuizen, D.: Evolutionary algorithms for solving multi-objective problems. Genetic and Evolutionary Computation. Springer (2007)
8. Coelho, C.A.: Constraint-handling tecniques used with evolutionary computation. In: Proc. 14th Annual Conference Companion on Genetic and Evolutionary Computation, GECCO (2012)
9. Dougherty, D., Cooper, D.J.: Tunning Guidelines of a Dynamic Matrix Controller for Integrating (Non-Self-Regulating) Processes. Ind. Eng. Chem. Res. 42 (2003)
10. Garriga, J.L., Soroush, M.: Model Predictive Control Tuning Methods: A Review Processes. Ind. Eng. Chem. Res. (2010)
11. Serapião, A.B.S.: Fundamentos de Otimização por Inteligência de Enxames: Uma Visão Geral. Revista de Controle & Automação 20 (2009)
12. Hsiao, Y.T., Chuang, C.L., Chien, C.C.: Ant Colony Optimization for Designing of PID Controllers. In: IEEE International Symposium on Computer Aided Control Systems Design (2004)
13. Nagaraj, B., Murugananth, N.: A Comparative Study of PID Controller Tuning Using GA, EP, PSO and ACO. In: IEEE International Conference on Communication Control and Computing Technologies, ICCCCT (2010)
14. Chiha, I., Liouane, N., Borne, P.: Tuning PID Controller Using Multiobjective Ant Colony Optimization. Applied Computacional Intelligence and Soft Computing (2012)
15. Rao, S.S.: Engineering Optimization Theory and Practice, 4th edn. John Wiley & Sons Inc. (2009)
16. Dorigo, M., Maniezzo, V., Colorni, A.: Ant System: Optimization by a Colony of Cooperating Agents. IEEE Transaction on System, Man, and Cybernetics Agents Part B 26 (1993)
17. Nasri, M., Nezamabadi-pour, H., Maghfoori, M.: A PSO-Based Optimum Design of PID Controller for a Linear Brushless DC Motor. World Academy of Science, Engineering and Technology 26 (2007)

Design of a Semantic Lexicon Affective Applied to an Analysis Model Emotions for Teaching Evaluation

Gutiérrez Guadalupe[1,2], Margain Lourdes[1], Padilla Alejandro[3],
Canúl-Reich Juana[2], and Ponce Julio[3]

[1] Universidad Politécnica de Aguascalientes, Mexico
gues.02@gmail.com, lourdes.margain@upa.edu.mx
[2] Universidad Juárez Autónoma de Tabasco, Mexico
juana.canul@ujat.mx
[3] Universidad Autónoma de Aguascalientes, Mexico
{padilla2004,julk_cpg}@hotmail.com

Abstract. There is an exponential interest by companies and researchers to identify the emotions that the users can express on their comments in different social media, for this reason the sentiment analysis has turned in one of the most researches on Natural Language Processing area with the purpose of creating resources as the lexical to do this homework. However, due to the variety of areas and contexts it is necessary to create or adequate lexical that allows to get the desired results. This work presents the basis for the development of Affective Semantic Lexicon. For the development of the lexicon is considered a grammatical lexicon and a graphic lexicon that will be evaluated then, be implement on a Sentiment Analysis Model applied to Professors' Assessments of Higher Education Institutions.

Keywords: Affective semantic lexicon, model, sentiment analysis.

1 Introduction

Internet is the global network where in recent years people have proliferated the social media, which have generated that the users feel fascinated by sharing opinions or comments, sentiment status, actions and experiences. This fascination has favored an especial interest on companies and researchers to analyze the users' opinions and can identify their emotions about services, products, purchases, sales, courses, among others. To process a lot of that data that there is in Internet, it is necessary the use of linguistic mechanisms with the finality is to define the correct sense of words. The Computational Linguistics Technology also called Natural Language Processing (NLP) combines the linguistics and the computing with the purpose to model human language under a computational scheme. The sentiment analysis, also called opinion mining, has be-come one of the most active areas of NLP; at first its application was directed mainly to trade, but its application has been extended to other areas such as education, medicine, politics and others. Bing [1] says that most of the time individuals

© Springer International Publishing Switzerland 2015
S. Omatu et al. (eds.), *Distributed Computing & Artificial Intelligence, 12th Int. Conference*,
Advances in Intelligent Systems and Computing 373, DOI: 10.1007/978-3-319-19638-1_25

seek opinions before making a decision, so the use of information resulting from a sentiment analysis may identify potential customers to a company, or the success of a product or service.

At the present for sentiment analysis processing have employed several techniques, tools or resources that have been created for specific tasks or applications. Lexical is one of the most important resources for carrying out the process of sentiment analyzing, there are some available on the Internet that can be used freely (e.g. WordNet), however due to the variety of study areas and contexts does not exist a standard lexical, whether it could be a grammatical lexicon or a graphic lexicon to implement in different researches, for this reason they have been created or adequate several lexical according to the context to consider. WordNet is a large lexical database designed specifically for English, it contains general terms on several topics; Hobbs [3] suggests that many of synsets on WordNet are rarely relevant for each language. Despite this WordNet is considered a basic tool for the development of complex applications of textual analysis and PLN. By this work it is proposed to perform a lexical analysis in two levels. The first is about a lexical-grammatical level and the second is about a graphic lexical level. The first will be understood as the dimension of text analysis in order to interpret the emotions of the issuers in the text. The second will be understood as the dimension of the analysis of the graph type mood which expresses an emotional state. This kind of lexicon, initially seeks to interpret the emotional state or complement the emotion expressed by the issuer.

In this order of ideas, this paper describes the design of a semantic affective lexicon to implement in a Sentiment Analysis Model applied to Professors' assessments. The organization of this paper begins with the art state about the development of affective lexicons for sentiment analysis and the classification of grammatical lexicon and graphic lexicon, the description of the process of lexical development, research problems, preliminary results, conclusions and future work.

2   Justification and State of the Art

In European literature some recent studies are reported under the line of in-affective lexical research using opinion mining to sentiment analysis and irony [4].Others investigators are realizing a study on the posterior-to-prior polarity issue i.e. the problem of computing words' prior polarity starting from their posterior polarities [5]. In Mexico, this line of research is relatively new, some studies are documented from the edge of Computer Science at doctoral level [like 6]. However further studies exist that support the study and understanding of Corpus-based lexicography (e.g. Dictionary of Current Spanish and Dictionary's Spanish in Mexico).

This work focuses on the design stage of an affective semantic lexicon to implement a model for identifying emotions that students express in teacher evaluation. This assessment is applied to students each semester and is considered as an indicator for the opening of courses on teaching methodologies, in order to support teachers to improve their classes they teach, to benefit student learning. One of the limitations of the proposed sentiment analysis model is that most of the resources and tools are designed to analyze texts in English, although in recent years Spain has been known for his research in the area of sentiment analysis nevertheless an adjustment is necessary to recognize the linguistic varieties of Mexico.

The model proposed in this paper is focused on the context of education, but the realization of this model may support future research in the country, to devote to other contexts such as: marketing, election politics or even measure the effectiveness of e-learning courses, which have grown exponentially for users who want to learn about various topics. Here are some related jobs using lexical and other techniques are discussed to analyze emotions in user reviews.

- The *"SentiSense: An easily scalable concept-based affective lexicon for sentiment analysis"* [8] implements an affective semantic lexicon in English under the methodology of WordNet Affect, is based on concepts and linked to WordNet to support semantic process. Is also designed to identify 16 emotional categories. However it is not focused on a context, so that their results may not be successful also not apply any technique of artificial intelligence.
- The *"Sentiment analysis in Facebook and its application to e-learning"* [14] present a method to extract information about the users' sentiment polarity (positive, neutral or negative) and to detect significant emotional changes in the messages in Facebook. This project can serve as feedback for teachers, especially in the case of online learning. It combines lexical-based and machine-learning techniques.

3   Design of Phases of the Proposed Model to Sentiment Analysis

The general architecture of the proposed model shown in Figure 1, consists of seven phases, where each of which performs a specific function except for the zero. Phases involved in the design of the lexicon would be referred hereinafter as: Semantic Affective Colloquial Lexical (LeSeAC) is from 0 to 4, those which are described below in Figure 1.

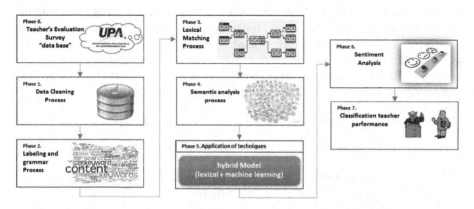

Fig. 1. Phases of the Proposed Model to Sentiment Analysis

3.1   Phase 0: Professor's Evaluation Data

Phase 0 represents the choice of the corpus or database to analyze, in this case it is called professor evaluation questionnaire, which allows students to evaluate the performance of each of their professors of the school year. For this work it is considered an educational assessment in the Polytechnic University of Aguascalientes in the period from September to December 2014. This evaluation is a survey composed by 21 questions applied to students of all academic programs. In the first 20 questions the student evaluates certain characteristics of the professor assigning a score of 1 to 10, where 1 is the lowest rating and 10 is the highest rating, on the other hand it is in question number 21 where students are allowed to give an opinion about the professor. The use of this database is permitted as long as the identities of the students and professors are protected.

3.2   Phase 1: Cleaning Comments Process

The fact of knowing that this is a database which mainly contains of real comments from students and considering that such information has not been processed, you must perform a cleaning process that involves the following tasks: correction chains, removing symbols or foreign characters, keyword recognition, among others. For example some of the errors to identify, eliminate or transform are: words like "veeeeeeeery" to "very" and possible misspellings like "fnid" to "find".

3.3   Phase 2. Grammatical Process and Labeling

During this phase the labeling process is performed, employing a POS-tagger (Part of Speech Tagger), which can be defined as a tool that seeks to convert an input text to a label text that is each input word assigned a role in the sentence. Also in this phase there is a recognizing entities process summarized in two tasks: to recognize proper names in text and disambiguation of names, the latter is when there are multiple ways of naming the same entity. Another process to do is to replace all the conjugations of verbs for infinitives and all plural for singular. It is equally important to identify the stopwords (prepositions, articles, etc.) because these words may subtract relevance to parts of speech. For example when is used stopword "but" in a sentence, the prefix idea that word is denied, which subtracts relevance and it is necessary focus on the later idea. The table 1 shows some of the stopwords and words used by students who subtracted importance at the beginning or middle of the sentence.

Table 1. Stopwords and words that remain relevant

Stopword	Subtracts importance
But	Due
Except that	Not suitable
Nonetheless	Missing
However	Never
Although	Then

3.4   Phase 3: Matching Process between Lexical

In this phase, a matching is performed between the lexical involved for analyzing emotions and the corpus of professor's evaluation. On one hand the lexicon of emotions is linked to WordNet [7] lexical and colloquialisms lexical to identify the meaning of the words that make up the comments. Use of WordNet is considered necessary because enables the relationship between similar terms.

3.5   Phase 4. Semantic Analysis Process

The LeSeAC's design and development are based on a lexicon called SentiSense [8], which was developed in the English language. The development methodology Senti-Sense is considered in this work because it has the advantage of being a lexicon based on concepts and emotional labeling and linked to WordNet for a semantic analysis.

For the analysis of emotions approach strategy coupled with labels emotions associated with a set of words is used. Such strategies have been used previously in [9, 10] extracting news headlines associating six types of emotions with an accuracy of 38%. By combining this approach with classification algorithms has come to obtain an accuracy of 61% [11]. For the process of Semantic Analysis of Emotions in this work, we have defined the type of parsing is simple analysis.

- Simple Syntactic Analysis: Refers to semantically associate a word or graphic.
- Double Syntactic Analysis: Refers to semantically associate a couple of words or graphic.
- Multiple Syntactic Analysis: Refers to semantically associate multiple words or graphic.

To identify the emotions that students often say in opinions about professors, a study called Proof of Concept was performed by Content Validity of Panel of Experts (Face Validity) [12], which was applied to professors in the educational institution with specialization in psychology. The positive emotions that experts consider more frequent in students evaluate their professors, are: like and admiration and the negative emotions are: dislike and anger. The table 2 presents some comments that the students have done about their professors, where clearly the emotions: like, admiration, dislike and anger are denoted.

Table 2. Examples of original emotional comments

Professor	Original Comment	Emotion
PROF02	*Es una maestra muy eficaz en la forma de explicar las clases.*	Like
PROF02	*La Maestra tiene un gran nivel y domina el tema con excelencia, con total seguridad volvería a tomar otro curso con ella si tuviera la oportunidad.*	Admiration
PROF01	*Le hace falta ética y tolerancia, aparte se duerme durante las clases.*	Dislike
PROF05	*El profesor es muy irresponsable y elitista. Se nota su favoritismo por ciertas personas en el salón.*	Anger

4   LeSeAC Design

On the comments that students make professors identify terms (keywords) that act as amplifiers and reducers as well as some of the negative colloquialisms commonly used by students to express their opinions; the table 3 presents some of these colloquialisms and its definition from the colloquial and dialectal terms of Spanish dictionary.[1]

Table 3. Examples of colloquial terms

Category	Colloquialism	Definition
Adjective	*Onda*	Negative or positive characteristic about specific person
Noun /adjective	*Sangrón*	Unfriendly person
Noun /adjective	*Naco*	Person with bad taste, without grace or style, without personality, ignorant.
Noun /adjective	*Gacho*	Evil, vile
Noun /adjective	*Fresa*	High class person (or appears to be), who usually disdains the lower classes.
Noun /adjective	*Farol*	Man usually make unsubstantiated statements that sustains vainly

After regularize the comments of students is applied a POS-tagger using Freeling tool [13]. Freeling is a system for linguistic analysis of text, like tagging, lemmatization, etc. Through the POS-tagger is possible to identify the grammatical category of each word as verbs, adjectives, adverbs and nouns. Similarly it is applied a lemmatization process (stemming) to identify and transform a word to its corresponding motto, such as word repentant can be reduced to the word or motto "repent". To form the LeSeAC lexicon is necessary to creating XML files then have communication with WordNet, so it is necessary to follow the EAGLES recommendations for do a morphosyntactic tagging. The following is an excerpt of the emotional categories to consider.

```
<LESEACEmoCat>
        <EmoCat name="gustar" antonym="disgustar"/>
        <EmoCat name="disgustar" antonym="gustar"/>
        <EmoCat name="admiración" antonym="ira"/>
        <EmoCat name="ira" antonym="admiracion"/>
</<LESEACEmoCat>
```

LeSeAC is formed by the professor evaluation corpus and is connected to Word-Net, it should be noted that it was necessary to create another lexicon following the standard of WordNet to form the Colloquialisms dictionary where the definitions were obtained based on the Colloquialisms and Terms Dialectal of Spanish dictionary. An excerpt from the LeSeAC corpus presented below.

---

[1] http://www.jergasdehablahispana.org/

```
<?xml version="1.0"?>
<LeSeACorpus>
   <Concept synset="ID-01785971-v" pos="verb" gloss="poner enojado" emo-
tion="ira"/>
   <Concept synset="ID-01789514-v" pos="verb" gloss="enojar continua o cróni-
camente="ira"/>
   <Concept synset="ID-01791911-a" pos="adjective" gloss="que no es elaborado o
con_mucho_detalle sencillo" emotion="disgust"/>
   <Concept synset="ID-01799235-v" pos="verb" gloss="desilusionar mostrarse
poco_responsable" emotion="disgust"/>
   <Concept synset="ID-01805889-a" pos="adjective" gloss="que no está contento
que_experimenta o manifiesta disgusto" emotion="disgust"/>
</LeSeACorpus>
```

The figure 2 represents the link between WordNet colloquialisms dictionary and LeSeAC developed by FreeLing and Java in order to analyze the professor evaluation and identify four emotions (admiration, anger, like and dislike).

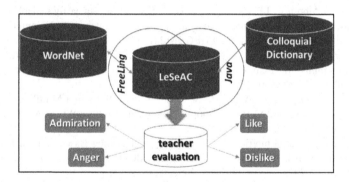

Fig. 2. LeSeAC linking with WordNet and colloquialisms dictionary

5   Conclusions and Future Work

Several researches are analyzing the preferences and needs of users for the purpose of sale; however the education area has also search to analyze emotions in order to improve distance education courses, classes, and others. LeSeAC lexicon is a proposal that seeks to implement a Sentiment Analysis Model applied to teaching evaluations in order to classify them according to their performance, however the aim is to offer to the professor constructive criticism chord comments from students to improve the impartation of classes and provide for students a quality education.

Future work is planned to apply two classification techniques (Support Vector Machine and neural networks) to automate the extension of LeSeAC lexicon through a tool developed using Java and FreeLing.

References

1. Bing, L.: Sentiment Analysis and Opinion Mining (Synthesis Lectures on Human Language Technologies). Morgan & Claypool Publishers (2012) ISBN-13: 978-1608458844
2. Fernández, D.: El nivel léxico-gramatical y su interacción con el nivel discursivo semántico en la elaboración de métodos de trabajo en el análisis del discurso. Boletín Millares Carlo, núm. 27. Centro Asociado UNED. Las Palmas de Gran Canaria (2008)
3. Hobbs, J., Gordon, A.: The Deep Lexical Semantics of Emotions. In: Affective Computing and Sentiment Analysis. Text, Speech and Language Technology, vol. 45. Springer Science+Business Media (2011)
4. Baca, Y.: Impacto de la ironía en la minería de opiniones basada en un léxico afectivo. Universidad Politénica de Valencia (2014)
5. Guerini, M., Gatti, L., Turchi, M.: Sentiment analysis: How to derive prior polarities from sentiwordnet. In: Proceedings of the Conference on Empirical Methods in Natural Language Processing (EMNLP), pp. 1259–1269 (2013)
6. Díaz, I.: Detección de afectividad en texto en español basada en el contexto lingüístico para síntesis de voz. Tesis Doctoral. Instituto Politécnico Nacional, México (2013)
7. Miller, G.A.: WordNet: A Lexical Database for English. Communications of the ACM 38(11), 39–41 (1995)
8. Carrillo de Albornoz, J., Plaza, L., Gervás, P.: SentiSense: An easily scalable conceptbased affective lexicon for Sentiment Analysis. In: The 8th International Conference on Language Resources and Evaluation, LREC 2012 (2012)
9. Strapparava, C., Mihalcea, R.: Annotating and Identifying Emotions in Text. In: Armano, G., de Gemmis, M., Semeraro, G., Vargiu, E. (eds.) Intelligent Information Access. SCI, vol. 301, pp. 21–38. Springer, Heidelberg (2010)
10. Banea, C., Mihalcea, R., Wiebe, J.: Sense-level Subjectivity in a Multilingual Setting. In: Proceedings of the IJCNLP Workshop on Sentiment Analysis Where AI Meets Psychology, Chiang Mai, Thailand (2011)
11. Sidorov, G., et al.: Empirical Study of Machine Learning Based Approach for Opinion Mining in Tweets. In: Batyrshin, I., González Mendoza, M. (eds.) MICAI 2012, Part I. LNCS, vol. 7629, pp. 1–14. Springer, Heidelberg (2013)
12. Mora, M.: Descripción del Método de Investigación Conceptual, Reporte técnico 2003-01. Universidad Autónoma de Aguascalientes (2003)
13. Padró, L., Stanilovsky, E.: FreeLing 3.0: Towards Wider Multilinguality. In: Proceedings fo the Language Resources and Evaluation Conference (LREC 2012). ELRA, Estambul (2012)
14. Ortigosa, A., Martín, J., Carro, R.: Sentiment analysis in Facebook and its application to e-learning. Computers in Human Behavior 31, 527–541 (2014)

The Effect of Training Data Selection on Face Recognition in Surveillance Application

Jamal Ahmad Dargham[1], Ali Chekima[1], Ervin Gubin Moung[1], and Sigeru Omatu[2]

[1] Fakulti Kejuruteraan, Universiti Malaysia Sabah, Jalan UMS, 88400 Kota Kinabalu, Sabah, Malaysia
[2] Faculty of Engineering, Department of Electronics, Information and Communication Engineering, Osaka Institute of Technology, 5-16-1, Omiya, Asahi-ku, Osaka, 535-8585, Japan
{jamalad,chekima}@ums.edu.my, menirva.com@gmail.com, omatu@rsh.oit.ac.jp

Abstract. Face recognition is an important biometric method because of its potential applications in many fields, such as access control and surveillance. In surveillance applications, the distance between the subject and the camera is changing. Thus, in this paper, the effect of the distance between the subject and the camera, distance class, the effect of the number of images per class, and also the effect of database used for training have been investigated. The images in the database were equally divided into three classes: CLOSE, MEDIUM, and FAR, according to the distance of the subject from the camera. It was found that using images from the FAR class for training gives better performance than using either the MEDIUM or the CLOSE class. In addition, it was also found that using one image from each class for training gives the same recognition performance as using three images from the FAR class for training. It was also found that as the number of images per class increases, the recognition performance also increases. Lastly, it was found that by using one image per class from all the available database sessions gives the best recognition performance.

Keywords: Principal Component Analysis, PCA, training data selection, face recognition, surveillance, distance from the camera.

1   Introduction

Face detection and face recognition is a growing research biometric field because of their potential application as an important tool for security surveillance and human-computer interaction. Many face detection and face recognition techniques have been proposed with satisfactory success [1]. However, face recognition for surveillance is difficult to tackle because CCTV cameras photograph people at tilted angles or in low light, thus capturing poor quality video images [2]. The low quality video coupled with the large variations in the subjects face orientations in the acquired video decreases the recognition accuracy. Various techniques have been proposed for face recognition for surveillance application [3, 4, 5], which can be classified into two

© Springer International Publishing Switzerland 2015
S. Omatu et al. (eds.), *Distributed Computing & Artificial Intelligence, 12th Int. Conference*,
Advances in Intelligent Systems and Computing 373, DOI: 10.1007/978-3-319-19638-1_26

groups: software and hardware based techniques. One of the problems with face recognition for surveillance is that data often acquired from different video sources and in different configurations. Vera-Rodriguez et al. [6] investigated the effect of distance between the camera and the subject on recognition rate. They categorized the images in the database into three classes, namely; close, medium, and far class according to the distance between the camera and the subject. They used mug-shot frontal face for their library and surveillance videos for their testing. Their work shows that Equal Error Rate increases (recognition performance dropped) when the person is further away from the camera regardless of the face recognition technique used. In Yuxi Peng [7] thesis, using mug-shot frontal face for training data and video camera images for testing, the author's work also shows that as the distance between a person and the camera increases, the recognition rate decreases regardless of the face recognition technique used. In this work, an investigation of the selection of training data on the performance of face recognition in surveillance application will carried out. In particular, the effect of images classes on face recognition, the effect of the number of image frames per class, and the effect of database used on face recognition will be investigated.

2   Surveillance Dataset

The ChokePoint Dataset [8] is a collection of surveillance videos of 25 persons in portal 1 and 29 persons in portal 2. The portal 2 dataset were used for database preparation since it has more subjects available compared to portal 1. Portal 2 dataset consist of two sub datasets; Portal 2 Entering scene dataset (P2E) and Portal 2 Leaving scene dataset (P2L). P2L dataset contain better quality frontal faces. Thus, P2L dataset were selected for database preparation for all experiments. The P2L dataset have 29 persons. Although the video images were captured in four sessions, only session 1 (S1), session 2 (S2), and session 3 (S3) were used. Session 4 video images were not used because the captured image of the face is not frontal as seen in Fig. 1. All the frontal face images are aligned so that their eyes are in the same horizontal line.

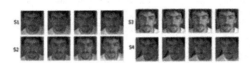

Fig. 1. Samples of the frontal face images. First row is taken from session 1, second row is from session 2, third row is from session 3 and fourth row is from session 4.

3   Database Preparation

The ChokePoint database is a series of video frames of a person walking towards the camera. Each person has different number of video images captured since some people walk faster than others thus having lesser video images captured by the camera. The smallest number of video frames per person is 44 frames. The video frames were then divided into three classes, namely; FAR, MEDIUM, and CLOSE class. Since the lowest number of frames available is 44, 42 frames are selected so that each class has the same number of frames, which is 14 frames per class. These 42 frames are divided equally into the three classes mentioned. From the 1st frame to the last 42th frame, each video image is showing the person getting closer to the camera location. The definitions of the classes are:

- FAR class - The first 14 frames from the 42 frames per person. The distance between the person and the camera is longer than MEDIUM and CLOSE class.
- MEDIUM class - The next 14 frames after the FAR class frames from the 42 frames per person. The distance between the person and the camera is shorter than FAR class and longer than CLOSE class.
- CLOSE class - The last 14 frames from the 42 frames per person. The distance between the person and the camera is shorter than FAR and MEDIUM class.

A total of 3528 video images with frontal face of persons were selected from the ChokePoint database. Three databases representing three different session containing 1176 video images each were created. They represent 28 different persons. 14 persons are used for training and testing, and the other 14 different persons are used for testing only. All the video images cover from chin to crown of the person and both eyes of the person are aligned in the same horizontal line. The image size is 96 by 96 pixels and all the images are in grey scale format.

3.1   Training Database

Four cases of training image selection were prepared. For case one, three images per person from the same class of the same database will be used. For case two, three images per person, one image from each class of the same database session will be used. For case three, nine images per person from the same database session with three images per class will be used. Lastly, for case four, nine images per person with one image per class from each database will be used. Table 1 shows the criteria for the selection of the training datasets for the four cases.

Table 1. Training database criteria

Case	Training database size	Number of persons in training database	Number of images per person	Number of classes per person	Number of images per class
1	42 images	14 persons	3 images	1 class	3 images
2	42 images	14 persons	3 images	3 classes	1 image
3	126 images	14 persons	9 images	3 classes	3 images
4	126 images	14 persons	9 images	3 classes	3 images

Case 1 and 2 are used to evaluate the effect of class on face recognition. While case 1 and 3 are used to study the effect of the number of images per class on face recognition. Lastly, the effect of database used will be evaluated using case 3 and 4.

3.2   Testing Database

Two testing databases were created. The first database, Client test database, has 588 images of the 14 persons for each session. This database will be used to test the recall capability of the face recognition system. The second database, Imposter test database, also has 588 images of 14 different persons for each session. This database will be used to test the rejection capability of the system.

4   The Face Recognition System

The face recognition system that is used in this work is the Principal Component Analysis (PCA). Fig. 2 shows the block diagram of the system. By using PCA, the set of training and testing images are projected to a lower dimensional feature space. The projection will produce two new sets of feature vectors; training and testing feature vectors with much smaller dimensions compared to the original image dimensions. These feature vectors, representing the images, are then used in the matching task between test images and training images.

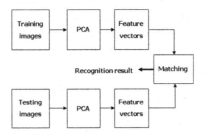

Fig. 2. Block diagram of PCA based face recognition system.

5   The Similarity Matching

For the matching task, the Euclidean distance is used. Each person has 42 frames, $T = \{F_1, F_2, ... F_{42}\}$ where T represents the test person and F represents each frame. Calculating the Euclidean distance for all the 42 frames belonging to the person T in the training database will produce a collection of Euclidean distances $E = \{e_1, e_2, ... e_{42}\}$ where E is a collection of the 42 Euclidean distances of the 42 frames, and e represents Euclidean distance for each frame. This will also produce a collection of likelihood of each frame belonging to which person from the training database, $L = \{l_1, l_2, ... l_{42}\}$ where L is a collection of likelihood for each frame, and l represents the person from training database which has the smallest distance e with test frame F. Let $P = mode\{L\}$, a person from the training database, be the highest occurrence from collection L, then the Euclidean distance average, represent by E_{av}, of all e from E that have the same likelihood, P, were calculated. If the average Euclidean distance between test frames y and frames x in the training database, $E_{av}(x,y)$ is smaller than a given threshold t, then frames y and x are assumed to be of the same person. The threshold t is the largest Euclidean distance between any two images in the training database, divided by a threshold tuning value (*Tcpara*) as given in equation (1).

$$t = \frac{max\{\|\Omega_j - \Omega_k\|\}}{Tcpara} \tag{1}$$

where $j, k = 1, 2, ..., M$. M is the total number of training images, and Ω is the feature vector of the images. To measure the performance of both systems, several performance metrics are used. These are:

1. For Recall

- **Correct Classification.** If a test frame y_i is correctly matched to an frame x_i of the same person in the training database.
- **False Acceptance.** If test frame y_i is incorrectly matched with frame x_j, where i and j are not the same person
- **False Rejection.** If test frame y_i is of a person i in the training database but rejected by the system.

2. For Reject

- **Correct Classification.** If test frame y_i, from the Imposter test database is rejected by the face recognition system.
- **False Acceptance.** If test frame y_i is accepted by the program.

For this work, the threshold tuning parameter, *Tcpara*, was set so that the system has equal correct classification rates for both recall and reject. This classification rate is defined as **Equal Correct Rate**.

6   Results and Discussion

Fig. 3, Fig. 4, and Fig. 5 shows the face recognition result using case 1 training image selection for FAR class, MEDIUM class, and CLOSE class respectively. As can be seen from Fig. 3, Fig. 4, and Fig. 5, using S2 database for training always gives the best average equal correct rate regardless of the class used. By comparing the best average equal correct rate from all three classes as shown in Fig. 6, it was found that the FAR class gives the best performance (64.29%) followed equally by the MEDIUM and CLOSE class (57.14%). Thus, it is better to use FAR class images for training. To measure the overall performance of a training image class, an *Average Equal Correct Rate*, defined as the average of equal correct rate from the three databases S1, S2, and S3, for the given class is used for comparison.

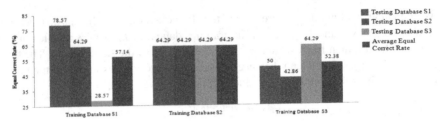

Fig. 3. Face recognition result using case 1 training image selection for FAR class

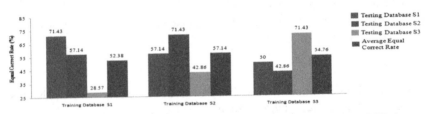

Fig. 4. Face recognition result using case 1 training image selection for MEDIUM class

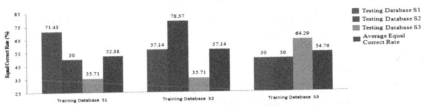

Fig. 5. Face recognition result using case 1 training image selection for CLOSE class

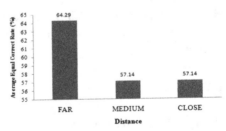

Fig. 6. Comparison of the best average equal correct rate from each class

The highest equal correct rate obtained using case 2 as shown in Fig. 7 is 64.29% which is the same obtained in case 1 as shown in Fig. 6. Thus, selecting all training images from the FAR class from session 2 gives the same performance as selecting one image from each class from the same database. The training database S2 gives the highest average equal correct rates of 64.29% for case 2 and 71.43% for case 3 as shown in Fig. 7 and Fig. 8 respectively. Comparing case 2 and case 3 recognition results shows an improvement in the performance from 64.29% to 71.43%. This shows that recognition performance improves as the number of images per class increases. Comparing the performances of case 3 and case 4, (Fig. 9(a)), shows that using one image per class from each database gives better recognition performance compared to using three images per class but from the same database session. As can be seen from Fig. 9(b), the best performance is achieved in case 4. Thus, the best choice for training data is to select them from all available classes and from all available databases.

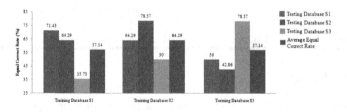

Fig. 7. Face recognition result using case 2 training image selection method

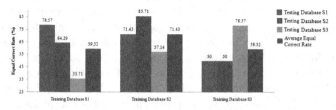

Fig. 8. Face recognition result using case 3 training image selection method

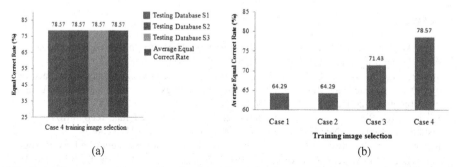

(a)   (b)

Fig. 9. (a) Face recognition result using case 4 training image selection method. (b) Comparison of the best average equal correct rate from using case 2 and case 3 training database criteria for training.

The performance of our system is also compared with several PCA based methods as shown in Table 2. Although direct comparison between our system and their system is not meaningful as different training and testing conditions are used. Nevertheless, our system performance of 78.57% is comparable to that of [6] with an average rate of 80.4%

Table 2. Comparison of several PCA based face recognition in surveillance application

Methods	Recognition rate
Vera-Rodriguez et al. [6]	CLOSE = 91.47% correct recall rate
	MEDIUM = 85.53% correct recall rate
	FAR = 64.31% correct recall rate
Yuxi Peng [7]	1.0 meter distance = 8.5% correct recall rate
Our proposed method (Case	78.57% correct recall and reject rate

7   Conclusion

In this work, the effect of the distance class, the number of images per class, and the effect of database used for training have been investigated. It was found that using three images from FAR class gives better performance than MEDIUM and CLOSE classes regardless of the database used. However, the same recognition performance can be achieved by using one image from each class. It was also found that as the number of images per class increases, the recognition performance also increases. Lastly, it was found that by using one image per class from all the databases gives the best recognition performance. Thus, the best choice for training data is to select them from all available classes from all available data-bases. Yoo et al. [9] reported that the YC_BC_R and $YC_G'C_R'$ colour spaces improve the performance of face recognition. Thus, in our further work, we intend to study whether using YC_BC_R colour space will improve the recognition rate for surveillance application.

References

1. Zhao, W., Chellappa, R., Phillips, P.J., Rosenfeld, A.: Face recognition: A literature survey. ACM Computing Surveys (CSUR) 35(4), 399–458 (2003)
2. Wiliem, A., Madasu, V.K., Boles, W.W., Yarlagadda, P.K.D.V.: A feature based face recognition technique using Zernike moments. In: Proceedings of the RNSA Security Technology Conference 2007, pp. 341–355 (2007)
3. Jillela, R.R., Ross, A.: Adaptive frame selection for improved face recognition in low-resolution videos. In: Neural Networks, IJCNN 2009, pp. 1439–1445 (2009)
4. Kamgar-Parsi, B., Lawson, W., Kamgar-Parsi, B.: Toward Development of a Face Recognition System for Watchlist Surveillance. IEEE Transactions on Pattern Analysis and Machine Intelligence 33, 1925–1937 (2011)
5. Davis, M., Popov, S., Surlea, C.: Real-Time Face Recognition from Surveillance Video. In: Zhang, J., Shao, L., Zhang, L., Jones, G.A. (eds.) Intelligent Video Event Analysis and Understanding. SCI, vol. 332, pp. 155–194. Springer, Heidelberg (2011)
6. Vera-Rodriguez, R., Fierrez, J., Tome, P., Ortega-Garcia, J.: Face Recognition at a Distance: Scenario Analysis and Applications. In: de Leon F. de Carvalho, A.P., Rodríguez-González, S., De Paz Santana, J.F., Rodríguez, J.M.C. (eds.) Distributed Computing and Artificial Intelligence. AISC, vol. 79, pp. 341–348. Springer, Heidelberg (2010)
7. Peng, Y.: Face Recognition at a Distance: a study of super resolution. Master thesis, University of Twente (2011)
8. National ICT Australia Limited (2014),
http://arma.sourceforge.net/chokepoint/
9. Yoo, S., Park, R., Sim, D.: Investigation of Color Spaces for Face Recognition. In: Proceedings of Machine Vision Application, pp. 106–109 (2007)

Combining Scheduling Heuristics to Improve e-mail Filtering Throughput

D. Ruano-Ordás, J. Fdez-Glez, F. Fdez-Riverola, and J.R. Méndez

Dept. Informática, University of Vigo, Escuela Superior de Ingeniería Informática, Edificio Politécnico, Campus Universitario As Lagoas s/n, 32004, Ourense, Spain
drordas@uvigo.es, jfgonzalez3@gmail.com,
{riverola,moncho.mendez}@uvigo.es

Abstract. In order to alleviate the massive increment of spam deliveries all over the world, spam-filtering service providers demand the development of new filtering schemes able to efficiently classify illegitimate content using less computational resources. As a consequence, several improvements have been introduced in rule-based filter platforms during the last years. In this context, different research works have shown the relevance of scheduling the evaluation of rules to improve filtering throughput, contributing some interesting heuristics to address this task. In this work, we introduce a novel scheduling approach that takes advantage of the combination of individual heuristics for globally improving filtering throughput.

Keywords: spam detection, rule optimization schedulers, prescheduling rules, filtering throughput, Wirebrush4SPAM framework.

1   Introduction and Motivation

Several years have passed since Bill Gates declared "Two years from now, spam will be solved". However, current statistics shows that more than 60% of the e-mail deliveries are spam [1]. This increment is supported by newest communication advances such as New Generation networks or 4G networks, which ensure a quick and easy Internet connection almost everywhere. In such a situation, the use of spam filtering services and products are the most effective way to fight against spam.

In this context, most filtering services are based on using rule-based classification engines such as the well-known Apache SpamAssassin [2]. These software products are able to successfully take advantage of combining individual classification approaches (e.g., regular expressions, Naïve Bayes [3], Sender Policy Framework [4], DomainKeys Identified Mail [5], etc.) into a single, more accurate, filter. In this way, any generic rule-based spam filter is always composed of a decision threshold (called *required_score*) and a set of scored rules. In addition, each individual rule contains a logical test (rule trigger) and a numeric score (positive or negative). Every time a rule is triggered during the classification of a new incoming e-mail, its associated score is added to the global e-mail score counter. After the execution of all the rules included

© Springer International Publishing Switzerland 2015

S. Omatu et al. (eds.), *Distributed Computing & Artificial Intelligence, 12th Int. Conference*,

Advances in Intelligent Systems and Computing 373, DOI: 10.1007/978-3-319-19638-1_27

in the filter, a message is finally classified as spam if its global counter is greater or equal than the configured threshold (*required_score*).

The main drawback of this approach is the time spent to filter an e-mail due to the extra workload needed to evaluate its confidence when using different classification techniques. However, recent works have shown how these drawbacks can be overcome by (*i*) using efficient parsing techniques (*ii*) taking advantage of the multithreading capabilities of current processors to execute rules concurrently and (*iii*) using some execution strategies such as smart filter evaluation (SFE), learning after report (LAR) and rule scheduling heuristics [6].

As filtering throughput achieved by using SFE and multi-threading capabilities is highly dependent on the rule execution order, we recently introduced five rule scheduling techniques (PFS, NFS, GDV, GAV and PSS) that were successfully used to achieve significant throughput improvements [7]. In the present contribution, we show how different heuristics can be successfully combined to increase global performance even more.

While the first section has motivated the work, the rest of the paper is structured as follows: Section 2 introduces the latest advances on rule-based filter scheduling. Section 3 presents the main contribution of the work: a multi heuristic ensemble. In section 4 we include a comparison showing the results of our experiments. Finally, section 5 presents the main conclusions and outlines future work.

2   Latest Advances on Rule-Based Spam Filters

After the successful introduction of the rule-based SpamAssassin filter as a revolutionary platform to fight against spam, some works have introduced different strategies and approaches to improve both filter performance and throughput [7, 8]. From them, we highlight (*i*) learn after report (LAR), (*ii*) smart filter evaluation (SFE), (*iii*) sufficient condition rules (SCR), (*iv*) use of efficient programming languages as well as flexible internal design and (*v*) optimization of rule scheduling.

In order to provide incremental learning capabilities using machine learning approaches, SpamAssassin needs to execute auto-learning tasks before delivering its final classification. Intuitively, this issue can generate significant increments in the time required to classify each new incoming e-mail. LAR execution scheme [6] (included in Wirebrush4SPAM) addresses this situation creating a new thread for each individual learner. As a consequence, knowledge updates are carried out in background whilst the classification report is sent to the mail transfer agent (MTA).

From another perspective, some filtering products are able to prematurely abort the filter evaluation without affecting the classification result. In this way, the SFE approach [6] defines a condition to safely stop the filter execution by combining *pending_add* and *pending_substract* variables, which should be continuously updated with the sum of positive and negative scores of all the unexecuted rules, respectively. SFE guarantees that the execution of a filter can be safely stopped when the current global message score is not included in the interval defined by Equation (1).

$$[required\_score - pending\_add_t, required\_score + pending\_substract_t) \quad (1)$$

Additionally, sufficient condition rules (SCR) [6] contain non-numerical score values ('+' or '-'). SCRs are executed before other rules because when a SCR matches the target e-mail, the evaluation is automatically aborted and the message is directly classified according to the specific score of the rule (spam if score is '+' or ham when score is '-').

Complementarily, there are other approaches that take into account design principles and programming language artifices with the goal of improving filtering throughput. In fact, the usage of ANSI/C language or the utilization of finite state machine parsers can also improve filtering speed.

Finally, different scheduling schemes can be used to achieve even more filtering time savings [7]. In this line, we have previously implemented five scheduling heuristics that can be successfully used as criteria to prioritise rules for execution (i.e., PFS, NFS, GDV, GAV and PSS). PFS, NFS, GDV and GAV have been designed to anticipate a situation in which the filter execution can be safely aborted (SFE), while PSS heuristic sorts rules with the goal of reducing the amount of locks arising from the concurrent execution of the rules. Table 1 shows a brief description of these heuristics.

Table 1. Description of the PFS, NFS GDV, GAV and PSS heuristics for rule scheduling

Name	Description	Heuristic type	
		Improve SFE	*Lock Prevention*
Positive First Scheduling	Descending ordering according to the score value of each rule	✓	
Negative First Scheduling	Ascending ordering according to the score value of each rule.	✓	
Greater ABS Value	Descending arrangement using the absolute value of the rule score as criterion.	✓	
Greater Distance Value	Descending ordering using the absolute value of the distance between the score of each rule and the *required_score* threshold.	✓	
Plugin Separation Scheduling	Sorting heuristic based on maximizing the separation between rules that belong to the same filtering technique.		✓

3 Multi-Heuristic Ensemble: A More Powerful Approach

In our previous experiments [7], we detected several situations where using only a single heuristic (from those presented in Table 1) we were not able to successfully discriminate the best rule for being executed first between two ones. These situations happens when the selected heuristic (h) assigns the same priority value for both rules (i.e., $h(r_1) = h(r_2)$). To cope with this situation, we present in this work a novel multi-heuristic ensemble (MHE), which aims to hybridise two individual heuristics to make the most of them. Therefore, the main difference with previous available approaches is that MHE is a heuristic combinatory scheme instead of a simple heuristic.

MHE combines a main heuristic (h) with an auxiliary one (h') for sorting rules in those situations in which h is not able to break the tie between two rules. Using the MHE criteria, rule r_1 will be executed before any other rule r_2 when Equation (2) gets true. Otherwise r_2 will be executed before r_1.

$$h(r_1) > h(r_2) \lor \left(h(r_1) = h(r_2) \land h'(r_1) > h'(r_2) \right) \qquad (2)$$

where h and h' stands for the main and auxiliary heuristics, respectively.

Due to the definition of the MHE criteria, the user must select both the main heuristic (GAV in most cases) as well as another one as auxiliary measure.

4   Experimental Results

In order to determine the utility of our approach, we have configured the Wirebrush4SPAM middleware for using the proposed ensemble method in conjunction with the five available scheduling strategies (showed in Table 1). For our experiments, we have selected the same corpus as [7]. The dataset is publicly available in [8] and composed by 200 e-mails written in multiple languages (i.e., Spanish, English and Portuguese) and equally distributed into ham and spam categories. E-mails included in the dataset are represented using the transmission format (RFC-2822 [9]). RFC-2822 specifies the syntax for messages sent between computer users. As long as our main goal is focused on evaluating the throughput (performance is not relevant), we applied a simple evaluation process (train and test) using 10% of the available messages for training purposes. Table 2 shows the final distribution of the selected corpus.

Table 2. Detailed description of the corpus used for the evaluation of the MHE approach

	Corpus ratio	HAM	SPAM	Σ
Train	1/10	10	10	20
Test	9/10	90	90	180
	Σ	100	100	200

For experimentation purposes we designed a spam filter (publicly available in [10]) with a global *required_score* value of 5 containing 185 rules structured as follows: (*i*) 5 Naïve Bayes, (*ii*) 169 regular expressions and (*iii*) 7 network tests (i.e., 2 belonging to RWL/RBL and 5 associated to SPF technique). All the experiments were executed using the latest version of the Wirebrush4SPAM platform running on an Intel 3.60 Mhz. Core i7 4-Duo CPU with 16 GB of RAM and Ubuntu 14.10 64bits GNU/Linux OS.

Time required for the execution of any rule performing network tests usually suffers variations across different executions due to its dependency on the network status (e.g., latency, congestion, overload, etc.). With the aim of minimizing this circumstance and achieve a more realistic result, we processed 10 times each message belonging to the the experimental dataset, simulating the execution of the experiments in a corpus of 1800 e-mails.

Although in real deployments of spam filters the amount of positive scores is usually greater than negative ones, in our experiments we tested two complementary alternatives having both positive and negative trends. In this context, a filter presents a positive trend when the global amount of positive scores is greater than the sum of negative ones. Otherwise, the filter trend is considered negative. The positive filter used in our experiments assigns scores in the range of [-89.36, 162.98], whilst the scores associated with the negative trend filter are defined in the interval [-162.98, 89.36] (i.e., our filters are completely symmetric).

To accurately compare the real throughput achieved by each heuristic, we use the following measures: (*i*) $\overline{X}_{SFE}$ represents the ratio of SFE executions in the whole experiment, (*ii*) $\overline{X}_{Avoided}$ stands for the mean of e-mail rules avoided for each message and (*iii*) $\overline{X}_{Time}$ symbolizes the average time used to classify each e-mail (in milliseconds). Table 3 shows the obtained results for the positive trend filter.

Table 3. Heuristic evaluation using a positive trend filter

	HAM			SPAM			SPAM + HAM		
	$\overline{X}_{SFE}$	$\overline{X}_{Avoided}$	$\overline{X}_{Time}$	$\overline{X}_{SFE}$	$\overline{X}_{Avoided}$	$\overline{X}_{Time}$	$\overline{X}_{SFE}$	$\overline{X}_{Avoided}$	$\overline{X}_{Time}$
NONE	0,6	7,011	9,827	0,6	6,448	12,614	0,6	13,46	11,221
NFS	0,504	6,726	9,237	0,595	17,966	9,961	0,55	24,693	9,599
PFS	0,889	9,191	9,166	0,695	17,32	9,415	0,792	26,511	9,291
GAV	0,933	12,531	9,121	0,893	16,506	10,633	0,913	25,037	9,877
GDV	0,508	13,635	8,994	0,591	18,933	8,673	0,55	32,568	8,834
PSS	0,512	6,626	9,270	0,737	9,7577	10,114	0,734	16,384	9,692
GDV + GAV	0,504	16,435	8,447	0,588	18,44	9,083	0,547	34,875	8,865
GDV+ PFS	0,511	6,668	9,717	0,597	18,242	10,437	0,553	24,911	10,077
GDV + NFS	0,504	6,695	9,025	0,591	18,211	11,201	0,551	24,906	10,113
GDV + PSS	0,504	6,733	8,556	0,591	18,206	9,598	0,547	24,94	9,027
GAV + PFS	0,975	13,313	9,093	0,906	11,511	10,672	0,941	24,824	9,883
GAV + NFS	0,931	5,615	9,091	0,862	12,616	10,034	0,896	18,228	9,562
GAV + GDV	0,926	12,168	9,078	0,86	12,582	11,122	0,893	24,751	10,101
GAV + PSS	0,933	12,446	9,403	0,86	12,406	11,987	0,896	24,853	10,695

As we can observe from Table 3, the execution of the filtering rules without a scheduling heuristic (NONE) achieves the worst results in terms of the number of rules avoided per message and filtering time. This fact evidences the relevance of the rule execution order inside a filtering platform.

Moreover, if we focus our attention only on the simple heuristics, we can highlight the good performance results achieved by GDV. In fact, the great amount of unexecuted rules per e-mail ($\overline{X}_{Avoided}$) obtained by this approach enables it to economize valuable computational resources and consequently, it saves time to filter each message ($\overline{X}_{Time}$ measure). From a scientific point of view, trying to avoid the execution of rules is the best effective way to save computational resource and therefore improve the filtering performance. Complementarily, another relevant aspect showed in Table 3 is related with the existing link between $\overline{X}_{SFE}$ and $\overline{X}_{Avoided}$. At first glance it may initially seem that both concepts are directly related (a high number of SFE executions implies a large amount of avoided rules). However, as showed in Table 3, GDV heuristic achieves an extremely low SFE value while the number of avoided rules is the best. This situation is caused by the inner nature of the GDV heuristic, which allows the sorting of rules in order to make the most of the SFE technique (see [7] for a deeper discussion about SFE).

Regarding the outcomes achieved by the combination of two different heuristics in a positive trend scenario, from Table 3 we can realize that the MHE approach together with GDV and GAV achieve the best throughput. This fact demonstrates that, a strategy designed to adequately balance the usage of I/O and computing operations can successfully complement the usage of SFE execution scheme. Moreover, we should also note that MHE heuristic combined with GDV and PSS also achieves a good filtering time, mainly due to the ability of PSS to sort rules keeping the maximum separation between them when they belong to the same filtering technique (i.e., it reduces the locks associated with the use of shared resources).

Additionally, with the goal of obtaining a complementary perspective of the proposed heuristic, we present in Table 4 the performance results achieved in a negative trend filter scenario.

As we can see from Table 4 (following the same behaviour as Table 3), the absence of a scheduling heuristic has a negative impact on the global filtering time. Analysing the results achieved by individual heuristics, we can observe that, although GAV is able to achieve the SFE finalization condition a large number of times, the best performance is achieved by the GDV heuristic when considering the number of avoided rules and filtering time.

Taking into account results achieved individually by GDV and GAV heuristics, we believe that a MHE combination of these heuristics would probably achieve a successful throughput evaluation. In fact, analyzing results achieved by all the hybridised heuristics, we can observe that using GDV as the main heuristic for MHE allows to obtain the best filtering throughput while minimizing the resource consumption (preventing the execution of a large number of rules per message).

Table 4. Heuristics evaluation using a negative trend filter

	HAM			SPAM			SPAM + HAM		
	$\overline{X}_{SFE}$	$\overline{X}_{Avoided}$	$\overline{X}_{Time}$	$\overline{X}_{SFE}$	$\overline{X}_{Avoided}$	$\overline{X}_{Time}$	$\overline{X}_{SFE}$	$\overline{X}_{Avoided}$	$\overline{X}_{Time}$
NONE	0,556	7,037	13,256	0,822	2,806	11,684	0,689	4,922	12,471
NFS	0,6	8,997	10,821	0,822	7,413	9,902	0,711	8,205	10,361
PFS	0,531	12,428	10,785	0,6	4,766	10,677	0,565	8,597	10,731
GAV	0,951	11,526	10,790	0,929	11,942	9,997	0,941	11,734	10,394
GDV	0,531	12,424	10,739	0,611	12,86	9,765	0,571	12,646	10,252
PSS	0,686	10,191	11,265	0,733	7,146	10,082	0,71	8,669	10,674
GDV + GAV	0,531	18,382	10,360	0,608	13,875	9,131	0,57	16,128	9,746
GDV + PFS	0,533	10,364	10,409	0,602	7,171	9,388	0,567	8,767	9,899
GDV + NFS	0,531	12,66	11,448	0,615	7,237	9,375	0,573	9,948	10,412
GDV + PSS	0,531	12,295	11,111	0,606	7,148	9,329	0,568	9,722	10,220
GAV + PFS	0,929	11,506	10,457	0,928	11,188	9,167	0,928	11,347	9,812
GAV + NFS	0,893	11,186	10,916	0,948	11,351	9,166	0,921	11,268	10,041
GAV + GDV	0,929	16,113	10,693	0,928	11,92	9,141	0,928	14,016	9,917
GAV + PSS	0,904	11,546	11,828	0,931	11,822	9,264	0,917	11,684	10,546

5   Conclusions and Future Work

In rule-based spam filters, schedulers allow to optimize rule execution plans by using different heuristics as sorting criteria with the goal of (*i*) improving the filtering throughput (minimize the time needed to classify each new incoming e-mail) and (*ii*) reducing the consumption of computational resources. In this work, we propose a novel multi-heuristic ensemble (MHE) method able to adequately combine the scheduling knowledge generated by using simple heuristics previously tested in [7]. For comparison purposes, we carried out a comparative empirical analysis specifically intended to measure the throughput differences between the usages of simple heuristics together with several MHE combinations.

Experimental results showed that the best heuristics are those focused on minimizing the number of executed rules. Thus, GDV and GAV heuristics generated the best rule execution plans when compared to other simple techniques based on avoiding locks caused by the usage of shared resources (PSS). Moreover, when combining heuristics using the proposed MHE algorithm, GDV as primary heuristic and GAV provided the best results. From a scientific point of view, these results demonstrate that the combination of two heuristics (GDV and GAV) allows the filtering platform to minimize the use of computational resource consumption while reducing the use of intensive I/O operations.

Regarding future work in filter throughput optimization, we believe that sorting rules based on the relation between its CPU consumption and I/O delays could provide better results. Moreover, we should also design novel algorithms and heuristics useful to sort META rules (see [2, 6, 7]). To support all this improvements, we will also need to address the development of new mechanisms able to obtain information about the execution behaviour of each rule (e.g., start time, end time, etc.). All this information is essential to find relevant constraints and, hence, design novel heuristics able to improve filtering throughput even more.

Acknowledgements. This work was partially funded by the following projects: [14VI05] Contract-Programme, from the University of Vigo, and [TIN2013-47153-C3-3-R] Platform of integration of intelligent techniques for analysis of biomedical information, from the Spanish Ministry of Economy and Competitiveness. D. Ruano-Ordás was supported by a pre-doctoral fellowship from the University of Vigo.

References

1. Spam and Phishing Statistics Report Q1-2014,
   http://usa.kaspersky.com/internet-security-center/threats/spam-statistics-report-q1-2014
   (accessed December 18, 2014)
2. Apache Software Foundation.: Spamassassin Spam Filter,
   http://spamassassin.apache.org (accessed December 18, 2014)
3. Metsis, V., Androutsopoulos, I., Paliouras, G.: Spam Filtering with Naïve Bayes - Which Naïve Bayes? In: 3rd Conference on E-mail and Anti Spam, California (2006)
4. Wong, M., Schlitt, W.: RFC4408 - Sender Policy Framework for Authorizing Use of Domains in E-Mail, Version 1, http://www.ietf.org/rfc/rfc4408.txt (accessed December 18, 2014)
5. DomainKeys Identified Mail (DKIM), http://www.dkim.org (accessed December 18, 2014)
6. Pérez-Díaz, N., Ruano-Ordás, D., Fdez-Riverola, F., Méndez, J.R.: Wirebrush4SPAM: a novel framework for improving efficiency on spam filtering services. Software Pract. Exper. 43(11), 1299–1318 (2013)
7. Ruano-Ordás, D., Fdez-Glez, J., Fdez-Riverola, F., Méndez, J.R.: Effective scheduling strategies for boosting performance on rule-based spam filtering frameworks. J. Syst. Software 86(12), 3151–3161 (2013)
8. Ruano-Ordas, D., Mendez, J.R.: Corpus of 200 E-mails,
   http://dx.doi.org/10.6084/m9.figshare.1326662 (accessed March 5, 2015)
9. Resnick, P.: RFC2822 - Internet Message Format,
   https://www.ietf.org/rfc/rfc2822.txt (accessed March 5, 2015)
10. Ruano-Ordas, D., Mendez, J.R.: Wirebrush4SPAM Trend Filters,
    http://dx.doi.org/10.6084/m9.figshare.1327581
    (accessed March 5, 2015)

Analyzing the Impact of Unbalanced Data on Web Spam Classification

J. Fdez-Glez, D. Ruano-Ordás, F. Fdez-Riverola, J.R. Méndez,
R. Pavón, and R. Laza

Dept. Informática, University of Vigo, Escuela Superior de Ingeniería Informática,
Edificio Politécnico, Campus Universitario As Lagoas s/n, 32004, Ourense, Spain
jfgonzalez3@gmail.com,
{drordas,riverola,moncho.mendez,pavon,rlaza}@uvigo.es

Abstract. Web spam is a serious problem which nowadays continues to threaten search engines because the quality of their results can be severely degraded by the presence of illegitimate pages. With the aim of fighting against web spam, several works have been carried out trying to reduce the impact of spam content. Regardless of the type of developed approaches, all the proposals have been faced with the difficulty of dealing with a corpus in which the difference between the amount of legitimate pages and the number of web sites with spam content is extremely high. Unbalanced data is a well-known common problem present in many practical applications of machine learning, having significant effects on the performance of standard classifiers. Focusing on web spam detection, the objective of this work is two-fold: to evaluate the effect of the class imbalance ratio over popular classifiers such as Naïve Bayes, SVM and C5.0, and to assess how their performance can be improved when different types of techniques are combined in an unbalanced scenario.

Keywords: web spam detection, unbalanced data, sampling techniques, ensemble of classifiers.

1 Introduction and Motivation

In recent years, communication networks have been indiscriminately exploited with business purposes because Internet is an economic and effective media to reach a large number of potential customers. More specially, the spam activity in the web (i.e., web spam or spamdexing) comprises the usage of any type of manipulative techniques to fraudulently promote web sites, achieving false high ranking positions in web search engines. Therefore, web spam has resulted in several obstacles such as reducing the quality of retrieved results provided by search engines, waste of valuable user time when discarding irrelevant documents and/or the increment in computational resources that most search engines have been forced to perform in order to fight against this problem.

In the context of web spam filtering, the dominant approach is based on the application of machine learning (ML) algorithms, in which a general inductive process

© Springer International Publishing Switzerland 2015

S. Omatu et al. (eds.), *Distributed Computing & Artificial Intelligence, 12th Int. Conference,*
Advances in Intelligent Systems and Computing 373, DOI: 10.1007/978-3-319-19638-1_28

automatically builds a classifier by learning the characteristics of the underlying predictive class. In this training process, most data mining algorithms usually assume that available datasets present the same class distribution. However, this assumption is not common in real world scenarios where one class (generally the less relevant category) is represented by a large number of examples, while the other (the relevant one) is composed of only a few. This situation is known as the class imbalance problem, and it represents the major obstacle in inducing accurate classification models [1]. WWW is not an exception, in which the asymmetric distribution between legitimate and spam pages represents an important inconvenient for web spam detection. Although in the last years there have been different attempts to deal with this situation, there is no simple solution and researchers seemed to agree that its negative effects depend on both the particular domain (together with its computational representation) and the specific family of the selected ML classifiers.

Taking into account the class imbalance problem inherent to web spam detection, and also the fact that in skewed distributions results heavily depend on the sensitivity of the classifier, in this work, in order to obtain relevant results, we evaluate and compare the performance (both individual and combined) of three popular algorithms commonly used for spam classification (SVM, C5.0 and Naïve Bayes). The rest of the paper is organized as follows: Section 2 presents a brief survey covering relevant web spam filtering techniques. Section 3 details the corpus used in our experiments and introduces common balancing strategies. Section 4 describes the experimental protocol carried out and comments achieved results. Finally, in Section 5 we point out main conclusions and outline future research work.

2   Web Spam Filtering Techniques

The purpose of this section is to concisely present the most relevant algorithms used for web spam detection in the last years. Research on web spam filtering has evolved from content-based methods to other type of approaches using specific link and user behaviour mining techniques. In this context, several approaches have focused on detecting content-based web spam by analysing content features in pages (e.g., popular terms, topics, keywords or anchor text) to identify illegitimate changes. Among the early content spam works, [2] statistically analysed content features of illegitimate pages whereas [3] used ML methods for filtering spam domains. Later, a specific study showed an exhaustive analysis about how various content features and ML models could contribute to the quality of a web spam detection algorithm [4]. On the other hand, other approaches were focused on link properties of web pages [5, 6] instead of dealing with content features.

Complementarily, [7] reported optimized results by the appropriate combinations of link-based techniques and content-based methods. Additionally, the work presented in [8] was focused on link and content-based features with the goal of constructing accurate classifiers using SVM. In the work of [9], the authors also combined link and content-based features using C4.5 in order to detect web spam. Finally, [10] proposed a new method based on expectation-maximization (EM) and Naïve Bayes classification to deal with the labelling problem. All these works confirmed that, regardless the specific data source used in their experiments, ML approaches are widely applied for web spam detection and classification.

3   Benchmark Corpus and Balancing Strategies

In order to analyse the particular consequences of dealing with unbalanced data in the specific case of web spam classification, the experiments carried out were focused on assessing the performance of different well-known classifiers when the difference between the amount of legitimate domains and the spam content was reduced. For this purpose we selected WEBSPAM-UK2007 corpus, a publicly available dataset widely used for web spam research. Additionally, it contains the original raw HTML of web pages, which makes this dataset particularly applicable for our purposes given the need of directly handling with web content.

The WEBSPAM-UK2007 dataset includes 6,479 hosts labelled using three different categories (i.e., spam, ham and undecided), but in our work we did not make use of those instances belonging to the latter category as we were mainly interested in studying a biclass problem. Moreover, in the Web Spam Challenge 2008 [11] existing labelled domains were separated into two complementary groups (i.e., train and test). As a result, Table 1 shows the final distribution of each group of entries used in our experiments.

Table 1. Structure of the corpus finally used for our analysis

	Spam domains	Ham domains	Σ
Train set	208	3,641	3,849
Test set	113	1,835	1,948
Σ	321	5,476	5,797

As we can observe from Table 1, classes are asymmetrically distributed with an imbalance rate of 1:17. Additionally, all the pages with empty or meaningless content (e.g., redirections to error pages) were removed before running our experiments.

Having regard to the obstacle that represents a high class-imbalance distribution in machine learning, in the last years several strategies have been successfully developed with the goal of improving the accuracy of standard classification methods that work with asymmetric distributions. These techniques mainly include: sampling algorithms, cost-sensitive methods, recognition-based strategies and active learning approaches [12].

Because of the good performance working with different ML techniques, and also taking into consideration the low computational resources required by sampling strategies, in this work we focused on it. Specifically, methods belonging to the group of *sampling* techniques have been used to alleviate the class imbalance problem by either eliminating some data from the majority class (under-sampling) or adding some artificially generated (or duplicated data) to the minority class (over-sampling). In this context, [13, 14, 15] reported the effect of sampling methods for unbalanced data distributions. Complementarily, the work of [16] compared random and focused sampling strategies for both over and under-sampling methods. Their results showed that random re-sampling, often regarded as a suboptimal strategy, gathered competitive results when compared to more complex methods.

4   Experimental Results

The experimental process carried out was divided in two stages: (*i*) the application of a web content resampling strategy and (*ii*) the execution of different anti-spam techniques (individually or combined) to analyse how sensitive they are when learning from no balanced data.

Therefore, in the first step a resampling strategy was applied in order to alleviate the effects of the class imbalance problem on the selected corpus (described in Table 1). With the goal of comprehensively evaluating the true impact of unbalanced data over the global performance of anti-spam classifiers, we used a random under-sampling strategy, which consists on randomly eliminating some web domains from the majority class in order to achieve the desired ratio between classes. To generate different balancing scenarios in a straightforward way, we executed our experiments using the original corpus (1:17) and four complementary reduced rates (1:8, 1:4, 1:2 and 1:1). Moreover, tests were repeated 10 times using random under-sampling for each ratio, guaranteeing that the training set was different in each round. The results presented in this work correspond to the average values obtained in the 10 independent iterations.

In the second stage of our experimental protocol, different representative ML techniques (i.e. C5.0, SVM and Naïve Bayes) were tested in order to analyse their sensitivity and global performance in a class imbalance scenario. Moreover, with the goal of assessing how the ensemble of simple classifiers can deal with unbalanced data, these three techniques were executed together in all possible combinations, comparing their output values against the individual performances obtained by each individual algorithm. In particular, for C5.0 and SVM classifiers, every web page was represented by a vector comprising 96 content features described in [11]. SVM uses the default settings of its LIBSVM implementation [17], and C5.0 [18] is performed without any advanced option. On the other hand, the visible content of all the pages belonging to each domain is used as the input of a Naïve Bayes text classifier. As with other techniques, Naïve Bayes algorithm is used with defaults settings without any variation.

In order to carry out the comparisons mentioned above, results obtained were presented using the area under the ROC curve (AUC), sensibility and specificity. This type of validation has been widely used by the spam filtering research community in contrast to specific measures more focused on accuracy (e.g. precision, recall, f-score, etc.) which are not suitable when dealing with unbalanced data since they do not consider the proportion of examples belonging to each class and, therefore, do not provide information about the real cost of the misclassified instances.

As previously stated, in order to directly compare the outcomes generated from our different configurations, our benchmarking protocol was structured in two different scenarios consisting in (*i*) individually running each ML technique and (*ii*) combining the techniques used in the previous step. Both scenarios were executed varying the class imbalance ratio.

Under the first scenario, each individual algorithm was executed independently. For a precise assessment of each classifier, Table 2 combines both sensitivity and specificity for the best cut-off threshold together with the AUC value for each test carried out.

Table 2. Sensitivity, specificity and AUC values obtained for each simple classifier

	Ratio	AUC	Sensitivity	Specificity
C5.0	1:17	0.562	14.2	98.4
	1:8	0.649	50.4	79.5
	1:4	0.651	66.4	65.9
	1:2	0.648	77.0	52.7
	1:1	0.573	82.3	32.4
SVM	1:17	0.534	7.1	99.7
	1:8	0.590	26.5	91.6
	1:4	0.602	29.2	91.2
	1:2	0.604	30.1	90.8
	1:1	0.624	34.5	90.3
BAYES	1:17	0.530	15.9	90.2
	1:8	0.541	17.7	90.6
	1:4	0.542	18.6	89.8
	1:2	0.481	32.0	71.7
	1:1	0.479	64.6	39.6

As can be appreciated on Table 2, all the classifiers achieved the best performance when the class imbalance ratio was considerably reduced. Specifically, SVM obtained its best results when the amount of legitimate domains was the same as the number of spam sites. If we focus on SVM values, we can realize that due to the imbalance elimination, its sensitivity has significantly increased and its specificity has barely changed. On the other hand, the best performance of C5.0 and Naïve Bayes algorithms are obtained when dealing with a rate of 1:4. Below this class imbalance ratio (1:4), we can see that both techniques showed a similar behaviour: its sensitivity has increased considerably, but this fact makes that their specificity experienced a significant slowdown.

In the next experiment we measured the global performance achieved by all the possible combinations of the three previous classifiers. In order to easily compare the performance of each scenario, Table 3 presents the AUC results obtained by all of them, in addition to their individual execution.

Table 3. AUC results of individual and combined approaches working with different class imbalance ratio

	Class imbalance ratio				
	1:17	**1:8**	**1:4**	**1:2**	**1:1**
C5.0	0.562	0.649	0.651	0.648	0.573
SVM	0.534	0.590	0.602	0.604	0.624
BAYES	0.530	0.541	0.542	0.481	0.479
C5.0+SVM	0.579	0.660	0.687	0.684	0.646
C5.0+BAYES	0.565	0.675	0.687	0.618	0.525
SVM+BAYES	0.54	0.611	0.623	0.558	0.534
C5.0+SVM+BAYES	0.601	0.659	0.692	0.673	0.612

From Table 3 we can observe that, as happened when using simple classifiers, the ensembles have also experienced an improvement with the reduction of the class imbalance ratio. All the combinations have reached their best result with a ratio of 1:4, clearly influenced by C5.0 and Naïve Bayes. Moreover, except the C5.0+BAYES ensemble, the rest of the combinations have improved their performance when compared to the individual execution. Specifically, it is worthwhile highlighting the combination of C5.0 with BAYES, which has achieved a good accuracy (0.687) in spite of the low value gathered by the second model (0.542). Finally, we can conclude that the best model in a class imbalance scenario is the ensemble formed by the three individual classifiers, because this approach achieved the highest accuracy (0.692) being less sensitive to class imbalanced data.

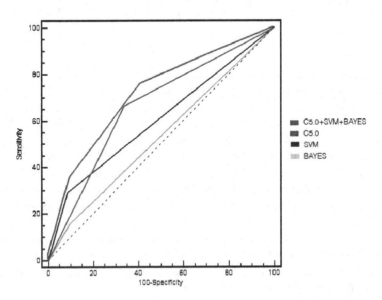

Fig. 1. Summary of ROC curves achieved by individual classifiers and the best ensemble model

In order to facilitate the visual comparison of the performance of each individual technique and the best ensemble, Figure 1 shows the ROC curves obtained in each experiment.

5   Conclusions and Future Work

The objective of this work was to carry out a straightforward analysis for assessing both (*i*) the true impact of the class imbalance distribution over the performance of popular classifiers in web spam detection and (*ii*) the behaviour of the combination of different classifiers working with unbalanced data.

From the experiments carried out we can conclude that, as seen in previous works [13, 14, 15], the application of a simple random undersampling method can improve significantly the performance of different well-known classifiers (i.e. C5.0, SVM and Naïve Bayes) when learning from imbalanced training data. Specifically, in the field of web spam detection, we can corroborate and extend to other classifiers that, as demonstrated in [7], spam-filtering area is a clear example of imbalanced data and the use of random undersampling strategies is an efficient approach that can help to overcome this obstacle. Moreover, the simple combination of different algorithms achieved a better accuracy than the one obtained by individual tests, even working with different class imbalance ratios. In particular, the combination of the three tested classifiers achieved the highest accuracy.

Regarding future work in the analysis of the class imbalance problem in web spam detection, the evaluation of other types of sampling strategies is undoubtedly an interesting field of study. In this work, we have used a random under-sampling method, however, the application of a more complex heuristic during the resampling process would avoid the elimination of valuable instances.

In addition, bearing in mind the appealing results obtained by the fusion of classifiers, it is worthwhile to focus on analysing the combination of algorithms using a more complex framework such as a rule-based system, where each technique has associated an individual weight according to its accuracy. Finally, another interesting aspect to consider is the possibility of training each algorithm of the ensemble using a different class distribution with the goal of obtaining better results.

Acknowledgements. This work was partially funded by the following projects: [14VI05] Contract-Programme, from the University of Vigo, and [TIN2013-47153-C3-3-R] Platform of integration of intelligent techniques for analysis of biomedical information, from the Spanish Ministry of Economy and Competitiveness. D. Ruano-Ordás was supported by a pre-doctoral fellowship from the University of Vigo.

References

1. García, S., Derrac, J., Triguero, I., Carmona, C.J., Herrera, F.: Evolutionary-based selection of generalized instances for imbalanced classification. Knowledge-Based Systems 25(1), 3–12 (2012)

2. Fetterly, D., Manasse, M., Najork, M.: Detecting phrase-level duplication on the World Wide Web. In: Proceedings of the 28th Annual International ACM SIGIR Conference on Research and Development in Information Retrieval, pp. 170–177 (2005)

3. Ntoulas, A., Najork, M., Manasse, M., Fetterly, D.: Detecting spam web pages through content analysis. In: Proceedings of the 15th International Conference on World Wide Web (WWW 2006), pp. 83–92 (2006)

4. Erdélyi, M., Garzó, A., Benczúr, A.A.: Web spam classification: a few features worth more. In: Proceedings of the 2011 Joint WICOW/AIRWeb Workshop on Web Quality (WebQuality 2011), New York, USA, pp. 27–34 (2011)

5. Gyöngyi, Z., Berkhin, P., Molina, H.G., Pedersen, J.: Link spam detection based on mass estimation. In: Proceedings of the 32nd International Conference on Very large data bases, VLDB, pp. 439–450. Endowment, Seoul (2006)

6. Benczur, A., Csalogany, K., Sarlos, T., Uher, M.: SpamRank–Fully Automatic Link Spam Detection. In: Proceedings of the 1st International Workshop on Adversarial Information Retrieval on the Web, Japan (2005)

7. Geng, G.G., Wang, C.H., Li, Q.D., Xu, L., Jin, X.B.: Boosting the performance of web spam detection with ensemble under-sampling classification. In: Proceedings of IEEE 4th International Conference on Fuzzy Systems and Knowledge Discovery, pp. 583–587 (2007)

8. Abernethy, J., Chapelle, O., Castillo, C.: Webspam identification through content and hyperlinks. In: Proceedings of the 4th International Workshop on Adversarial Information Retrieval on the Web (2008)

9. Becchetti, L., Castillo, C., Donato, D., Leonardi, S., Baeza-Yates, R.: Web spam detection: link-based and content-based techniques. In: Proceedings of the European Integrated Project Dynamically Evolving, Large Scale Information Systems, pp. 99–113. Heinz-Nixdorf-Institut. (2008)

10. Karimpour, J., Noroozi, A.A., Alizadeh, S.: Web Spam Detection by Learning from Small Labelled Samples. International Journal of Computer Applications 50(21), 1–5 (2012)

11. Castillo, C., Chellapilla, K., Denoyer, L.: Web spam challenge 2008. In: Proceedings of the 4th International Workshop on Adversarial Information Retrieval on the Web, AIRWeb 2008 (2008)

12. He, H., Garcia, E.A.: Learning from imbalanced data. IEEE Transactions on Knowledge and Data Engineering 21(9), 1263–1284 (2009)

13. Drummond, C., Holte, R.C.: C4.5, class imbalance, and cost sensitivity: Why under-sampling beats over-sampling. In: Proceedings of the International Conference on Machine Learning (2003)

14. Laza, R., Pavón, R., Reboiro-Jato, M., Fdez-Riverola, F.: Assessing the suitability of mesh ontology for classifying medline documents. In: Proceedings of the 5th International Conference on Practical Applications of Computational Biology & Bioinformatics, PACBB 2011, pp. 337–344 (2011)

15. Van Hulse, J., Khoshgoftaar, T.M., Napolitano, A.: Experimental perspectives on learning from imbalanced data. In: Proceedings of the 24th International Conference on Machine Learning, Corvallis, OR, USA, pp. 935–942 (2007)

16. Batista, G.E.A.P.A., Prati, R.C., Monard, M.C.: A study of the behavior of several methods for balancing machine learning training data. ACM SIGKDD Explorations Newsletter - Special Issue on Learning from Imbalanced Datasets 6(1), 20–29 (2004)

17. Chih-Chung, C., Chih-Jen, L.: LIBSVM: a library for support vector machines. ACM Transactions on Intelligent Systems and Technology 2, 27:1–27:27 (2011), http://www.csie.ntu.edu.tw/~cjlin/libsvm

18. Data Mining Tools C5.0, Rulequest Research (2013), http://www.rulequest.com (accessed December 19, 2014)

Implementing an IEEE802.15.7 Physical Layer Simulation Model with OMNET++

Carlos Ley-Bosch, Roberto Medina-Sosa, Itziar Alonso-González,
and David Sánchez-Rodríguez

Institute for Technological Development and Innovation in Communications (IDeTIC)
University of Las Palmas de Gran Canaria, Las Palmas de Gran Canaria, Spain
{carlos.ley,itziar.alonso,david.sanchez}@ulpgc.es

Abstract. Visible Light Communications (VLC) uses visible light spectrum as transmission medium for communications. VLC has gained recent interest as a favorable complement to radio frequency (RF) wireless communications systems due to the ubiquity and wide variety of applications. In 2011 the Institute of Electrical and Electronic Engineers published the standard IEEE 802.15.7 [1]. Nowadays, simulation tools are widely used to study, understand and achieve better network performance. This paper describes the design and implementation of a physical layer model based in IEEE802.15.7 standard using OMNET++ simulation tool [2]. This software is a popular tool for building networks' and modeling their behavior. The main goal of this paper is to introduce the developing and implementing of a software module to simulate the Physical Layer (PHY) based on IEEE802.15.7. The developed module, called *simVLC* will let researchers and students to study and simulate different scenarios in this standard.

Keywords: VLC, IEEE802.15.7, Physical Layer, MAC, OMNET Simulator.

1   Introduction

In recent years, different research lines have focused on energy saving techniques for illuminating sources, since traditional lamps present very low energy efficiency. This way, new more efficient lighting devices are being developed to replace incandescent and fluorescent lights, see [3] as an example. Some of these new devices are based on LED technology, which provides great savings in energy consumption and have a high lifespan, between 10,000 and 50,000 hours.

The use of illumination systems for data transmission is based on adding communication capabilities to lamps, while maintaining their main function as light source, [4]. Transmitted data signals should not alter the illumination perception to the human eye. The most important requirement is to have light sources ON/OFF switching times faster than the persistence of the human eye. VLC transmits data via intensity modulation using light emitting diodes LEDs. Hence LEDs are used for both lighting and communications purposes.

© Springer International Publishing Switzerland 2015

S. Omatu et al. (eds.), *Distributed Computing & Artificial Intelligence, 12th Int. Conference*,
Advances in Intelligent Systems and Computing 373, DOI: 10.1007/978-3-319-19638-1_29

A wide range of communications applications including wireless local network, personal area networks, vehicular networks and machine-to-machine communications among many others can be developed [5]. The Institute of Electrical and Electronic Engineers (IEEE) has published IEEE Standard 802.15.7 [1] in 2011, which defines a Physical Layer (PHY) and Medium Access Control Layer (MAC) for wireless optical communications using visible light.

There is a growing interest in researching and developing simulation tools for VLC communications. For example, in [6] a simulation model has been developed for ns-2 simulation platform, but they focus mainly on MAC layer modeling, not on physical layer.

The aim of our work is developing a simulation tool to simulate and evaluate IEEE 802.15.7 networks, with a dependable physical layer model to rely on.

The remainder of this paper is organized as follows: Section 2 introduces an overview of the IEEE802.15.7 physical layer (PHY). Next, in section 3, we describe the *simVLC* simulation module. In section 4, we evaluate our module *simVLC*, some scenarios are simulated and some results are presented. Finally, we sum up the conclusions and we present future work.

2   Overview IEEE802.15.7 Physical Layer

IEEE 802.15.7 defines Physical Layer (PHY) and Medium Access Control Layer (MAC) for short-range wireless optical communications using visible light. This standard supports tree types of network topologies: peer-to peer, star and broadcast, see figure 1(a).

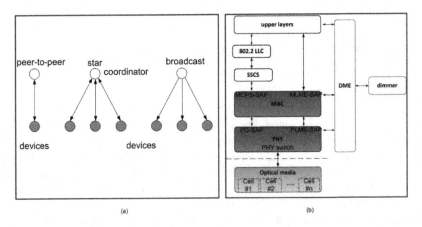

Fig. 1. (a) Topologies defined in IEEE802.15.7. (b) VPAN device architecture.

A visible-light personal area network (VPAN) communication device is defined by a PHY layer (including an optical transceiver) and a MAC layer that provides access to the optical channel. Figure 1 (b) shows this device architecture.

IEEE 802.15.7 standard defines three PHY layer operational modes. Data rates in each PHY mode support a variety of optical sources, detectors, and environments.

- PHY I: Outdoor usage with low data rates, which vary from 11.67 to 266.6 kb/s.
- PHY II: Indoor usage with moderate data rates, from 1.25 to 96 Mb/s.
- PHY III: Intended for communications using color-shift keying (CSK). Operates from 12 to 96 Mb/s.

PHY-I and PHY-II systems make use of OOK and variable pulse position (VPPM) modulations. PHY-III uses color shift keying (CSK) modulation, where multiple optical sources are combined to produce white light [1][7].

Indoor wireless visible light optical channel can be characterized by three main elements: an emitting source (LEDs), an enclosed area, and an optical receiver [8].

3   Module *simVLC* Using OMNET++

OMNET++ [2] is an extensible, modular, component based simulation library and framework for building network simulators. It is free for academic use and widely used by researchers to simulate systems where the discrete event approach is suitable.

Many simulation frameworks have been developed for OMNET++, amongst which we highlight: INET, which provides a base protocol model library, INETMANET, which adds experimental support for mobile ad-hoc networks, and MiXiM which offers detailed models for radio propagation. However, none of the existing frameworks offered support to simulate the characteristics of physical layer used in optical wireless communications.

We decided to use INETMANET framework as the base for our model because it included support for simulating IEEE 802.15.4 Wireless Personal Area Networks (WPANs) [9]. Similarities between IEEE 802.15.4 and 802.15.7 MAC layers led us to decide the use of INETMANET 802.15.4 implementation as starting point. Nevertheless, it was necessary to make extensive changes to INETMANET implementation to build a new Physical Layer model to support VLC. We made all code modifications with the aim of integrating our module into INETMANET framework in the future.

3.1   Implementing of the Module *SimVLC*

SimVLC is designed to simulate networks with multiple nodes exchanging data through an optical wireless channel. We have focused on star network topology: with one node - the Coordinator - acting as the MAC layer master, and the rest of the nodes which are the so-called Devices. We have designed a propagation model based on line-of-sight (LOS) channel path between nodes according to [10]. We have used a typical application scenario of VPAN: indoor and short-range distance. Network nodes are enclosed and arranged in a three-dimensional area, see figure 2. Each node can move on its own way, according to different mobility patterns.

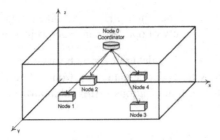

Fig. 2. Three-dimensional enclosed area with nodes arranged VPAN star network topology

Optical Channel

This entity in our model is responsible for modeling the physical channel environment. Its main functions are: managing position and orientation of nodes, managing transmitted frames through the optical channel, and calculating propagation delay.

Nodes

Each node is modeled by means of several entities or modules, which model protocol stack based in IEEE802.15.7. There are other modules to model additional node features like energy consumption (battery) and mobility patterns. Figure 3 shows node modules arranged in a graphical representation.

IEEE802.15.7 PHY

This module is the core of our simulation model. The main implemented functions are:

— Transmission and reception of frames (PHY-PDUs).
— Modeling of optical transceiver characteristics, such as: transmission clocks and data rates, defined according the so-called MCS-Ids (modulation and coding scheme identifiers).
— Calculating optical received power, based on LOS propagation model, and Signal-to-interference-plus-noise ratio (SNIR).
— Implementation of PHY layer and optical transceiver states.

All the functionality needed to implement our simulation model can be summarized on three main modules, (see figure 3):

• *IEEE802157StarNet:* module for network simulation modeling.
• *IEEE802157Node:* compound module for a node.
• *IEEE802157Nic:* compound module for a network interface.

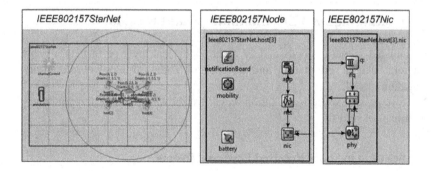

Fig. 3. Representation of three main modules developed in OMNET++ for VLC

Mobility of Nodes

A specific module is also needed for modeling nodes positional movement and orientation. INETMANET's original mobility models did not contemplate the orientation of nodes, so we included new mobility models to add the management of position and orientation of nodes. For example, our circular mobility pattern models the movement of a node around a circle at a certain speed, but also manages node orientation depending on its position. Some other enhanced mobility patterns have been developed, such as: random, linear, rectangular or stationary.

This modular structure is completed with some other additional modules. Figure 4 shows a detailed view of all the elements involved in the modeling of a VLC network composed of two nodes: one coordinator and one device.

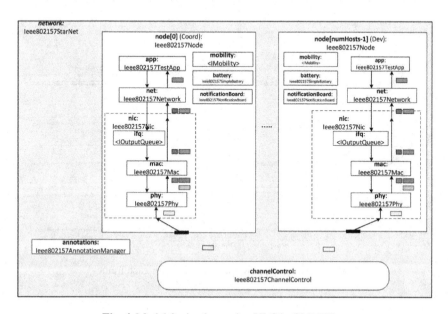

Fig. 4. Model for implementing VLC in OMNET++

4   Evaluating *simVLC*. Simulations and Results

Our simulated scenarios focus mainly on how physical layer performance is affected by factors such as orientation and distance of nodes, transceiver directivity, and propagation model.

Next we present several simulation scenarios, showing some of the potential of our simulation tool *simVLC*.

4.1   Scenario to Study Received Power vs. Distance

This scenario is designed to study the received optical power vs. distance. In this scenario there is one coordinator (transmitter) and one mobile device (receiver), which moves back and forth linearly at constant speed.

In figure 5, we can see how received power varies inversely to distance squared $(1/d^2)$. At t = 0 seconds both nodes are at closest distance and received power is maximum. At t = 6 seg. nodes are at the farthest distance and received power is minimum. Since there are only two nodes in our scenario, SNIR is only affected by received power, because there are no other nodes transmitting. In figure 5 we can see a zoom of figure (a), showing on detail the behavior when the nodes are perfectly aligned (blue) or with a small alignment error (red).

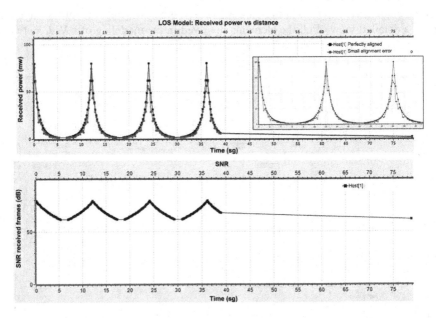

Fig. 5. (a) Received power and (b) SNIR for each frame

4.2 Scenario to Study Radiated Power Patterns

This scenario evaluates radiated power patterns, according to lambertian optical sources, and depending on the *Modal Index* (*m*), which determines the directivity of optical transmitters such as LEDs. The greater *m*, the greater the transmission directivity is, with more radiated power concentrated close to the transmitter orientation vector. In this scenario we have a static transmitter node, and a mobile receiving node which moves around the transmitter according to a circular mobility pattern. See figure 6 for received power values depending on *m*.

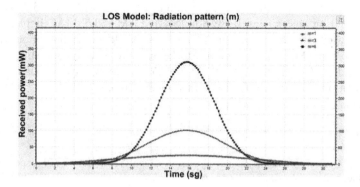

Fig. 6. Scenario to study radiation patterns

4.3 Scenario "*heatmapper*"

The last scenario we present in this paper is aimed to study coverage maps in an enclosed area. In figure 7, colors and values shown represent received power data at coordinate points (x,y) where receiving nodes are located. There is one coordinator at the ceiling which transmits to many receiving devices arranged in a grid pattern, located on the same plane XY. Device nodes are oriented to coordinator. Distance and orientation angles are different for each node. See figure 7.

(X,Y)	0.25	0.75	1.25	1.75	2.25	2.75	3.25	3.75	4.25	4.75	5.25	5.75	6.25	6.75	7.25	7.75	8.25	8.75	9.25	9.75
0.125						0.150	0.184	0.324	0.527	0.755	0.916	0.916	0.755	0.527	0.324	0.184	0.150			
0.375					0.150	0.250	0.464	0.792	1.182	1.469	1.469	1.182	0.792	0.464	0.250	0.150				
0.625				0.082	0.184	0.337	0.657	1.182	1.846	2.353	2.353	1.846	1.182	0.657	0.337	0.184	0.082			
0.875				0.150	0.250	0.445	0.916	1.741	2.852	3.742	3.742	2.852	1.741	0.916	0.445	0.250	0.150			
1.125				0.118	0.259	0.575	1.247	2.506	4.318	5.844	5.844	4.318	2.506	1.247	0.575	0.259	0.118			
1.375				0.138	0.312	0.721	1.643	3.491	6.327	8.837	8.837	6.327	3.491	1.643	0.721	0.312	0.138			
1.625			0.070	0.176	0.364	0.872	2.080	4.648	8.837	12.726	12.726	8.837	4.648	2.080	0.872	0.364	0.176	0.070		
1.875			0.076	0.172	0.410	1.012	2.506	5.844	11.580	17.139	17.139	11.580	5.844	2.506	1.012	0.410	0.172	0.076		
2.125			0.085	0.184	0.445	1.112	2.852	6.860	14.017	21.155	21.155	14.017	6.860	2.852	1.112	0.445	0.184	0.085		
2.375		0.052	0.095	0.464	1.182	3.047	7.451	15.475	19.812	70.812	10.678	7.451	3.047	1.182	0.464	0.250	0.092			
2.625		0.062	0.190	0.464	1.182	3.047	7.451	15.476	21.012	29.012	15.476	7.451	3.047	1.182	0.464	0.190	0.062			
2.875		0.065	0.184	0.445	1.112	2.852	6.860	14.017	21.155	21.155	14.017	6.860	2.852	1.112	0.445	0.184	0.065			
3.125		0.076	0.172	0.410	1.012	2.506	5.844	11.580	17.139	17.139	11.580	5.844	2.506	1.012	0.410	0.172	0.076			
3.375		0.070	0.176	0.364	0.872	2.080	4.648	8.837	12.726	12.726	8.837	4.648	2.080	0.872	0.364	0.176	0.070			
3.625				0.138	0.312	0.721	1.643	3.491	6.327	8.837	8.837	6.327	3.491	1.643	0.721	0.312	0.138			
3.875				0.118	0.259	0.575	1.247	2.506	4.318	5.844	5.844	4.318	2.506	1.247	0.575	0.259	0.118			
4.125				0.150	0.250	0.445	0.916	1.741	2.852	3.742	3.742	2.852	1.741	0.916	0.445	0.250	0.150			
4.375				0.082	0.184	0.337	0.657	1.182	1.846	2.353	2.353	1.846	1.182	0.657	0.337	0.190	0.082			
4.625					0.150	0.250	0.464	0.792	1.182	1.469	1.469	1.182	0.792	0.464	0.250	0.150				
4.875						0.150	0.184	0.324	0.527	0.755	0.916	0.916	0.755	0.527	0.324	0.184	0.150			

Fig. 7. Coverage map

5   Conclusion

The work we present here has been developed to model some of the main characteristics of the physical layer of IEEE802.15.7 standard. Our designed simulation module, *simVLC*, has been developed with OMNET++ simulation platform. We have shown only a few simulation scenarios in this paper, but many others could be easily defined depending on user requirements. In the future, we plan to incorporate non direct path reflections to our propagation model, and also models for modulation schemes and calculations of Bit Error Rates (BER).

Nowadays we are working on the development of the IEEE 802.15.7 MAC layer module, based on the existing IEEE 802.15.4 INETMANET MAC implementation. At the time we finish this module, we plan to propose the integration of our simulator *simVLC* into INETMANET framework, to facilitate researchers and students the simulation and evaluation of IEEE 802.15.7 PHY and MAC layers. Our future work will focus on studying improvements to this standard too.

References

1. Wireless Optical Communication using Visible Light, IEEE802.15.7-2011, specifications
2. Website, http://www.omnetpp.org
3. Philips Lumileds Lighting Company: Case Study: Beyond Compact Fluorescent Lighting (2008), http://www.philipslumileds.com/pdfs/CS20.pdf
4. Deicke, F., Fisher, W., Faulwaßer, M.: Optical Wireless Communication to Eco-System. In: Future Network & MobileSummit Conference 2012 (July 2012)
5. Bhalerao, M., Sonavane, S.: Visible light communication: A smart way towards wireless communication. In: IEEE International Conference on Advances in Computing, Communications and Informatics (ICACCI 2014), pp. 1370–1375 (September 2014)
6. Musa, A., Baba, M., Mansor, H., Asri, H.: The Design and Implementation of IEEE 802.15.7 Module with ns-2 Simulator. In: 2014 IEEE International Conference on Computer, Communications and Control Technology (I4CT 2014), September 2-4 (2014)
7. Roberts, R.D., Rajagopal, S., Sang-Kyu, L.: IEEE 802.15.7 Physical Layer Summary. In: 2nd IEEE Workshop on Optical Wireless Communications, pp. 772–776 (2011)
8. González, O., Guerra, M., Martín, R.: Multi-User Visible Light Communications. In: Advances in Optical Communications, ch. 2, pp. 36–63. Intech (2014)
9. Chen, F., Wang, N., German, R., Dressler, F.: Performance Evaluation of IEEE 802.15.4 LR-WPAN for Industrial Applications. In: Fifth Annual Conference on Wireless on Demand Network Systems and Services, pp. 89–96 (January 2008)
10. Kahn, M., Barry, J.: Wireless Infrared Communications. Proceedings of the IEEE 85(2), 265–298 (1997)

P2P Self-organizing Agent System: GRID Resource Management Case

Oleg Bukhvalov and Vladimir Gorodetsky

St. Petersburg Institute for Informatics and Automation, Russian Academy of Science
39, 14-th Liniya, 199178 St. Petersburg, Russia
psychoveter@gmail.com, gor@iias.spb.su

Abstract. The three decade history of multi–agent system technology has provided, for IT developers, a lot of promises. However, it has exhibited much less successful practice and about zero industrial applications. Nevertheless, the big potential of multi-agent paradigm and technology is far from the exhaust. Indeed, distributed large scale networked systems, in particular, self-organizing systems remain to be its very promising application area. Indeed, models of self-organization and their requirements to the needed implementation technology, on the one side, and capabilities of the multi-agent technology, on the other side, well meet each other. The paper demonstrates this fact by example of real time open distributed GRID resource management self-organizing system intended to provide for both, GRID node load balancing and equalization of the input task waiting times. The presented software implementation of an open GRID resource management system uses P2P multi-agent architecture and platform supporting for self-organizing policy and P2P agent communication. The lessons learnt are outlined and generalized.

1 Introduction

The three decade history of multi–agent system (MAS) technology has provided, for IT developers, a lot of promises. However, it has exhibited much less successful practice and about zero industrial applications. Nevertheless, the big potential of MAS paradigm and technology is far from the exhaust, and this potential is especially valuable to conceptualize and implement management and control of the large scale computer networked objects like GRID, global scale air traffic control systems, production and logistics B2B networks, Smart Grid, and many others analogous applications. These networked objects comprise thousands and more nodes that can usually freely join and leave the network and change neighbors to optimize the network configuration. In such networks, it is practically impossible to on-line monitor states and interactions of its components, to support job allocation and synchronized scheduling of their networked resources, search for data and services available, etc. As a rule, complexity of the networked object on-line management and control considerably exceeds the possibilities of the existing methods and means developed for these purposes. This very heavy challenge forces to reconsider the existing on-line management principles and architectures of networked object distributed control, while entrusting these tasks with interaction of local management functions of self-organizing nature.

© Springer International Publishing Switzerland 2015

S. Omatu et al. (eds.), *Distributed Computing & Artificial Intelligence, 12th Int. Conference*,
Advances in Intelligent Systems and Computing 373, DOI: 10.1007/978-3-319-19638-1_30

The existing experience has proved that self-organization model and its requirements to the implementation technology, on the one side, and capabilities of the MAS technology, on the other side, well meet each other. There exist a number of successful MAS self-organizing applications [5] which persuasively demonstrate the advantages and efficiency of MAS technology for self-organizing management of large scale networked objects [1, 4, 5, 9, etc.]. Examples can be found in adaptive planning and scheduling for B2B production networks and logistics [9], road traffic control [2], P2P routing protocols [5], software infrastructures of computer networks [1, 3] supporting operation of large-scale open networked systems including mobile ones, etc.

The paper objective is to demonstrate that future perspectives of MAS self-organizing control are very promising, and the existing achievements in MAS technology are capable to support the development and implementation of on-line management of really complex large scale open networked objects of dynamic configuration using simple agents and efficient self-organizing policy. Due to limited paper space, this technology is explained based on software implementation of an open GRID resource management system that uses P2P multi-agent architecture and platform supporting for self-organizing policy and P2P agent communication.

In the rest of the paper, section 2 describes the statement of application problem that is adaptive GRID resource management. The MAS architecture and P2P infrastructure supporting implementation technology for this application are sketched in section 3. Section 4 outlines experimental settings and results. In Conclusion, the lessons learnt through this research and development are outlined.

2   Problem Statement

Let us consider GRID of an ad-hoc structure with the nodes corresponding to particular computing resources. An assumption is that the network has no centralized control and solves the task allocation problem in distributed mode. The task flow is distributed too: every network node can be an input of the incoming task flow. The incoming tasks can be of various complexities unknown to GRID computers. The GRID computers can be of different processing power unknown to other network nodes. The GRID nodes can randomly leave and join the network. Every node of the network "knows" only its network neighbors and can exchange messages with them on P2P basis. Thus, the GRID is a dynamically altering P2P networked object.

Processing power of every GRID node is given as a proportion coefficient compared to an "etalon" node with power equal to 1. Any task complexity is also given in relative units compared to the etalon task processing time on GRID node. No node "knows" input task complexity that is a task attribute used for simulation purposes only. Each task is mapped input time when it arrives into a GRID node. It is assumed that, at any time instant, every node processes only a task and solves it to the very end. The node buffer is loaded with queue of the tasks waiting for processing. Since different tasks are of various complexity and nodes are of various processing power, some computers can be overloaded whereas other ones can be idle. The load-balancing task is solved to improve the total GRID productivity.

The load balancing is achieved due to task exchange among the GRID neighbor nodes on regular basis. As a result, after several such actions, every task entering at the beginning a node can be found out in the queue of any other node achievable due to the GRID connectivity. Thus, load balancing actuating mechanism is task exchange among the GRID nodes according to a self-organization policy. The policy objective is to provide load balancing for GRID nodes, and to equalize the waiting times of the tasks in the node queue.

3   Multi-agent Architecture and Self-organization Policy

Let us set up a software agent at each GRID node and assume that, at any time, it "knows" the own task queue length and its neighbors in the GRID. No other information is available for this agent. It can exchange messages with the agents of neighbor GRID nodes. The GRID holders interested in maximal load of their computers, but task holders interested in minimization of the waiting time in queues of the GRID. These requirements determine two criteria of the load balancing self-organization policy determining the strategy of task exchange among the GRID neighbors.

The general idea of the self-organizing load balancing policy used was inspired by one described in [10] but the latter was significantly transformed. In this policy, each agent periodically transmits a number of the tasks from its queue to their neighbors. This number is proportional to the own queue length, on the one hand, and to the total number of the node neighbors, on the other one:

$$N = \alpha \, Q + \beta \, M \;, \text{ if } \; N > Q \; \text{ then } \; N = Q \,, \tag{1}$$

where N – the total task number to be sent to the neighbors, Q – own task queue length and M – the current number of the neighbors. The proportionality coefficients α and β are selected empirically. This part of the policy is responsible for load balancing only.

To equalize the waiting times of the tasks, a heuristics is used: the local (node-based) policy orders the tasks of the queue according to their input times and send to the neighbors the computed number of task from the queue begin.

The GRID self-organizing control architecture is given in Fig. 1. In it, the agents realizing the self-organizing policy constitute the P2P overlay network set up over the distributed P2P agent platform [6, 7], which instances, in their turn, form overlay network too. In this distributed MAS, all the agents execute the same role named *GridAgent*. This role can perform two scenarios: (1) cyclic proactive behavior and (2) receiving the messages from the neighbor agents with the list of tasks transferred to it. The agents operate and interact in asynchronous mode.

The distributed P2P agent platform was developed and implemented according to the FIFA service-oriented reference architecture [7] supporting the distributed tables of White and Yellow pages and services, as well as corresponding agent-based mechanisms for distributed gossip protocol-based search for agents and services. The overlay network of P2P providers supports the message exchange. In the implementation software, emulator of distributed P2P communication network [8] simulates P2P

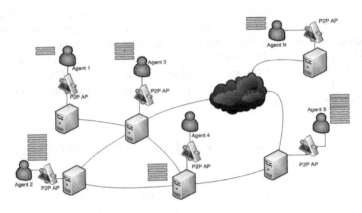

Fig. 1. Architecture of self-organizing GRID load balancing system

provider interactions thus making it possible to simulate the distributed GRID and self-organizing control software on a single computer. In other words, thanks to emulator, the prototype operates as if it was installed on the distributed GRID computers.

The basic software components of the developed prototype of self-organizing distributed MAS infrastructure, i.e. FIPA-compliant distributed P2P agent platform and emulator of distributed communication network were developed within the earlier projects [6–8]. The generator of GRID network configuration was developed specifically for the application in question. Together they constitute basic tool kits supporting P2P MAS simulation technology have been used in some other applications [5].

Within the MAS in question, each GRID node is mapped P2P communication provider (peer, for short), each peer is mapped the instance of the distributed P2P agent platform, and each instance of P2P agent platform is mapped the single application agent, *GridAgent*. Thus, topology of the agent network is the same as GRID topology, in the developed architecture. All tool kits and other software components are coded in C++ using the *STL, POCO, wxWidgets, Boost* libraries. Agent software and software simulating the external environment uses *wxWidgets* library whereas simulation of random flows of tasks is based on *Boost* library. The details of software implementation of the developed GRID load balancing system can be found in [11].

GridAgent executes its role in three use cases (scenarios):

1. *Initialization* – registers the *GridAgent* service in the own Yellow page of the P2P agent platform instance. This use case initializes the agent and its variables.

2. *Message receipt* – extracts (from the message) the names of the tasks sent and put them into the own task queue according to the increase of the task arrival time.

3. *Proactive behavior* – it is the scenario controlling the task queue and task run.

The external environment is simulated by the *GridControlModule* component [11]. *The life cycle* of the GRID simulation system comprises the following stages:

1. *Initialization.* The *VirtualEnvironment* interface module [11] initializes peers, P2P agent platform instances and agents using configuration file specifying virtual environment that is generated automatically using GRID topology specification.

2. *Generation of the experimental settings* comprising average task complexity, experiment duration, attributes of the task flow Poisson probability distributions, probabilities defining the presence/absence of GRID nodes, in the network. These settings are used to generate experiment scenario, GRID nodes and task flows initial states. The scenario is run by the set of events ordered according to global time scale.

3. *Experiment* launching the proactive agent behavior scenario according to the initial system configuration. The system is logging in time the dynamics of the agent states, monitors the message exchange, load of communication channels, etc. [11].

4   Experiment Objectives, Settings and Results

The objective of the experiments is to validate operability of the developed self-organizing MAS from several viewpoints and in various conditions, in particular, to study the GRID load balancing system performance when GRID nodes can freely leave it and join GRID again. Let us remind that the nodes can have various processing powers, and node agents have no such information about own neighbors.

The self-organizing MAS *productivity* is understood as its capability to balance, at average, the loads of the GRID nodes and equalize waiting times of all the tasks incoming into GRID. The variations of task waiting times are measured through *variances* of the latter. This metric is computed for each GRID node as well as for GRID as a whole. Let us note that the averaged waiting time value itself is not a characteristic of the load balancing control system, because it depends on the of the intensities of total input task flows and processing powers of the GRID computers. The GRID network topology used in experimentations contains in total 30 nodes assigned task flows incoming into each of them.

It total, five different experiments with different settings were run. Within each experiment except one of them, the node fails and corresponding time failure interval durations were simulated at random, so that practically all the nodes, at least once, were absent in the network during some random time interval. Analogously, the computer processing powers of all the nodes were simulated at random, for each experiment. The variations of it for the network computers were set in the interval [0.5, 1.5].

Every experiment scenario is the time-ordered set of three types of events:

(1) adding a new task for processing to some randomly selected GRID node,
(2) leaving the network by a GRID node, and
(3) joining a node (previously existed or new one) to the GRID network.

Each event is assigned the time of occurrence (in milliseconds) on the global time axis. The external task flow does not input in the idle node. On the other hand, if a node is getting to leave the network it passes all the tasks from its queue to the neighbors selecting them at random according to the uniform probability distribution. The time scale of the real system and the one of the simulated system are connected through the ratio of the length of the proactive behavior cycle duration in the real system to the corresponding length in the simulated one. In the real system, this cycle is set equal to 5 minutes whereas in simulated one it is equal to 5 seconds. Thus, the one day (24 hours) of GRID performance is simulated during 24 minutes.

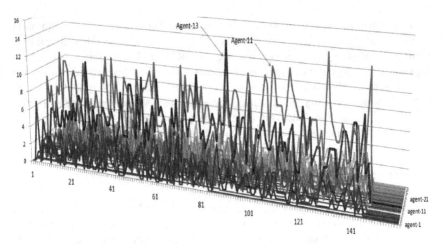

Fig. 2. Experiment 1: Plot of input task flows for particular GRID nodes (agents)

Other settings of each experiment concern with the following attributes:

- the total number of GRID nodes (computers) – N ;
- topology (node connectivity) of the GRID;
- experiment duration D (in milliseconds);
- duration of the proactive agent behavior cycle in real system, C_r ;
- duration of the proactive agent behavior cycle in simulated system, C_m ;
- attributes λ_i of Poisson probability distribution, for each GRID node; their aver-
  aged value over all GRID nodes is denoted as $M[\lambda_i]|_{i=1}^N$;
- processing power p_i of i -th node that is presented as proportionality coefficient
  with regard to this value of etalon computer; their averaged value over all GRID
  nodes is denoted as $M[p_i]|_{i=1}^N$;
- probability of the presence, for every GRID node at the start time, $P_{o,i}$,
- total number of fails of every GRID node S_i ; this value indicates how many
  times the i -the node changes its status of presence, in the computer network;
- averaged value of the processing time for l -th task on etalon node $M[t_e]$;
- attributes α and β of the function (1) defining the self-organization policy
  used.

In total, five different experiments were carried out. In this section, the experiments #1, and #5 are briefly analyzed only, although conclusion section takes into account the results of all the five experiments. The experiment settings differ in the values of α and β coefficients in (1) and also in particular settings of other attributes listed in previous section.

Experiment #1: $\alpha=0.5$, $\beta=0.5$. In fig.2 the plots of input task fows of different GRID nodes are depicted. In this and other analogous figures, X (horizontal) – the axis indicating simulated system time (in seconds with quantization of 5 sec. according to the sinulation cycle); vertical axis indicats of tasks incoming duting sequential 5-second time intervals. Z–axis in horizontal plane indicates the name of the GRID node (agent), which incoming task flow is indicated along the vertical axis.

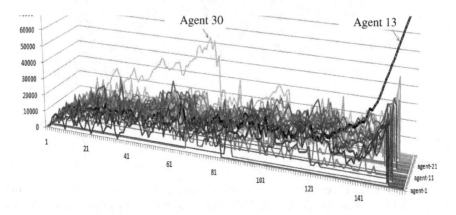

Fig. 3. Experiment 1: Plot for averaged waiting time of tasks, for various GRID nodes (agents)

One can mention that the nodes #11 and #13 have considerably more heavy task load than other ones. Fig. 3 shows the plot of avaraged waiting times of the tasks in the node queues in particulat time instants. This plot shows that variances of waiting times of different tasks for particular GRID nodes are more or less close. The peacks observed in these plots for nodes #13 and #20 are caused by the fact that their neighbors leave the network within corresponding time intervals. Indeed, the node #20 is the neighbor of the node #13, and the latter has the most intensive input task flow. Correspondingly, if the node #20 neighbors leave the network the waiting times of the tasks of its queue increase. This increase takes place up to 80-th interval when its neighbors join the network again. Later on, within the time interval #140, the neighbors of the node #13 leave the network again, what naturally leads to the increase of the corresponding waiting times. Fig. 7 demonstrates these facts.

3. Experiment #5: $\alpha = 0.8$, $\beta = 0.5$. In it, the entire input task flow incomes into node #8 having 5 neighbors. For this case, the plot of averaged waiting times is given in Fig. 4. Its self-organizing strategy also performs well: the averaged waiting time is balanced among the GRID computers. The same fact is peculiar to the lengths of the queues in various GRID nodes.

Table 1. Statistical properties of the waiting times in Experiment #5

Averaged waiting time	17479.78
Maximal waiting time	51676
Variance of waiting times	5.08E7

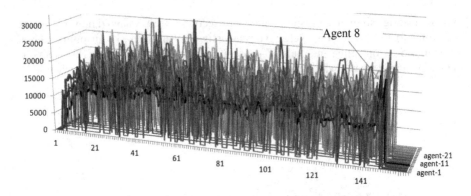

Fig. 4. Experiment 5: Plot of the averaged waiting times of tasks in the GRID nodes

Conclusion: Lessons Learnt

The developed and experimentally investigated application makes it possible to infer several facts about perspectives of MAS technology in conceptualization, design and implementation of self-organizing systems managing large scale networked objects.

Indeed, it was surprising for the authors how efficiently and with high quality can perform very simply organized MAS while solving not very simple problem: very primitive self-organizing policy provides for acceptable dynamic load balancing, while consuming vanishingly small processing and memory resources as overhead. The simple self-organization policy provides for fault tolerant performance while operating steady although with natural performance degradation in the situations when some GRID nodes leave it. The control system reacts on this fact as if a decrease of total capacity of GRID processing resources occurs. Experiments proved a guess that such MAS will not overload the communication channels with many messages [11]. The communication channel overhead depends only on the number of the neighbors of each GRID node. The investigated self-organization mechanism and main MAS-related conclusions can be extended to other classes of applications, e.g. to balancing of electrical grid load in case of multiple generators, etc.

Some specific observations concern with the most perspective application niche for MAS technology. A class of applications where the advantages of the MAS paradigm and the MAS technology are apparent is large-scale networked objects of complex dynamic topology operating in unpredictable environment. MAS can cope with such challenging tasks using multiple simple agents where the main role belongs to the agent interactions but not to the single agent intelligence. However, the most of other applications that, until the recent times, were the subjects of MAS developer interests evidently failed. This lesson exhibits that super intelligent BDI agent systems built according to basic FIPA standards like FIPA platform reference architecture and ACL have practically no industrial perspectives. The experience of y Magenta Technology [12] and Smart Solution Ltd. [13] is a convincible evidence, for this statement.

Acknowledgment. This work is funded by the Project 14-07-00493 of the Russian Foundation for Basic Research.

References

1. Bernon, C., Chevrier, V., Hilaire, V., Marrow, P.: Applications of Self-Organising Multi-Agent Systems: An Initial Framework for Comparison. Informatica 30, 73–82 (2006)
2. Camurri, M., Mamei, M., Zambonelli, F.: Urban Traffic Control with Co-Fields. In: Weyns, D., Van Dyke Parunak, H., Michel, F. (eds.) E4MAS 2006. LNCS (LNAI), vol. 4389, pp. 239–253. Springer, Heidelberg (2007)
3. Clarke, I., Sandberg, O., Wiley, B., Hong, T.W.: Freenet: A Distributed Anonymous Information Storage and Retrieval System. In: Federrath, H. (ed.) Designing Privacy Enhancing Technologies. LNCS, vol. 2009, pp. 46–66. Springer, Heidelberg (2001)
4. Di Marso Serugendo, G., Gleizes, M.-P., Karageorgos, A.: Self-organisation and emergence in multi-agent systems: An overview. Informatica 30(1), 45–54 (2006)
5. Gorodetski, V.: Self-Organization and Multiagent Systems. Journal of Computer and Systems Sciences International 51(2), 256–281, (3), 391–409 (2012)
6. Gorodetski, V., Karsaev, O., Samoylov, V., Serebryakov, S.: Development Tools for Open Agent Networks. Journal of Computer and Systems Sciences International 47(3), 429–446 (2008)
7. Gorodetsky, V., Karsaev, O., Samoylov, V., Serebryakov, S.: P2P Agent Platform: Implementation and Testing. In: Joseph, S.R.H., Despotovic, Z., Moro, G., Bergamaschi, S. (eds.) AP2PC 2007. LNCS, vol. 5319, pp. 41–54. Springer, Heidelberg (2010)
8. Gorodetsky, V., Karsaev, O., Samoylov, V., Serebryakov, S., Balandin, S., Leppanen, S., Turunen, M.: Virtual P2P Environment for Testing and Evaluation of Mobile P2P Agents Networks. In: UBICOMM 2008, pp. 422–429. IEEE Computer Press, Valencia (2008)
9. Karuna, H., Valckenaers, P., Saint-Germain, B., Verstraete, P., Zamfirescu, C.B., Van Brussel, H.H.: Emergent Forecasting Using a Stigmergy Approach in Manufacturing Coordination and Control. In: Brueckner, S.A., Di Marzo Serugendo, G., Karageorgos, A., Nagpal, R. (eds.) ESOA 2005. LNCS (LNAI), vol. 3464, pp. 210–226. Springer, Heidelberg (2005)
10. Omicini, A., Gardelli, L.: Self-Organisation and MAS: An Introduction, http://unibo.lgardelli.com/teaching/2007-selforg-mas.pdf
11. http://space.iias.spb.su/pub/grid.pdf
12. Magenta Technology, http://magenta-technology.com/RU/about/who-we-are.html
13. Smart Solution Ltd., http://smartsolutions-123.ru/en/

Awakening Decentralised Real-Time Collaboration: Re-engineering Apache Wave into a General-Purpose Federated and Collaborative Platform

Pablo Ojanguren-Menendez, Antonio Tenorio-Fornés, and Samer Hassan

GRASIA: Grupo de Agentes Software, Ingeniería y Aplicaciones,
Departamento de Ingeniería del Software e Inteligencia Artificial,
Universidad Complutense de Madrid, Madrid, 28040, Spain
{pablojan,antonio.tenorio,samer}@ucm.es

Abstract. Real-time collaboration is being offered by plenty of libraries and APIs (Google Drive Real-time API, Microsoft Real-Time Communications API, TogetherJS, ShareJS), rapidly becoming a mainstream option for web-services developers. However, they are offered as centralised services running in a single server, regardless if they are free/open source or proprietary software. After re-engineering Apache Wave (former Google Wave), we can now provide the first decentralised and federated free/open source alternative. The new API allows to develop new real-time collaborative web applications in both JavaScript and Java environments.

Keywords: Apache Wave, API, Collaborative Edition, Federation, Operational Transformation, Real-time.

1 Introduction

Since the early 2000s, with the release and growth of Wikipedia, collaborative text editing increasingly gained relevance in the Web [1]. Writing texts in a collaborative manner implies multiple issues, especially those concerning the management and resolution of conflicting changes: those performed by different participants over the same part of the document. These are usually handled with asynchronous techniques as in version control systems for software development [2] (e.g. SVN, GIT), resembled by the popular wikis.

However, some synchronous services for collaborative text editing have arisen during the past decade. These allow users to write the same document in real-time, as in Google Docs and Etherpad. They sort out the conflict resolution issue through the Operational Transformation technology [3].

These services are typically centralised: users editing the same content must belong to the same service provider. However, if these services were federated, users from different providers would be able to edit contents simultaneously. Federated architectures provide multiple advantages concerning privacy and power

© Springer International Publishing Switzerland 2015 269
S. Omatu et al. (eds.), *Distributed Computing & Artificial Intelligence, 12th Int. Conference,*
Advances in Intelligent Systems and Computing 373, DOI: 10.1007/978-3-319-19638-1_31

distribution between users and owners, and avoid the isolation of both users and information in silos [4].

The rest of this paper is organised as follows: first, Operational Transformation frameworks' state of the art is outlined in Section 2. Section 3 depicts the reengineering approach and used technologies and tools. Concepts of the Wave Platform and changes made are explained in Section 4. Afterwards, the results are discussed in Section 5. Finally, conclusions and next steps are presented in Section 6.

2   State of the Art of Real-Time Collaboration

The development of Operational Transformation algorithms started in 1989 with the GROVE System [5]. During the next decade many improvements were added to the original work and a *International Special Interest Group on Collaborative Editing* (SIGCE) was set up in 1998. During the 2000s, OT algorithms were improved as long as mainstream applications started using them [6].

In 2009, Google announced the launch of Wave [7] as a new service for live collaboration where people could participate in conversation threads with collaborative edition based on the Jupiter OT system [8]. The Wave platform also included a federation protocol [9] and extension capabilities with robots and gadgets. In 2010 Google shutted down the Wave service and released the main portions of the source code to the Free/Open Source community. Since then, the project belongs to the Apache Incubator program and it is referred as Apache Wave. Eventually, Google has included Wave's technology on some other products, such as Google Docs. Despite its huge technological potential, the final product had a very constrained purpose and hardly reusable implementation.

Other applications became relevant during that time, such as the Free Libre Open Source Software (FLOSS) Etherpad. However, it was mostly after the Google Wave period when several FLOSS OT client libraries appeared, allowing integration of real-time collaborative edition of text and data in applications. The most relevant examples are outlined as follows.

TogetherJS [10] is a Mozilla project that uses the WebRTC protocol for peer-to-peer communication between Web browsers in addition to OTs for concurrency control of text fields. It does not provide storage and it needs a server in order to establish communications. It is a JavaScript library and uses JSON notation for messages.

ShareJS [11], is a server-client platform for collaborative edition of JSON objects as well as plain text fields. It provides a client API through a JavaScript library.

Goodow [12], is a recent FLOSS framework copying the Google Drive Real-Time API with additional clients for Android and iOS, while providing its own server implementation.

On the other hand, Google provides a *Real-Time API* as part of its Google Drive SDK [13]. It is a centralised service handling simple data structures and plain text.

In general, these solutions are centralized. Despite their claim of focusing in collaboration, users from different servers cannot work or share content. They just provide concurrency control features without added value services like storage and content management. They mostly allow collaborative editing of simple plain text format.

3   Reengineering: Technologies and Tools

This section summarises the procedure followed to re-engineer and build a generic Wave-based collaborative platform, together with the technologies used.

Wave in a Box [14] is the FLOSS reference implementation of the Apache Wave platform, which supports all former Google Wave protocols and specifications [15] and includes both implementations of the Server and the Client user interface. Most of its source code is original from Google Wave and was provided by Google, although it was complemented with parts developed by community contributors.

In particular, the Client part has been used as ground to develop the new API, with same technologies: Java and the Google Web Toolkit (GWT) FLOSS framework [16]. The Client is written in Java but is compiled and translated into JavaScript by GWT in order to be executed in a web-browser.

The lack of technical documentation forced to perform a preliminar extensive source code analysis outcoming documentation and UML diagrams. Then, initial developments within the Wave client were performed to assess whether the Apache Wave implementation could be used to develop new applications within fair parameters of quality and cost.

New general functionality was added in separated components, on top of underneath layers such as the federation protocol and server storage system. This has proved the feasibility of reusing the original code and Wave core features. The new source code is GWT-agnostic in order to be reusable in Java platforms. GWT is used to generate just the top JavaScript layer.

Concerning software testing, the JavaScript framework Jasmine [17] was used in addition to existing unit tests. The developed test suite for the API attacks the public API functions in a web-browser environment testing new layers together with the rest of the architecture stack.

The development has been tracked and released in a public source code repository [18]. It includes documentation and examples on how to use the API. Besides, during the development process, several contributions have been made to the Apache Wave Open Source community, in the form of source code patches, documentation and diagrams.

4   Generalising the Wave Federated Collaborative Platform

This section shows the fundamentals of the Wave platform and how they have been used to turn Wave into a general-purpose platform unlike the former conversation-based one.

4.1 Conversations: Wave Data Models and Architecture

This subsection exposes the conversation approach of Apache Wave, its data models and general architecture. From a logical point of view, the Wave platform handles two data models: the *Wave Data Model* [19] and the *Wave Conversational Model* [20]. First, the Wave Data Model defines general data entities used within the platform:

- **Participant**: user of the platform. It may be a human or a robot [7].
- **Document**: recipient of collaborative real-time data.
- **Wavelet**: set of Documents shared by a set of Participants.
- **Wave**: set of Wavelets sharing the same unique identifier.

Documents are the smallest entity that can store data which can be edited in a collaborative way. Documents are logically grouped in Wavelets. In addition, a Wavelet has a set of participants, which are able to access –read and edit– those Documents. Finally, the Wave Data Model defines the Wave concept as just a group of Wavelets sharing the same Wave identifier.

Data is represented in XML and the Wave Operational Transformation (OT) system [21] provides the concurrency control and consistency maintenance for editing this XML in a Document by multiple users at the same time. It also generates events to notify changes to other parts of the system, locally or remotely. XML is used to represent two types of data: rich *Structured Text* in a HTML-like format and *Abstract Data Types* (ADTs) like maps, lists, sets, etc.

On the other hand, the Wave Conversational Model was defined to manage *Conversations*, the major concept of the Wave product. A Conversation is a Wavelet having a set of participants, and a set of Documents supporting the Conversation Thread. A Conversation Thread is compound of Documents storing paragraphs as Structured Text and a Document storing the tree-structure of those paragraphs using ADTs. Conversation Metadata is also stored as a Document using ADTs. This schema is summarised below in Figure 2.

Those data models are implemented in separated layers of the Wave Client architecture as it is shown in Figure 1. All the components of this architecture are developed, packaged and deployed as an unique Java/GWT application.

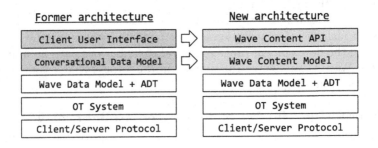

Fig. 1. Wave Client architecture

4.2   General-Purpose Collaboration: Generalising the Wave Data Model and Architecture

Last section outlined the Wave's general data model that could be used in alternative ways. This section introduces a general approach to use it (the Wave Content Model) and a mechanism to consume it (the Wave Content API).

The Wave Content Model. This is a new general-purpose and dynamic data model replacing the former Wave Conversational Model (see Figure 2). It allows to edit Abstract Data Types collaboratively on real-time by different users.

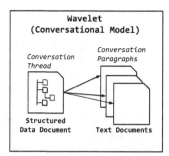 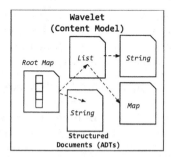

Fig. 2. Wave Data models

The main task was to develop a suitable layer that allows to dinamically create and handle ADTs within Documents of a Wavelet. ADTs are Java classes managing part of the Document content in a particular way. They can be combined declaring new compound types. The Conversation Thread implementation is an example of inmutable compound type as long as inner data structure can't be change on execution.

However, to provide a dynamic composition of ADTs, a Composite pattern [22] is applied. Such pattern defines a hierarchy of data types that can be combined and nested: map, list and string values. Each type is backed by the matching ADT; these new data types control where and how to create and handle ADTs instances within Documents:

This dynamic model is named the *Wave Content Model*. For the shake of clarity, a Wave is now called *Content Instance*, and it provides a main Wavelet where arbitrary data types can be stored dynamically starting from a provided root map. Applications can add new instances of lists and maps to this root or their nested lists and maps, and eventually store string values.

From the architecture perspective, all existing components related to Conversations have been discarded. In particular, the two top layers of the architecture have been replaced (see Figure 1). First, the Wave Conversational Model by the new Wave Content Model. Second, in order to consume the new model in a general way -not just by one single application- the old client is replaced by an API as it is depicted below.

The Wave Content API. With the new Wave Content Model any application could use collaborative data structures. However, according to the technology used in the Apache Wave implementation, just new Java or GWT Web Applications could use them directly. With the aim of delivering these new capabilities to any Web Application developed in any technology, a JavaScript API has been built.

Although GWT eventually translates Java code into JavaScript, this is not suitable to be consumed directly by non-GWT JavaScript code in a web-browser environment, for several reasons: the exception handling is not understood by outer code, and GWT-generated JavaScript syntax is obfuscated.

JSNI and Overlay Types [16] are features of GWT allowing to write arbitrary native JavaScript code and objects integrated transparently with Java code. These features have been used to develop a native JavaScript layer, following the Proxy pattern, which exposes the Wave Content Model functionality as an API. A summary of the features provided by the API follows:

Session management: controls user authentication and life cycle of content instances and

Content Instance management: Maps, lists and strings are created through a provided factory and a root map is provided as a hook.

Data types management: exposes type-specific operations such as the addition of an element to a list or getting map keys.

5   Discussion

This paper introduces the only federated platform for real-time collaboration available nowadays. However, using Wave as its starting point involves some issues.

There are several critiques concerning the complexity of the Wave OT system [11]. Its highly complex implementation –together with the lack of good documentation– causes the maintenance of the source code to be a hard task. However, OT systems are inherently complex and to design OT-based languages and control algorithms require knowledgeable people.

Some existing OT implementations are simpler, using the JSON language and a smaller set of OT operations [11] [12]. In contrast, Wave uses XML dialects that supports both, *rich text* edition straight away and structured data, instead of just plain text and JSON. Wave is the only open OT system providing full rich text and text annotations.

Regarding the API design, it works with data structures (map, list) –as the Google Drive Real-Time API–, in contrast with direct JSON objects. It is hard to conclude which approach is more appropriate for third-party developers since the lack of information about the adoption level and critics in both cases.

Java/GWT as implementation language and Jetty as the HTTP server [23], could be seen as a pitfall as long as nowadays trends are to develop using JavaScript directly and to use high-performance servers. However, GWT is still a highly adopted and mature project which a strong community. And from the server perspective, it would be easy to adapt the code to run in non-blocking IO servers [24], extending the life of the original source code.

6 Concluding Remarks

A federated platform to develop web applications with real-time collaborative editing capabilities has been presented in the previous sections. It has been developed as a generalisation of the Apache Wave platform, the FLOSS project formerly known as Google Wave.

Nowadays there is no other federated (or distributed) platform for real-time collaboration. Moreover, this work takes the Wave Federation Protocol further, making it a general protocol. Thus, now on top of the Wave Content Model anyone can define new inter-operable collaborative data formats for text documents, spreadsheets, drawings, games, social media, social activity, etc. New applications could adopt them using an existing provider or becoming a new one. Providers can scale on interoperability since OT storage system is agnostic from underlaying content. Clients just need to be aware of data formats.

The provided API is a functional alternative to existing collaborative platforms. It provides a full-stack of software ready to be deployed, with functionalities only comparable with the proprietary Google Drive Real-Time API. Features such as the participation model, content storage and capabilities to search and manage contents, are already included in the Apache Wave platform but not implemented in any alternative.

The API is offered in JavaScript, to be integrated in web applications. Besides, a Java version will be soon released, in order to allow also Android and Java applications to have collaborative capabilities.

This work shows the unexplored high potentials of Google's original development, in spite of its complexity and lack of documentation. Thus, this work steps out engineering challenges for reusing Apache Wave and we hope it paves the way for other researchers and developers.

Acknowledgments. This work was partially supported by the Framework programme FP7-ICT-2013-10 of the European Commission through project P2Pvalue (grant no.: 610961).

References

1. West, J.A., West, M.L.: Using Wikis for Online Collaboration: The Power of the Read-Write Web. John Wiley & Sons (2008)
2. Berliner, B.: CVS II: Parallelizing software development. In: USENIX Winter 1990 Technical Conference, pp. 341–352. USENIX, Berkeley (1990)
3. Sun, C., Ellis, C.: Operational transformation in real-time group editors: Issues, algorithms, and achievements. In: Proceedings of the 1998 ACM Conference on Computer Supported Cooperative Work, pp. 59–68. ACM, New York (1998)

4. Yeung, C., Liccardi, I., Lu, K., Seneviratne, O., Berners-Lee, T.: Decentralization: The future of online social networking. In: W3C Workshop on the Future of Social Networking Position Papers, W3C (2009)
5. Ellis, C.A., Gibbs, S.J.: Concurrency control in groupware systems. In: Proceedings of the 1989 ACM SIGMOD International Conference on Management of Data, SIGMOD 1989, pp. 399–407. ACM, New York (1989)
6. Bigler, M., Raess, S., Zbinden, L.: ACE - a collaborative editor, http://sourceforge.net/projects/ace/
7. Ferrate, A.: Google Wave: Up and Running. O'Reilly Media, Inc. (2010)
8. Nichols, D.A., Curtis, P., Dixon, M., Lamping, J.: High-latency, low-bandwidth windowing in the jupiter collaboration system. In: Proceedings of 8th ACM Symposium on User Interface and Software Technology, pp. 111–120. ACM, New York (1995)
9. Baxter, A., Bekmann, J., Berlin, D., Gregorio, J., Lassen, S., Thorogood, S.: Google Wave Federation Protocol Over XMPP (2009), http://wave-protocol.googlecode.com/hg/spec/federation/wavespec.html
10. Mozilla Labs: Togetherjs, https://togetherjs.com/
11. Joseph, G.: ShareJS, http://sharejs.org/
12. Chuanwu, T.: Google docs–style collaboration via the use of operational transforms, https://github.com/goodow
13. Google Inc.: Google Drive SDK: Realtime API, https://developers.google.com/drive/realtime/
14. North, A.: Google Wave Developer Blog: Wave open source next steps: Wave in a Box, http://googlewavedev.blogspot.com.es/2010/09/wave-open-source-next-steps-wave-in-box.html (2010)
15. Google Inc.: Google Wave Protocol, http://www.waveprotocol.org/
16. Cooper, R., Collins, C.: GWT in Practice. Manning Publications (2008)
17. Pivotal Labs: Jasmine, Behavior-Driven JavaScript, http://jasmine.github.io/
18. Ojanguren-Menendez, P.: Real-time collaboration API for Wave, https://github.com/P2Pvalue/incubator-wave
19. North, A.: Wave model deep dive (2010), https://cwiki.apache.org/confluence/display/WAVE/Wave+Summit+Talks
20. Gregorio, J., North, A.: Google Wave Conversation Model (2009), http://wave-protocol.googlecode.com/hg/spec/conversation/convspec.html
21. Lassen, S., Mah, A., Wang, D.: Google Wave Operational Transformation (2010), http://wave-protocol.googlecode.com/hg/whitepapers/operational-transform/operational-transform.html
22. Gamma, E., Helm, R., Johnson, R., Vlissides, J.: Design Patterns: Elements of Reusable Object-Oriented Software. Pearson Education (1994)
23. The Jetty Project: Jetty, http://www.eclipse.org/jetty/
24. Roth, G.: Architecture of a highly scalable nio-based server (2007), https://today.java.net/pub/a/today/2007/02/13/architecture-of-highly-scalable-nio-server.html

Efficient P2P Approach for a Better Automation of the Distributed Discovery of SWs

Adel Boukhadra, Karima Benatchba, and Amar Balla

Ecole National Supérieur d'Informatique, BP 68M, Oued-Smar, Alger, Algérie
{a_boukhadra,k_benatchba,a_balla}@esi.dz

Abstract. In this paper, we present a scalable approach for visualizing and browsing the search space of available Web services to effectively and efficiently resolve the problem of distributed discovery for Semantic Web services (SWs). We investigate the use of matching technique of ontologies OWL-S, an approach to provide a collaborative mechanism to discover basic SWs distributed among all peers in a purely distributed and heterogeneous P2P network. Our scalable approach is based on the matching technique of OWL-S in order to reduce the time complexity of the distributed discovery for SWs with respect to their semantic similarity, to simplify the management of the P2P network, to optimize the ratio of service exchange and to ensure the quality of service. The network peers offer their SWs to other ones in a distributed and heterogeneous P2P computing, and are able to use distant services to improve system responses to requests by the system that are asked. The experimental results show that the proposed approach enhances the network scalability while providing good overall performances. Also, we show that our approach can perform more effective and efficient distributed discovery of SWs with low cost in P2P computing.

Keywords: SWs, Distributed Discovery, P2P Computing, Matching of Ontology, Similarity Measure, OWL-S process model.

1 Introduction

In SOC[1], SWs can be viewed as an elementary unit of the whole distributed application. That is, a distributed application consists of some number of distributed SWs that are discovered together, which may be unavailable or inaccessible in order to provide the required functionality. Web service discovery is the process of finding multiple distributed SWs for a given task that can meet the needs of increasingly complex requests of user to ensure high availability of services [4] [5] [9].

Actually, it is not always easy to find Web services that match with users queries in an open and highly dynamic environment, such as Internet for enhancing the computational performance of automated service discovery. There is increasingly need for automatic discovery of service. This Web service discovery

---

[1] **SOC:** Service-Oriented Computing.

© Springer International Publishing Switzerland 2015
S. Omatu et al. (eds.), *Distributed Computing & Artificial Intelligence, 12th Int. Conference,*
Advances in Intelligent Systems and Computing 373, DOI: 10.1007/978-3-319-19638-1_32

satisfying the query becomes even more complex. This is due to the unavoidable limitations of todays standard specifications, e.g., WSDL[2] , which do not encompass such aspects. WSDL descriptions are limited to problems related to connectivity and functional properties. They cannot, by themselves, guarantee the success of the distributed discovery of service. The semantic of stateful services represents a very important issue to be considered during discovery, to provide users with an additional means to refine the search in such a diverse environment [2] [3].

Currently UDDI[3] registries are the dominating technological basis for Web services discovery. They allow business compliance and reuse, and in perspective they could provide control over data and facilitate Web services lifecycle management. But existing registries are still small and mostly private; there is no control over provided information, qualitative characteristics of Web services and ability of quality-based retrieval. The discovery supported by UDDI is inaccurate as services retrieved may be inadequate due to low precision and low recall [2] [5] [6].

Web services discovery based on P2P computing require special attention from collaboration and interoperability in a large-scale distributed computing. One of the most promising applications of the Web services discovery based in the P2P computing is to optimize the bandwidth of peers to distribute a large amount of SWs through the network. It has often been known as the ultimate solution for scalability [3] [9] [11] [12].

In this paper, we will focus on the contribution an efficient approach for an automatic and a distributed discovery of SWs, which address some aspects, related to problems the time complexity of collaboration in the process of automatic discovery for SWs in a large-scale distributed computing. For this purpose, our approach is based on P2P computing that proved to be scalable, efficient and robust solutions for distributed discovery of SWs. Indeed, we propose a matching technique of ontology OWL-S process model, in order to achieve semantic interoperability in a distributed and heterogeneous P2P network, minimize false responses, and improve the overall quality of results. The proposed approach aims to optimize considerably the service exchange rate and the network traffic, while providing good overall performances. Moreover, it enhances significantly the network scalability, where a variety of peers can communicate with each other on a variety of large-scale P2P networks.

The rest of the paper is organized as follows. We first study in Section 2 some existing research works in our field. Section 3 describes in detail our P2P approach and presents our main contributions in order to strengthen the power of collaboration for peers using the network resources effectively and supporting scalability in an inte-grated and convenient way. In Section 4, we report the encouraging results of our experimental evaluation of the introduced approach. Finally, the last Section concludes and sketches avenues of future work.

---

[2] **WSDL:** Web Services Description Language.
[3] **UDDI:** Universal Description Discovery and Integration.

2   Related Works

Recently, several works improve the decentralized discovering methods of SWs in the context of P2P networks. In [12], authors presented a model system which combines the advantages of centralized and decentralized structures. It distributes the functions of UDDI to the local Web peers, and groups Web peers and public Web peers. In [11], a discovery framework in P2P environment is proposed to replace traditional UDDI, by distributing the functions of the UDDI among all the peers in the P2P network. In [2], propose the P2P-based Semantic Driven Service Discovery (P2P-SDSD) framework to enable cooperation and communication based on a semantic overlay that organizes semantically the P2P-integrated knowledge space and emerges from local interactions between peers. In [6], present an intelligent service matchmaker, called iSeM, for adaptive and hybrid semantic service selection that exploits the full semantic profile in terms of signature annotations in description logic SH and functional specifications in SWRL. In [9], present a distributed approach to SWs publication and discovery by leveraging structured P2P network. In this work, the computers concerned constitute a P2P network to maintain the sharable domain and service ontologies to facilitate SWs discovery.

3   Proposed Approach for Distributed Discovery of SWs

The principal goal of our work is to develop a distributed approach allowing the implementation of automatic discovery of SWs in a large-scale P2P network. In this architecture, each peer implements a set of SWs described semantically with OWL-S process model. Figure 1 shows the architecture of our approach.

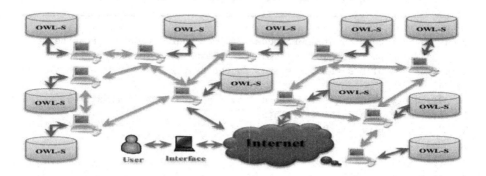

Fig. 1. A P2P Approach for Distributed Discovery of SWs

The distributed discovery of SWs became active research topics, thus P2P networks have been proposed based on the ultimate success of scalability to share and discover Web services effectively in a distributed and heterogeneous network. But, the introduction of P2P computing in distributed discovery of

SWs without considering its requirements and constraints may result in poor performance.

For this reason, we combine hybrid P2P system with a pure P2P system which reinforces the power of collaboration for the peers by the means of an effective use of the resources of P2P network and the management of scalability in an integrated and practical way. Our P2P approach must be able to minimize the traffic of peers to reduce the amount of message transferred by the peers and optimized bandwidth, and must also be able to achieve an acceptable performance (see Figure 1).

In this approach, the network peers are well structured. We can distinguish two types of peers: Child Peers and Super Peers. A Child Peer is connected and communicates with only Super Peer. The Super Peers, in addition to their role which is the same as that of Child Peers, act as directories of their kind. Super Peers manage two lists: one with information about the nodes that their Child is connected, and the other records information about the great neighboring peers. Child Peers provides a pure P2P network (see Figure 1).

In this first stage of our work, we describe a strategy for discovering SWs. This phase is to provide a distributed discovery algorithm of SWs. We assume that our epidemic discovery algorithm relies on the matching technique of OWL-S to discover new personalized SWs distributed on all peers in the P2P network. The main idea of our distributed discovery algorithm is to avoid the problem of overhead generated by control messages, to adapt existing OWL-S process model implementation to P2P systems and ensures the operation progress of the discovery in the P2P network. Each peer runs this algorithm, when it receives a user request for discovering SWs. Our main algorithm is defined as:

Algorithm 1. Distributed Discovery of SWs ();

```
Input: Upon reception of user request (R) at peer such as Search the
       SWs:(Input, Output, Precondition, Result and TextDescription).
Output: A set of SWs which responds to the user request.
Begin
1:  Discovery-Matching (); // to discover a local SWs in this peer
2:  If (There is a local SWs) Then:
3:  Send the Favorable Response; and Go to End;
4:  End If;
5:  Calculate the TTL;
6:  If (TTL > 0) Then:
7:  TTL: = TTL - 1;
8:  P2P-Discovery (); // Send the request to all the peers in the
9:  Else // if TTL=0; // Pure P2P networks
10: Drop the request; and Go to End;
11: End If;
End.
```

Each Super Peer in the P2P network executes the following main algorithm when receiving a request. Initially, each Super Peer executes the main algorithm from step 1 to step 3 where it tries to response to the request locally (local basic SWs). If there is not a possibility to answer the request locally, the Super Peer starts a new P2P Discovery (step 8 of the main algorithm).

To send a request that is not satisfied to relevant sources of data, the Super Peer initiating the request sends identical messages to its Child Peers. Each message contains the identifier of the sending peer and a parameter TTL[4], which for example is a counter that is initialized with a positive integer. This parameter represents the maximum number of hops. That is to say, the path length of each message in terms of number of Child Peers crossed in a pure P2P network.

After having received the message by a Super Peer, the Child Peer executes the request locally and sends the response to the transmitting Super Peer. In addition, if the Child Peer has no positive results, it decreases the TTL value by 1 (step 7 of the main algorithm). If the TTL value becomes 0, than destroy the message. The Child Peer forward the message contains the new TTL value to its neighbors.

By repeating this process, among all peers situated at a distance having a value less than the initial value of TTL will receive the message. Other Child Peer in the pure P2P system does not receive the message even if they have valid responses. If all Child Peers fail, the request will be forwarded to the next Super Peer where the process is repeated until all peers are accessed or a positive result is given. If no positive result is found, the result sent by the last peer consultation will be negative.

Once the goal described in the request is matched, a result is sent back to the client. The SWs can then be executed on the respective peer and results are sent back to the Service Requester. In the event that a peer has a positive answer, a message is returned to the client. The SWs can be run on the platform of the peer that holds and the result is sent to the applicant Web service.

In a pure P2P system, if all remote peers sends the query by flooding, than the network becomes overloaded reducing the performance of the P2P system as well, and usually a peer does not have incentives to send query to remote peer, it focuses on the peers in its neighborhood. To optimize the flooding, some criteria have to be defined to select the peer for which the request will be sent. Thus in this work, the mechanism is based on constrained flooding, where requests are passed between neighbors in the pure P2P system with a TTL. Each peer maintains a list of neighbors in the pure P2P system. One peer must limit the number of simultaneous transfers of request. The restriction of the neighborhood can reduce the routing overhead, however, that causes a lack of messages diversity in the pure P2P network. However, to achieve this objective, we propose to use a hops peers and group peers. In our solution, we assume that hops peers (ρ_{Hops}) are the only ones responsible for sending request to group peers (ρ_{Group}).

---

[4] **TTL:** Time To Live

The recent study by Stutzbach et al in [10], explains how a Weibull distribution can be used to better describe the session times of peers in a pure P2P networks. In a network of N peers with D being the average node degree. The shape α and scale λ parameters can be used to describe exponential distributions when $\alpha = 1$, every t seconds. We compute in what follows the maximum network size of supported peers supported by the hops peers :

$$\rho_{Hops} = 1/\exp^{-\lambda t(N/D)^{\alpha}} \tag{1}$$

The group peers (ρ_{Group}) is computed by considering the number of groups participated by the forwarding peer. The (ρ_{Group}) between two randomly chosen peers on any pure P2P network is approximated as follows [8]:

$$\rho_{Group} = \frac{\ln[(N-1)(Z_2 - Z_1) + Z_1^2] - \ln(Z_1)^2}{\ln(Z_2/Z_1)} \tag{2}$$

Where Z_i is the number of neighbors which are i hops away from the originator peer, and N is an estimation of the total number of peers available in the P2P network. By taking into account these requirements, we have defined the following evaluating function for:

$$\mathbf{TTL} = \ln(\rho_{Hops} + \rho_{Group}) \tag{3}$$

The procedure (Discovery-Matching ()) made in step 1 of Algorithm 1, it used for the automatic discovery of SWs [3].

4   Experimental Evaluation

In this section, we conduct the different experiments in order to evaluate and verify the performance and efficiency of the proposed approach. The experiments were conducted on a desktop PC with an Intel core i5-3570 3.5 GHZ processor, 4G RAM, Windows 7, and Java Standard Edition V1.7.0. To verify the performances of our approach, we must run our algorithm in a large scale network and experimentally evaluated it. For this reason, we are using PeerSim simulator [7] to simulate a P2P network. To evaluate the efficiency and scalability of our approach, we must compare our approach with a flooding protocol that is Gnutella P2P protocol [1].

In order to validate our proposed approach, we performed simulations with the following parameters: The hybrid P2P network of machines that we used in our experiments consists of twenty peers. The experimentation has been performed with the number of peers that varies in the range [100-1500], with increments of 200 in the pure P2P network. Experimentation results will be analyzed in the following: The analysis of generated requests has been performed on a series of 30 simulations. In each simulation we ran, a request has been submitted and we have collected the results obtained with both our approach and Gnutella P2P protocol on the same simulated network.

The results of the hit ratio are shown in the graphic of Figure 2 by setting different TTL values. We also observed that the value of hit ratio slightly increases when the TTL increases. For instance, the hit ratio obtained is about 80% of our approach better than the value of 20% obtained by the Gnutella P2P protocol when TTL = 10. As shown by this graphic, the results obtained by our approach are very satisfactory.

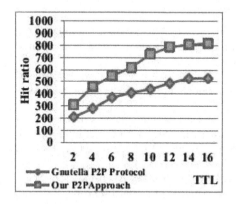

Fig. 2. Hit ratio vs the TTL

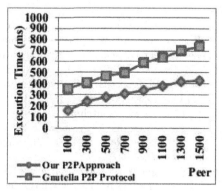

Fig. 3. Execution Time vs the Peer

To better demonstrate the effectiveness of our approach, a performance analysis has been performed to measure the variation in the execution time according to the number of peer. The execution time has been measured using The Eclipse Test and Performance Tools Platform (http://www.eclipse.org/tptp/).

The graph in Figure 3 summarizes some of the performance test results. We notice that the execution time increases linearly with the increasing number of peers. The results reported in this figure represent the average execution times found for each given request. More clearly shown in Figure 3, our approach performs better than the Gnutella. The results are very promising.

5   Conclusion and Future Work

In this paper, we presented a P2P purely decentralized approach that supports the distributed and scalable discovery runtime environments in which SWs are distributed among all peers of the network. We described a distributed cooperative strategy to provide a distributed algorithm that resolves the problem of flooding-based methods, while reducing the network traffic, and accelerates the distributed discovery of SWs. One of the core advantages to this approach is its applicability in high-scale environ-ments, while minimizing the over-provisioning of resources. This will allow designing a discovery approach of a large positive, satisfying two criteria that are important: the accuracy and completeness.

In a future work, we intend to extend the approach to propose strategies for the selection and composition of Web services by supporting a tool for information retrieval. We also plan to apply the proposed solution in a larger network of peers, and we think it will be more appropriate to develop a P2P system of disaster management.

References

1. Barjini, H., Othman, M., Ibrahim, H., Udzir, N.I.: Hybridflood: minimizing the effects of redundant messages and maximizing search efficiency of unstructured peer-to-peer networks. Cluster Computing 17(2), 551–568 (2014)
2. Bianchini, D., De Antonellis, V., Melchiori, M.: P2p-sdsd: on-the-fly service-based collaboration in distributed systems. International Journal of Metadata, Semantics and Ontologies 5(3), 222–237 (2010)
3. Boukhadra, A., Benatchba, K., Balla, A.: Ranked matching of owl-s process model for distributed discovery of sws in p2p systems. In: 2014 17th IEEE International Conference on Network-Based Information Systems (NBiS), pp. 106–113. IEEE (2014)
4. Fragopoulou, P., Mastroianni, C., Montero, R., Andrjezak, A., Kondo, D.: Self-* and adaptive mechanisms for large scale distributed systems. In: Grids, P2P and Services Computing, pp. 147–156. Springer (2010)
5. Furno, A., Zimeo, E.: Efficient cooperative discovery of service compositions in unstructured p2p networks. In: 2013 21st Euromicro International Conference on Parallel, Distributed and Network-Based Processing (PDP), pp. 58–67. IEEE (2013)
6. Klusch, M., Kapahnke, P.: iSeM: Approximated reasoning for adaptive hybrid selection of semantic services. In: Aroyo, L., Antoniou, G., Hyvönen, E., ten Teije, A., Stuckenschmidt, H., Cabral, L., Tudorache, T. (eds.) ESWC 2010, Part II. LNCS, vol. 6089, pp. 30–44. Springer, Heidelberg (2010)
7. Montresor, A., Jelasity, M.: Peersim: A scalable p2p simulator. In: IEEE Ninth International Conference on Peer-to-Peer Computing, P2P 2009, pp. 99–100. IEEE (2009)
8. Newman, M.E.J., Strogatz, S.H., Watts, D.J.: Random graphs with arbitrary degree distributions and their applications. Physical Review E 64(2), 026118 (2001)
9. Si, H., Chen, Z., Deng, Y., Yu, L.: Semantic web services publication and based discovery in structured p2p network. Service Oriented Computing and Applications 7(3), 169–180 (2013)
10. Stutzbach, D., Rejaie, R.: Understanding churn in peer-to-peer networks. In: Proceedings of the 6th ACM SIGCOMM Conference on Internet Measurement, pp. 189–202. ACM (2006)
11. Sun, F., Hao, S.: A discovery framework for semantic web services in p2p environment. In: 2010 International Conference on Electrical and Control Engineering (ICECE), pp. 44–47. IEEE (2010)
12. Zhu, Z., Hu, Y., Lan, R., Wu, W., Li, Z.: A p2p-based semantic web services composition architecture. In: IEEE International Conference on e-Business Engineering, ICEBE 2009, pp. 403–408. IEEE (2009)

Exploration of Unknown Territory via DALI Agents and ASP Modules

Stefania Costantini, Giovanni De Gasperis, and Giulio Nazzicone

Dipartimento di Ingegneria e Scienze dell'Informazione e Matematica (DISIM),
Universitá degli Studi dell'Aquila, Italy
{stefania.costantini,giovanni.degasperis}@univaq.it,
giulio.nazzicone@graduate.univaq.it

Abstract. This paper describes a new development related to the DALI logic agent-oriented programming language/framework, and discusses a case-study. In particular, DALI agents have been equipped with ASP (Answer Set programming) modules, which allow for affordable and flexible planning capabilities, and with the possibility to select plans according to a suite of possible preferences. The case-study concerns a DALI MAS (Multi-Agent System) whose composing agents cooperate for achieving the task of exploring an unknown territory.

1 Introduction

Adaptive autonomous agents are capable of adapting to unknown and potentially changing behavior. This requires agents to be capable of various forms of commonsense reasoning and planning. Since [1], we advocated agent architectures capable of smooth integration of several modules/components representing different behaviors/forms of reasoning, possibly based upon different formalisms. Therefore, the overall agent's behavior can be seen as the result of dynamic combination of these behaviors, also in consequence of the evolution of the agent's environment. We proposed in particular to adopt Answer Set Programming (ASP) modules, where ASP (cf., among many, [2,3,4] and the references therein) is a successful logic programming paradigm suitable for planning and reasoning with affordable complexity and many efficient implementation freely available [5]. We implemented this solution in the DALI [6,7,8,9] agent-oriented language. We have recently enhanced the integration by adopting ASP modules for planning purposes, allowing an agent to choose among the various plans that can be obtained by means of suitable preferences.

In this paper, we show the effectiveness of this solution by means of a case-study where DALI agents cooperate in order to explore an unknown territory upon occurrence of some kind of catastrophic event (earthquake, fire, flooding, terrorist attack, ect.). For simplicity, we represent the territory (also called "area") as a $N * N$ chessboard, where some squares are marked as unreachable/forbidden, and are therefore considered as "holes" in the chessboard. This represents the fact that the agents may be notified by an authority of the actual impossibility of traversing that location because of some kind of obstruction. The robot that each agent employs for exploration is represented as a chess' knight piece, which performs knight leaps.

© Springer International Publishing Switzerland 2015
S. Omatu et al. (eds.), *Distributed Computing & Artificial Intelligence, 12th Int. Conference,*
Advances in Intelligent Systems and Computing 373, DOI: 10.1007/978-3-319-19638-1_33

In this way, the problem can be modeled as a variant of the well-known "knight tour with holes" problem, for which solutions exist, also in ASP. As however the Hamiltonian path option results too heavy with reasonable instance size (actually, it is too heavy for size more than 8), we resorted to sub-optimal solutions which adopt soft constraints in order to visit each square as few times as possible.

We propose a solution based upon a MAS (Multi-Agent System), instead of a monolithic software solution. This because we consider important that each software component, that we implement as an agent, should partially keep its autonomy during asynchronous event processing. In fact, in this way each agent can be enriched with higher level reasoning/control behaviors that can coexists with the planning/executing activity. The MAS solution also permits to distribute the computational effort and increase overall robustness by means of advanced features such as self-monitoring and self-diagnostic, as shown in [10]. Even more importantly, as discussed below the MAS can be based upon a controller agent which, if necessary, is able to partition the territory to be explored into possibly overlapping sections of reasonable size, so that: (i) the sub-plan for each section's exploration is devised within a reasonable deadline, and can be chosen according to preferences related to that specific area's features; (ii) sections can be then explored in parallel by dedicated agents, embedded in mobile devices such as robots, drones or similar. The qualitative aspects of the proposed solution therefore consists in: (1) the interaction between the MAS and the ASP module(s), in terms of preferences for choosing among possible plans; (2) the flexible MAS architecture, that can be customized in order to cope with real-world problems rather than toy instances.

2   The DALI Language and Architecture

DALI [7,8] (cf. [9] for a comprehensive list of references) is an Agent-Oriented Logic Programming language. DALI agent is triggered by several kinds of events: external events, internal, present and past events.

External events are syntactically indicated by the postfix E. Reaction to each such event is defined by a reactive rule, where the special token $:>$. The agent remembers to have reacted by converting an external event into a *past event* (postfix P). An event perceived but not yet reacted to is called "present event" and is indicated by the postfix N.

In DALI, **actions** (indicated with postfix A) may have or not preconditions: in the former case, the actions are defined by actions rules, in the latter case they are just action atoms. An action rule is characterized by the new token $:<$. Similarly to events, actions are recorded as past actions.

Internal events is what make a DALI agent agent proactive. An internal event is syntactically indicated by the postfix I, and its description is composed of two rules. The first one contains the conditions (knowledge, past events, procedures, etc.) that must be true so that the reaction (in the second rule) may happen. Thus, a DALI agent is able to react to its own conclusions. Internal events are automatically attempted with a default frequency customizable by means of directives in the initialization file.

The DALI communication architecture implements the DALI/FIPA protocol, which consists of the main FIPA primitives, plus few new primitives which are particular to DALI. The architecture also includes a filter on communication based on ontologies and forms of commonsense reasoning.

The DALI programming environment at current stage of development [6] offers a multi-platform folder environment, built upon Sicstus Prolog programs, shells scripts, Python scripts to integrate external applications, a JSON/HTML5/jQuery web interface to integrate into DALI applications, with a Python/Twisted/Flask web server capable to interact with A DALI MAS at the backend.

3   Answer Set Programming in a Nutshell

"Answer set programming" (ASP) is a well-established logic programming paradigm adopting logic programs with default negation under the *answer set semantics*, which [11,12] is a view of logic programs as sets of inference rules (more precisely, default inference rules). In fact, one can see an answer set program as a set of constraints on the solution of a problem, where each answer set represents a solution compatible with the constraints expressed by the program. For the applications of ASP, the reader can refer for instance to [2,3,4]. Several well-developed answer set solvers [5] that compute the answer sets of a given program can be freely downloaded by potential users.

Syntactically, a program (or, for short, just "program") Π is a collection of *rules* of the form $H \leftarrow L_1, \ldots, L_m, not\ L_{m+1}, \ldots, not\ L_{m+n}$
where H is an atom, $m \geqslant 0$ and $n \geqslant 0$, and each L_i is an atom. Symbol $\leftarrow$ is usually indicated with :- in practical systems. An atom L_i and its negative counterpart $not\ L_i$ are called *literals*. The left-hand side and the right-hand side of the clause are called *head* and *body*, respectively. A rule with empty body is called a *fact*. A rule with empty head is a *constraint*, where a constraint of the form $\leftarrow L_1, ..., L_n$. states that literals $L_1, \ldots, L_n$ cannot be simultaneously true in any answer set.

Unlike other paradigms, a program may have several answer sets, each of which represent a solution to given problem which is consistent w.r.t. the given problem description and constraints, or may have no answer set, which means that no such solution can be found. Whenever a program has no answer sets, it is said to be say that the program is *inconsistent* (w.r.t. *consistent*).

All solvers provide a number of additional features useful for practical programming, that we will introduce only whenever needed. Solvers are periodically checked and compared over well-established benchmarks, and over challenging sample applications proposed at the yearly ASP competition (cf. [13] for a recent report).

4   The MAS Architecture

In this section we illustrate the specific features of the proposed architecture. The MAS is intended to fulfill the *bounded rationality principle* [14], by which an exploration plan for the unknown territory/area has to be devised and executed in a timely manner before a ultimate T_{max} deadline. Consequently, there is a second deadline $T_{PlanMax} < T_{Max}$ by which the exploration plan has to be devised, set so that the remaining time should be sufficient for plan execution, i.e., for actual area exploration. Also for logistic reasons, a requirement of the exploration is the *coverage* $C_\%$ parameter, which establishes the percentage of cells of the area which are required to be covered the exploration

path. The Knight Tour problem is known to be solved for a set of sizes [15] resulting in a Hamiltonian path, i.e. a path where each cell is visited exactly once. However, as it is well-known, the problem and its derivatives (including the one with holes
that we consider here [16]), is NP-complete and becomes therefore soon practically
intractable with the increasing instance size. Moreover, a Hamiltonian path may not
exist. Therefore in practice when exploring an area, especially in critical circumstances
such as the aftermath of a catastrophic event, one may compromise upon completeness. However, in order to make the exploration effective, we have introduced *coverage*
as a problem parameter. In this way, we accept sub-optimal paths that maximize the
number of explored cells w.r.t. the coverage parameter. Such solutions can be obtained
by introducing weak constraints in the general ASP solution. Thus, given the input set
$T_{PlanMax}, T_{Max}, C_\%, N, F$, where N is the size of the problem and F is the set of
forbidden cells, the MAS operates via the following steps.

1. Generate an exploration path within the $T_{PlanMax}$ deadline; in case of failure, return a trivial path of maximum possible length (always possible).
2. Explore the unknown territory at $C_\%$ coverage within the T_{Max} deadline in case of
   failure (insufficient time), maximize the length of the partially executed plan.

Since the ASP solver may possibly find more than one answer set, i.e., more than
one plan, it is useful to define a metric by which a plan could be preferred to an other
one. Reasonable metrics could measure a plan in terms of: (i) number of cells that have
to be visited when using coverage, (ii) length of the path, (iii) presence of loops (when
the hamiltonian constraint is released); (iv) plan cost, in case there is a specific cost
associated to each cell. Preference criteria can then be defined by selecting one metric,
or by combining different metrics: for instance, a criterium may consist in preferring
the shortest path, if it does not exceed a certain cost.

The proposed DALI MAS architecture is shown in Fig.1 and the agent behaviors are
the following.

- **COORDINATOR** agent: this agent synchronizes all the actions of the MAS and
  updates the global state of exploration. Its task are the following. (a) Ensure the
  proper activation of the MAS; (b) Communicate with the external world and whenever needed set new objectives for the MAS. Objectives can include: set a new
  territory to be explored, with its size; set a new preference criterium. Initialize the
  $T_{PlanMax}$ and T_{Max} deadlines. (c) Activate the **META-PLANNER** agent providing as input the preference criterium for plan selection. (d) Wait to receive from the
  META-PLANNER agent the exploration plan to be executed up to $T_{PlanMax}$; (e)
  deliver the found plan to the **EXPLORER** agent, which is in charge of executing
  within maximum time $T_{Max} - T_{PlanMax}$. If time elapses, it cancels the current
  running exploration. It also logs all events to a log server.
- **META-PLANNER** agent, whose tasks are the following. (a) Receive the triggering event from the **COORDINATOR** to start to search for a new plan. (b) Generate
  input for the **PLANNER** agent while monitoring its performances. If **PLANNER**
  agent does not deliver before $T_{PlanMax}$, cancel the plan request and ask **PLAN-
  NER** to generate a trivial plan. It also exploits the given preference criterium in

order to select the plan which is closer to present preferences whenever the **PLANNER** returns more than one answer.

- **PLANNER** agent, which receives as input the time constraints $T_{PlanMax}, T_{Max}, C_\%, N, F$ from **META-PLANNER** to generate the best exploration plan using the ASP solver module, if possible within the $T_{PlanMax}$ deadline. If more than a single answer is produced by the ASP solver, it returns all available plans to the **META-PLANNER**. If no solution exists, it generates a trivial maximized path.
- **EXPLORER**: it puts into action the plan provided by the **COORDINATOR**, if possible within the T_{Max} deadline, and notifies the **COORDINATOR** upon completion

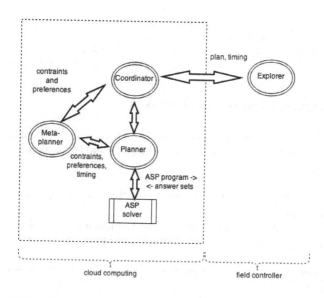

Fig. 1. DALI MAS architecture: coordinator, meta-planner, planner, explorer agents. Most part of the MAS can be deployed over a cloud computing, distributing and balancing the required computational resources. The ASP module is an external solver, configurable depending on the required capabilities. The EXPLORER agent is supposed to work in the field, embedded in a mobile robots or some other facility.

4.1 ASP Problem Definition

The Knight Tour with holes problem has constituted a benchmark in recent ASP competitions, aimed at comparing ASP solvers performances. We performed a number of modifications to the original version [17] concerning: the representation of holes; the objective of devising a path which, though not Hamiltonian, guarantees the required degree of coverage with the minimum number of multiple-traversals; simple forms of loop-checking for avoiding at least trivial loops. Our solution is formulated for the DLV ASP solver [18], though it might be easily reformulated for other solvers. The key modifications are the following.

- We modified the **reached** constraint, and transformed it into a soft constraint, so as not to be forced to finding a Hamiltonian path.

```
reached(X,Y) :- move(1,1,X,Y).
reached(X2,Y2) :- reached(X1,Y1), move(X1,Y1,X2,Y2).
:~ cell(X,Y), not forbidden(X,Y), not reached(X,Y).
```

- We added a new coverage-satisfaction rule, where *coverage* denotes the required degree of coverage and *number_forbidden* the number of holes, and V is the instance size, i.e., the chessboard edge. The maximum possible coverage is 100% of the available cells, i.e., $M = V * V$, while the minimum coverage N is computed in terms of *coverage*, considering the holes. Suitable application of the *count* DLV constraint [18] guarantees the desired coverage.

```
coverage(95).
number_forbidden(5).
cov(N) :- N <= #count{X,Y : reached(X,Y)} <= M,
    size(V), coverage(Z), number_forbidden(F),
    M = V * V, N2 = M * Z, N3 = N2 /100, N = N3 - F.
```

5 Experimental Results and Discussion

We have performed a number of experiments on our solution, adopting the DLV ASP solver [18] and configuring the MAS to use it. We considered in particular the following three scenarios.

1. Instance size 6, 8, 10, percentage of forbidden cells $< 5\%$, T_{Max} large and constant, 90 minutes (Table 1).
2. Instance size fixed to 8, high percentage of forbidden cells: 12%, 20%, 60%, no time constraint for ASP solver, T_{Max} 60 minutes (Table 2).

The performed tests have been then classified according to the achieved result, in particular as follows.

1. Result 1: Closed circuit, Hamiltonian path.
2. Result 2: Weak constraints over reached cells, which produces a Hamiltonian path with given target % of coverage: i.e, each cell is traversed only once, but some cells reamain unreached.
3. Result 3: Trivial solution (longest path, always feasible).

The experimental setup consists in a DALI MAS and an ASP-DLV module, where experiments have been performed on a laptop with a Intel(r) core i5-3337u, clock 1.80Ghz, CPU, 16Gb of DDR3 1666 MHz RAM. The execution results and time measures for configurations with relatively few holes are shown in Tables 1 and 2. In all experimental cases, the MAS found a Hamiltonian path only for small sizes (6,8), which is coherent with existing literature. In most cases however, the planner could found a partial coverage solution, thanks to the soft constraints that have been introduced in the ASP program. In case the percentage of forbidden cells increases, no Hamiltonian path

Table 1. Size parameter span, with $T_{PlanMax} = 90min$, H_{path} Hamiltonian path existence check. First and second lines differ concerning the placement of holes. With the first configuration Hamiltonian paths can be found for sizes up to 8, with the second one none can be found. The weak constraint formulation finds a solution for all given instance sizes and hole placement.

Instance size:	6	8	10
H_{path}, Res. 1:	yes(2-6),5s	yes(8),10s	–
H_{path}, Res. 3:	no($>$ 6),6s	no($>$ 8),10s	no,10s
Weak constr., Res. 2:	5min	10min	60min

Table 2. Instance size fixed to 8, forbidden cell % span, with $T_{PlanMax} = 60min$, H_{path} Hamiltonian path existence check. Note: ($*$) The META-PLANNER agent stopped the process: however, a solution can be computed in most cases, though requiring some hours.

Forbidden %:	12%(8)	20 %	60 %
H_{path}, Res. 1 1:	yes, 5s	–	–
H_{path}, Res. 3:	–	no, 10s	no,10s
Weak constr. Res. 2:	50min	50min(1 ad hoc)	–
Weak constr. Res. 2,3:	–	60min (*)	60min (*)

can be found in a reasonable time regardless the instance size. Therefore, in such cases it is more convenient to search directly for approximate solutions.

Experimental results demonstrate the usefulness of the proposed MAS architecture, that we have in fact enhanced so as to effectively cope with real-world instance sizes. The generalized architecture works as follows. The COORDINATOR agent has been modified so as to: (a) partition the territory that must be explored into a number of (possibly overlapping) sections (chessboards) of reasonable size, each one to be assigned to a META-PLANNER instance; (b) each solution is assigned to a separate EXPLORER agent, specifically assigned to that territory section. Each instance of the META-PLANNER agent relies upon its own associated copy of the planner agent. Moreover, different preference policies can possibly be associated with different sections of the territory to be explored, according to directions provided by the user/environment. A possible future development may consist in dynamic update of plan selection criteria, which can be seen as a form of partial re-planning.

References

1. Costantini, S.: Answer set modules for logical agents. In: de Moor, O., Gottlob, G., Furche, T., Sellers, A. (eds.) Datalog 2010. LNCS, vol. 6702, pp. 37–58. Springer, Heidelberg (2011); Revised selected papers
2. Baral, C.: Knowledge representation, reasoning and declarative problem solving. Cambridge University Press (2003)
3. Leone, N.: Logic programming and nonmonotonic reasoning: From theory to systems and applications. In: Baral, C., Brewka, G., Schlipf, J. (eds.) LPNMR 2007. LNCS (LNAI), vol. 4483, pp. 1–1. Springer, Heidelberg (2007)
4. Truszczyński, M.: Logic programming for knowledge representation. In: Dahl, V., Niemelä, I. (eds.) ICLP 2007. LNCS, vol. 4670, pp. 76–88. Springer, Heidelberg (2007)

5. Web-references: Some ASP solvers Clasp, `potassco.sourceforge.net`; Cmodels, `www.cs.utexas.edu/users/tag/cmodels`; DLV, `www.dbai.tuwien.ac.at/proj/dlv`; Smodels, `www.tcs.hut.fi/Software/smodels`

6. De Gasperis, G., Costantini, S., Nazzicone, G.: Dali multi agent systems framework, doi 10.5281/zenodo.11042. DALI GitHub Software Repository (July 2014) DALI, `http://github.com/AAAI-DISIM-UnivAQ/DALI`

7. Costantini, S., Tocchio, A.: A logic programming language for multi-agent systems. In: Flesca, S., Greco, S., Leone, N., Ianni, G. (eds.) JELIA 2002. LNCS (LNAI), vol. 2424, p. 1. Springer, Heidelberg (2002)

8. Costantini, S., Tocchio, A.: The DALI logic programming agent-oriented language. In: Alferes, J.J., Leite, J. (eds.) JELIA 2004. LNCS (LNAI), vol. 3229, pp. 685–688. Springer, Heidelberg (2004)

9. Costantini, S.: The DALI agent-oriented logic programming language: Summary and references 2015 (2015)

10. Bevar, V., Costantini, S., Tocchio, A., De Gasperis, G.: A multi-agent system for industrial fault detection and repair. In: Demazeau, Y., Müller, J.P., Rodríguez, J.M.C., Pérez, J.B. (eds.) Advances on PAAMS. AISC, vol. 155, pp. 47–56. Springer, Heidelberg (2012)

11. Gelfond, M., Lifschitz, V.: The stable model semantics for logic programming. In: Kowalski, R., Bowen, K. (eds.) Proceedings of the 5th International Conference and Symposium on Logic Programming (ICLP/SLP 1988), pp. 1070–1080. The MIT Press (1988)

12. Gelfond, M., Lifschitz, V.: Classical negation in logic programs and disjunctive databases. New Generation Computing 99, 365–385 (1991)

13. Calimeri, F., Ianni, G., Krennwallner, T., Ricca, F.: The answer set programming competition. AI Magazine 33(4), 114–118 (2012)

14. Gigerenzer, G., Selten, R.: Bounded Rationality, The Adaptive Toolbox. The MIT Press (2002)

15. Schwenk, A.J.: Which rectangular chessboards have a knight's tour. Mathematics Magazine, 325–332 (1991)

16. Delei, J.B.S., Wenming, D.: An ant colony optimization algorithm for knight's tour problem on the chessboard with holes. In: IEEE (ed.) First International Workshop on ETCS 2009, vol. 1, pp. 292–296 (2009)

17. Calimeri, F., Zhou, N.F.: Knight tour with holes ASP encoding (2014), `http://www.mat.unical.it/aspcomp2013/files/links/benchmarks/encodings/aspcore-2/22-Knight-Tour-with-holes/encoding.asp`

18. Leone, N., Pfeifer, G., Faber, W., Eiter, T., Gottlob, G., Perri, S., Scarcello, F.: The dlv system for knowledge representation and reasoning. ACM Transactions on Computational Logic 7(3), 499–562 (2006)

A Bloom Filter-Based Index for Distributed Storage Systems

Zhu Wang[1], Chenxi Luo[2], Tiejian Luo[2,*], Xia Chen[3], and Jinzhong Hou[2]

[1] Xingtang Telecommunications Technology Co., Ltd.,
40 Xueyuan Rd., Haidian District, Beijing 100083, China
[2] University of Chinese Academy of Sciences (UCAS),
19(A) Yuquan Rd., Shijingshan District, Beijing 100049, China
[3] SAPPRFT,
40 Xuanwumenwai Street, Xicheng District, Beijing 100866, China
{wangzhu09,houjinzhong13}@mails.ucas.ac.cn, cxluo91@yahoo.com,
tjluo@ucas.ac.cn, chenxia2000@sina.cn

Abstract. The indexing technique, which is capable of locating an item, is a key component in distributed storage systems. There have been many solutions for the index in distributed systems. One of the problems is the large number of items and the (relatively) low space available for the index. In this paper we propose a bloom filter based schema for the representation and lookup of items in the distributed systems. In each node, the method selects items and inserts them into a probabilistic data structure. After gathering all the data structures, the index node is in possess of all objects information and is capable of locating items in the system. To reduce the false checking times of the index, we choose items to be recorded in reference with the Internet user behavior pattern. We further use theoretical and experimental analysis to test our proposal. Results show that our method can achieve high performance with limited index space.

1 Introduction

Indexing algorithm which maps items of the system onto its storage node is a key component in large scale distributed systems. In many cases, the indexing data structure has to be deployed in many (even all) nodes in order to support fast lookup and high scalability. The index has to be very compact in size to be stored in the index nodes' memory. Therefore, we need a high performance indexing technique with limited space consumption in distributed systems.

Bloom filter[1] is a space-efficient probabilistic data structure for item representation and lookup in a set. When indexing space is limited, i.e. in the memory, the mathematical format offers fast item lookup with a low false rate. Many distributed systems that emphasize time efficiency are using bloom filters as their indexing technique when a small false rate is tolerable[2]. Our previous works[3,4] have adopted the algorithm in privacy index. In this paper we

---

* Corresponding author.

© Springer International Publishing Switzerland 2015   293
S. Omatu et al. (eds.), *Distributed Computing & Artificial Intelligence, 12th Int. Conference*,
Advances in Intelligent Systems and Computing 373, DOI: 10.1007/978-3-319-19638-1_34

optimize the lookup procedure and use it in distributed item representation and lookup.

The observation of user behavior indicates that in many applications, a small number of items attract a major part of user access[5,6]. The phenomenon inspires us to be selective in index construction when space is limited, which is shown in this paper. Further, we analyse the impact of access frequency on the performance of the index.

The rest of the paper is organized as follows. Section 2 describes the related work of our research. In Section 3, we give the new bloom filter construction and item lookup procedures. Then, the theoretical analysis of the system performance is deducted in section 4. Experimental results are shown in Section 5. Finally, we present the conclusion and future work in Section 6.

2   Related Work

2.1   Bloom Filter

Bloom filter[1] works as an probabilistic index which records all elements of a set. Given a set with n elements, the bloom filter with length m and hash number k has a false rate[7]:

$$f_{FP} = (1 - e^{-\frac{kn}{m}})^k \tag{1}$$

f_{FP} reaches minimal value when

$$k = \frac{m}{n} ln2 \tag{2}$$

Then the false positive is minimized

$$f_{FP} = 0.6185^{\frac{m}{n}} \tag{3}$$

Due to its simple structure and smooth integration characteristic, the mathematical format allows considerable potential improvement for system designers to develop new variations for their identical application requirements. Counting bloom filters[8] can be used to improve network router performance[9]. Other variations are adopted in state machines[10], Internet video[11] and publish/subscribe networks[12], etc. In this paper, we focus on fast object lookup in distributed storage systems.

2.2   Pure Bloom Filter Array (PBA)

Many distributed systems use Pure Bloom filter Array (PBA) to support item index and lookup[13]. Bloom filters are loaded with all the items in the nodes. Therefore all queries will find at least one match in one of the bloom filters. The system first calculates the bloom filter of each node one by one and collects result. Then it verifies the positive nodes. If the queried item does exist on the node, the lookup process stops immediately. The false checking time comes from the false positive rate of the bloom filters. In the first node, Q/s queries are already loaded in its bloom filter because of the uniform item distribution between nodes. So the queries that can cause false positive rate are $Q-Q/s$. The average false checking per query of PBA is

$$F_{PBA}/Q = \frac{(Q - Q/s)s}{2Q}f_{FP} = \frac{(s-1)f_{FP}}{2} \qquad (4)$$

Since the bloom filters have an $O(C)$ time complexity, the method can reduce lookup time remarkably.

3   Algorithm and Data Structure

In distributed storage systems, user access frequency varies from one item to another. Popular items attract a majority of visits. In our index, we build one bloom filter for one storage node. Only the popular items are inserted into the data structure. The index node collects all the bloom filters of the storage nodes and forms a bloom filter array, which records the popular items of the system. Unpopular items will not find a match in the index. In those cases, we use traditional method to find the items in the system. In item lookup, once we confirm an item's location, we stop the lookup process immediately. Since most queries are for popular items, the overall false positive rate can be reduced because of the small number of popular items stored.

Index Building. The system sets a load factor β, which is the ratio of the item number loaded by the bloom filter. Each node orders the items by their access time and inserts the items one by one until it reaches the load threshold. For a system of totally N items, the bloom filter arrays hold βN items.

Item Locating. When a query for a certain item arrives, the index node first calculates with the bloom filter of each node and collects the results. One possible situation after the calculation is that there is at least one positive result, it means that the queried item exists on the node with a probability of $1 - f_{FP}$. Then the system first queries the actual node whose bloom filter check result is positive to verify whether the queried item exists in the node. If it does exist on one of the positive nodes, the lookup procedure stops immediately; otherwise it continues to check on the remaining negative nodes until it finds the queried item. The other possible situation after bloom filter calculation is that there is no positive match. Under that circumstance, the system checks each unsearched node directly until it finds the item.

4   Theoretical Analysis

We first define the system environment and parameters. Let N be the total number of items. s is the node number. Those objects are unique items and their distribution on the nodes is the uniform distribution, so each node has approximately N/s items. Let Q be the total number of queries for the existing items. The load factor is β. The vector length of each bloom filter is m and the hash function number is k. Each bloom filter stores $n = N\beta/s$ items. When hash number k reaches its optimal in (2), the false positive rate is

$$f_p = 0.6185^{\frac{ms}{N\beta}} \tag{5}$$

Q is the query number. According to the access pattern, a small number of items attract the majority of user access (in the experiment we assume that the user access for items follows Zipfian distribution[14]). Given a bloom filter with load ratio β, the hit rate of the bloom filter is $r = r(\beta) > \beta$.

In the two steps of the lookup procedure, the bloom filter calculation takes place in index nodes' memory. The time consumption is rather low considering the O(C) time complexity of bloom filters. The vast majority of time cost comes from the node lookup process, in which the index node communicates with the storage node and the storage node check in its disk for queried items. So the time consumption is approximately linear to the average checking times of a query. In the many check in nodes, only one of the checking procedure can find the needed item. The rest nodes checking end without a match and waste a lot of time and system resource. Those redundant (false) checking are the key factor that lowers system performance. Now we try to deduct how much the false checking time is.

In the bloom filter calculation process, queries whose wanted item are indexed by the bloom filter (indexed queries) are $Q_{idx} = rQ$.

When Q queries arrive at the first node, 1/s of them already exists on the node. Only Q(1-1/s) queries can cause false positive nodes, which will result in $Q(1 - 1/s)f_p$ false checking times. While the queries that find its location stops querying, the total false checking times caused by bloom filter lookup process is

$$F_{BF} = \frac{Q(1 - 1/s)f_p \times s}{2} = \frac{Qf_p(s - 1)}{2}$$

It needs to mention that there exists a possibility that in a node, a query that is not loaded in its bloom filter receives a false positive response; but the queried item happens to be in the node outside the bloom filter content. That is to say, the bloom filter false positive rate miss-guides the lookup procedure and happens to find the item in the node. The remaining queries do not find a match through the bloom filter lookup process and have to proceed to the direct lookup step. The query number is

$$Q_{rmd} = (Q - Q_{idx})(1 - f_p) = Q(1 - r)(1 - f_p)$$

Those queries will go through the node checking process. The false checking time in that step is

$$F_{CN} = \frac{Q_{rmd}(1 - f_p)(s - 1)}{2} = \frac{Q(1 - r)(1 - f_p)^2(s - 1)}{2}$$

In sum, the distribution of false checking time when Q queries arrive is

$$F_{BA}/Q = F_{BF}/Q + F_{CN}/Q = \frac{f_p(s-1) + (1-r)(1-f_p)^2(s-1)}{2} \quad (6)$$

Compare (4) and (6), we can see that when $\beta = 1$, $F_{BA}/Q = F_{PBA}/Q$, which indicates that PBA is a special case of our method when load factor equals to one. The multiplier s-1 means that the false checking time has an approximately direct ratio to node number.

5 Experimental Evaluations

After the theoretical analysis, we use experiments to verify our deduction. In all experiments we set node number s=100, the total number of items $N = 10^6$. The items are scattered randomly among s nodes, so each node has approximately $n = 10^4$ items. The probability that an object be allocated in one node is identical among all servers. The total query number $Q = 10^4$.

5.1 Queries and Node Construction

Observations show that item access on the Internet follows Zipf's or Zipf-like distribution[5,6], in which the access time ratio of the 1st item to the nth item is $1 : n^{-\alpha}, 0 < \alpha \leq 1$. The queries follow the same distribution as the corpus.

In actually processing of bloom filter construction in storage node, we first order all items in the node by its popularity rank and pick the first βn items. Then we insert those items into its bloom filter. For easy deployment in the memory of the node, the bloom filter needs to be compact in size. In the experiment we set the length of the bloom filter vector m=40000. The hash number of the bloom filters reaches the nearest integer of its optimal value in (2).

5.2 Experiments with Different α

We first find the influence of α in the bloom filter based index. The range of α is 0.6, 0.7, 0.8, 0.9, 0.95 and 1. For each α, we generate the queries and run the experiment with β ranges from 0.1 to 1. Then we calculate the total false checking times of each α and β. In order to find if our proposed method have a positive effect, we use the PBA approach and the direct lookup approach for a comparison. The result is given in Fig. 1.

We can see that with the increase of α, the overall false checking decreases. The more the queries concentrate on popular items, the better the system performance is.

5.3   Experiments with Different β

In the experiment we want to analyze the impact of β. In the following experiment we repeat the indexing and querying procedure separately with the same system environment. Then we count the false checking of each method. The false checking times in each experiment is plotted in Fig.2.

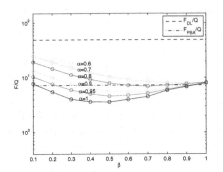

Fig. 1. Overall false checking in node

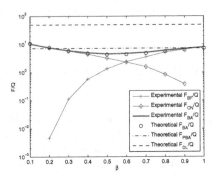

Fig. 2. False checking times(α=0.95)

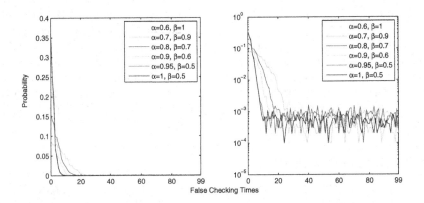

Fig. 3. False checking distribution, linear and log scale

The experimental result is in good accordance with the theoretical one. Our algorithm reaches its optimal false checking times when $\beta = 0.5$. It shows that our method can achieve better performance than PBA with proper parameters.

5.4   Distribution of False Checking Times

In the experiment, we record the false checking time of each query. Fig.3 shows the experimental false checking distribution when α ranges from 0.6 to 1 and

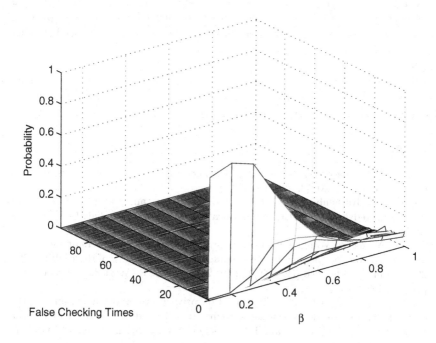

Fig. 4. False checking distribution (α=0.95)

β reaches the optimal value in Fig. 1. We see that the most queries find their residents within 20 node checking times. With the increase of α, the false checking times becomes concentrated on smaller numbers. That means when the access time of the items gets concentrated, the queries are more likely to find their resident node with lower false checking times. That is also the reason why the overall false checking times becomes lower with the increase of α, as shown in Fig.1.

Fig.4 shows the theoretical false check distribution with different β when α is 0.95. We can see that for every β, the false checking time concentrates on smaller integers. That shows that our method can find the resident node of the queries quickly.

6 Conclusion

In the paper we have presented a new bloom filter-based index for distributed storage systems. We have modified the object representation and lookup process of the index. We showed that the false checking times is a key indicator of the index system performance. We have further analyzed the false checking times of the new method with respect to two important variables - the load factor and the concentration level of the dataset. Then the experimental evaluation has been conducted and it was in good accordance with the theoretical analysis. The experiments showed that with proper setting of the load factor, our proposal can have better performance than traditional ones. Further we proved that the more concentrated the queries are, the better performance our algorithm can achieve.

In the future, we are going to change the algorithm to accommodate to more storage systems such as heterogeneous platforms. Also we will find the relationship between different query distribution patterns and the system performance.

References

1. Bloom, B.H.: Space/time trade-offs in hash coding with allowable errors. Communications of the ACM 13, 422–426 (1970)
2. Tarkoma, S., Rothenberg, C.E., Lagerspetz, E.: Theory and practice of bloom filters for distributed systems. IEEE Communications Surveys & Tutorials 14(1), 131–155 (2012)
3. Luo, T., Wang, Z., Wang, X.: A privacy preserving model for ownership indexing in distributed storage systems. In: Proceedings of the Workshops of the EDBT/ICDT 2014 Joint Conference (EDBT/ICDT 2014). CEUR Workshop Proceedings, pp. 416–420 (2014)
4. Luo, C., Wang, Z., Chen, X., Luo, T.: Lba privacy preserving index and its theoretical analysis in cloud storage systems. In: Proceedings of the 2nd International Workshop on Security and Forensics in Communication Systems, SFCS 2014, pp. 37–44. ACM, New York (2014)
5. Chierichetti, F., Kumar, R., Raghavan, P.: Compressed web indexes. In: Proceedings of the 18th International Conference on World Wide Web, pp. 451–460. ACM (2009)
6. Breslau, L., Cao, P., Fan, L., Phillips, G., Shenker, S.: Web Caching and Zipf-like Distributions: Evidence and Implications. In: IEEE INFOCOM, vol. 1, pp. 126–134 (1999)
7. Mullin, J.K.: A second look at bloom filters. Communications of the ACM 26(8), 570–571 (1983)
8. Ficara, D., Giordano, S., Procissi, G., Vitucci, F.: Multilayer compressed counting bloom filters. In: INFOCOM 2008. The 27th Conference on Computer Communications, pp. 311–315. IEEE (2008)
9. Song, H., Dharmapurikar, S., Turner, J., Lockwood, J.: Fast hash table lookup using extended bloom filter: an aid to network processing. In: Proceedings of the 2005 Conference on Applications, Technologies, Architectures, and Protocols for Computer Communications, SIGCOMM 2005, pp. 181–192. ACM, New York (2005)
10. Bonomi, F., Mitzenmacher, M., Panigrah, R., Singh, S., Varghese, G.: Beyond bloom filters: from approximate membership checks to approximate state machines. In: Proceedings of the 2006 Conference on Applications, Technologies, Architectures, and Protocols for Computer Communications, SIGCOMM 2006, pp. 315–326. ACM, New York (2006)
11. Wang, Z., Luo, T.: Intelligent video content routing in a direct access network. In: 2011 3rd Symposium on Web Society (SWS), pp. 147–152. IEEE (2011)
12. Jokela, P., Zahemszky, A., Esteve Rothenberg, C., Arianfar, S., Nikander, P.: Lipsin: line speed publish/subscribe inter-networking. In: Proceedings of the ACM SIGCOMM 2009 Conference on Data Communication, SIGCOMM 2009, pp. 195–206. ACM, New York (2009)

13. Zhu, Y., Jiang, H., Wang, J., Xian, F.: Hba: Distributed metadata management for large cluster-based storage systems. IEEE Transactions on Parallel and Distributed Systems 19(6), 750–763 (2008)
14. Kotera, I., Egawa, R., Takizawa, H., Kobayashi, H.: Modeling of cache access behavior based on zipf's law. In: Proceedings of the 9th Workshop on MEmory Performance: DEaling with Applications, Systems and Architecture, MEDEA 2008, pp. 9–15. ACM, New York (2008)

Construction of Functional Brain Connectivity Networks

Ricardo Magalhães[1,2,3], Paulo Marques[1,2,3], Telma Veloso[4], José Miguel Soares[1,2,3],
Nuno Sousa[1,2,3], and Victor Alves[4]

[1] Life and Health Sciences Research Institute (ICVS), School of Health Sciences,
University of Minho, Campus Gualtar, 4710-057, Portugal
[2] ICVS/3B's - PT Government Associate Laboratory, Braga/Guimarães, Portugal
[3] Clinical Academic Center – Braga, Braga, Portugal
[4] Department of Informatics, University of Minho, Portugal
{ricardomagalhaes,paulo.c.g.marques,josesoares,
njcsousa}@ecsaude.uminho.pt,
valves@di.uminho.pt

Abstract. Graph theory and the study of complex networks have, over the last decade, received increasing attention from the neuroscience research community. It allows for the description of the brain as a full network of connections, a connectome, as well as for the quantitative characterization of its topological properties. Still, there is a clear lack of standard procedures for building these networks. In this work we describe a specifically designed full workflow for the pre-processing of resting state functional Magnetic Resonance Imaging (rs-fMRI) data and connectome. The proposed workflow focuses on the removal of confound data, the minimization of resampling effects and increasing subject specificity. It is implemented using open source software and libraries through shell and python scripting, allowing its easy integration into other systems such as BrainCAT. With this work we provide the neuroscience research community with a standardized framework for the construction of functional connectomes, simplifying the interpretation and comparison of different studies.

Keywords: Rs-fMRI, Graph Theory, Connectome, Python, Shell scripting.

1   Introduction

The world of neuroscience research is constantly marked by the development of new methodologies that seek to deliver new tools to allow researchers to better study the structure and function of the brain. While these developments are key to the success of the field, their translation into practice is neither necessarily direct nor simple in terms of application and interpretation. There exists a gap between the highly subject focused and specialized research in charge of developing these methodologies and the more broadly competent research who needs to apply them in the laboratories. To complete this process it is then necessary the creation of tools and frameworks that will allow the application of these new methodologies to every day research situations.

One such development that has received a great deal of attention over the last decade is the application of graph theory and the principle of complex networks to the

© Springer International Publishing Switzerland 2015   303
S. Omatu et al. (eds.), *Distributed Computing & Artificial Intelligence, 12th Int. Conference*,
Advances in Intelligent Systems and Computing 373, DOI: 10.1007/978-3-319-19638-1_35

study of the brain through neuroimaging data [1]. A graph is a mathematical representation of a network, constituted by nodes, that can represent different brain areas, and edges that represent some connection or relationships between them, and allows the study of how the brain behaves as a network [2]. It is commonly applied using resting state functional Magnetic Resonance Imaging (rs-fMRI) data [3].

Despite the strong potential of this application, nowadays there is still a clear lack of free guidelines and standardized tools to build these graphs [4]. Difficulties range from the very large amounts of data to be processed to the lack of a standardized pre-processing and graph building workflows.

In this work we present a complete framework for the pre-processing of rs-fMRI data and the construction of functional connectivity graphs, using open source tools, through python and shell scripting, capable of parallel processing and ready for implementation on any research center.

2   The Complete Framework

The necessary data processing for the construction of a connectivity graph can be divided into two main phases (Figure 1): the pre-processing of MRI data, meant to improve the signal to noise ratio, remove confound data and define the brain areas that will represent the graph nodes (Figure 1 a)); the conversion and extraction of the rs-fMRI data, with the definition of the graph edges and construction of connectivity graphs (Figure 1 b)).

As much as possible the software chosen for the implementation of each block should be com-

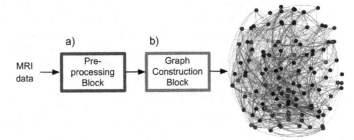

Fig. 1. Block design of the framework. Divided in two main componenets meant to properly prepare the data (a), and build the connectivity graphs (b).

patible and easily integrated with each other, improving the ease of implementation and workflow design. Furthermore, for the choice of software to be implemented in a data processing framework in research centers, there are several requisites that need to be taken into consideration:

- Robustness and validity proved by the specialized scientific community;
- A modular structure that allows for the easy re-adaptation of data processing workflows;
- Compatibility with the main operative systems used in the neuroimaging community;

- Possibility of integration in computational clusters and performance optimization through parallel processing;
- Open source and freely available software that can be implemented at no extra costs;

An overview of the software commonly reported in neuroimaging articles easily reveals three main software packages used in the pre-processing and analysis of rs-fMRI data: Statistical Parametric Mapping (SPM, http://www.fil.ion.ucl.ac.uk/spm/, [5]), Analysis of Functional NeuroImage (AFNI, http://afni.nimh.nih.gov/, [6,7]) and FMRIB's Software Library (FSL, http://fsl.fmrib.ox.ac.uk/fsl/, [8]). While SPM does present several free and well-tested tools for the pre-processing of rs-fMRI and the construction of functional connectivity graphs, such as REST [9], it is presented as a Matlab (http://www.mathworks.com) toolbox, with its use conditioned to the acquisition of a paid license AFNI is developed in C, compatible with Mac OS, Linux, Solaris and SGI, with open source code. Finally, FSL is freely available for Mac OS and Linux in C++ and shell scripts. Both AFNI and FSL present the necessary tools for the pre-processing and extraction of rs-fMRI data, while FSL presents more tools for the analysis of this kind of data, as well as better compatibility with other software. Furthermore, we have previously developed and tested workflows for the general processing of MRI data [10,11] using FSL, making it an ideal choice for the current implementation. Another very useful set of tools in the pre-processing of MRI data is FreeSurfer [12] developed by the Martinos Center for Biomedical Imaging. It is a unique tool in its ability to segment individual brains using different atlas. It is developed in C while also using shell scripts and is compatible with Mac OS and Linux.

For the construction of the connectivity graphs the options are far scarcer. While there are several options for the analysis and visualization of such graphs, only a few present the necessary tools to build them. Two of the most common options, REST and Conn [13], are once again developed in MATLAB, making its implementation in the desired framework impossible. An alternative strategy is the use of freely available libraries for a language compatible with the desired framework. Nipy [14] is a library developed in Python for the processing and analysis of neuroimaging data, meant to give an integrative platform for the easy implementation of new methods, with compatibility with FSL and SPM output data. In its nitime package it presents

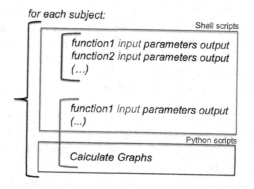

Fig. 2. Detailed structure of the framework implementation. Data pre-processing on the red bracket, graph construction block on gray bracket. Blue block represents the shell scripts and green block represents python scripts.

the necessary tools for the construction of connectivity graphs from rs-fMRI time series, and as such was chosen for the implementation in the present framework.

The present framework will then consist of a series of shell and python scripts, capable of carrying state of the art pre-processing of rs-fMRI data and constructing connectivity graphs using FSL and nipy. It can be easily integrated in the data processing workflows of any research center, or more user-friendly software such as BrainCAT. The use of shell scripts allows for an easy automatization of the all process and the processing of large amounts of data with minimal interaction from the user. The overall structure of the scripts can be seen in Figure 2. The common syntax for the use of FSL functions requires the indication of the input data, the parameters and of the output name. The use of nitime follows the common python language structure.

3   Pre-processing Tools

We have previously described pre-processing workflows for the analysis of rs-fMRI data [10], and the construction of connectivity graphs [11]. In the current framework we propose the full application of the referred workflows using FSL commands implemented in shell scripts. Using as inputs a list of subject reference codes, the pre-processing shell script then calls each of the necessary FSL functions, properly naming the output files.

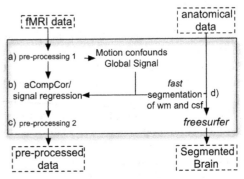

The proposed pre-processing workflow is constituted by the steps of format conversion, initial volumes removal, slice timing correction, motion correction and brain tissue extraction (Figure 3 a)), and normalization to the functional native space and temporal filtering (Figure 3 c)).

Fig. 3. Data pre-processing workflow. Input and output data in dashed lines. fMRI pre-processing described in three phases: a), b) and c); Anatomical segmentation in two: fast segmentation and freesurfer processing.

Furthermore we propose the addition of a step of removal of confound data: removal of white matter (wm) and cerebrospinal fluid (csf) through an aCompCor strategy [15], the removal of motion confounds [16] from the motion correction step and of the global signal (Figure 3 b)).

Additionally, we include in this block the processing of anatomical data (aMRI), used in the removal of confound data in the aCompCor strategy. The anatomical data is later segmented using the Freesurfer workflow (Figure 3 d)) and the result of the segmentation will be used in the connectivity graph construction block.

4   Implementation of Graph Construction Tools

As previously described, a graph is a mathematical representation of a network constituted by nodes, that represent the units under study, and edges that represent some
connection or relation between them. When constructing brain connectivity graphs,
the most natural interpretation of nodes is as the anatomical units of the brain, such as
the ones obtained from the FreeSurfer segmentation. For the edges, when dealing with
functional data, the Pearson correlation between the different anatomical areas is the
most common option. As such, we suggest the use of FSL's fslmeans (Figure 3 a))
function for the extraction of the time series, using the output from FreeSurfer as the
label option. This function extracts, for each different label in the brain segmentation,
the average time-series across all voxels of a brain region and saves it in a text file. To
do this it is necessary to convert the output from FreeSurfer into the FSL compatible
nifti format, which can be
done using the FreeSurfer
command   line   tool
mri_convert.

The final step necessary
to build the connectivity
graphs is to calculate the
correlations between each
pair of time series, creating
an adjacency matrix. This is
done using the python libraries nipy and nitime. Using
the TimeSeries function the
timeseries for each region is
imported into an object.
Finally, the function Corre-
lationAnalyzer (Figure 4 b))
can be used to calculate all
the Pearson correlations
between the time series and
create a correlation matrix
object, which can then be
saved as a text-file, or a
correlation matrix image.

The entire workflow was

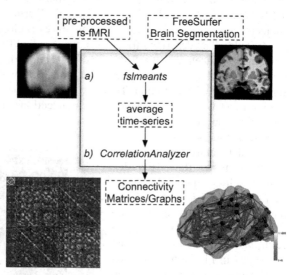

Fig. 4. Construction of the connectivity matrices. Inputs
and outputs surrounded by dashed lines. This step in comprised by two stages: a) extraction of the time series using
the segmented brains and the fMRI data; b) calculation of
the connectivity matrices;

implemented using shell and python scripts. For each subject code the script sequentially calls all the Freesurfer and FSL functions as well as the python scripts necessary
to calculate the adjacency matrices (Figure 3).

5   Application of the Framework to Real Data

To test the developed framework, a dataset of 54 healthy subjects from the
SWITCHBOX project were used in this study (30 males and 24 females aged between
51 and 82 years old and with a mean age of 64.85 – 8.82 years). The goals and tests of
the study were explained to all participants, who provided informed written signed
consent. For each subject two different acquisitions were used in the framework: a T1
Magnetization-Prepared Rapid Gradient Echo (MPRAGE) with repetition time
(TR)=2730 msec, echo time (TE)=3.5 msec, field of view (FoV)=256·256mm, flip
angle=7°, in-plane resolution of 1·1mm, and 1mm slice thickness, used as a structural
acquisition; a T2* Echo-Planar Imaging (EPI) acquisition with 180 volumes,
TR=2000 msec, TE=30 msec, FoV=224·224mm, flip angle=90°, in-plane resolution
of 3.5·3.5mm, and 4.5mm slice thickness, as a resting state acquisition. All acquisitions were performed on a clinically approved Siemens Magnetom Avanto 1.5 T
(Siemens Medical Solutions, Elangen, Germany) scanner, in Hospital de Braga, using
a 12-channel receive-only Siemens head coil. During the resting-state acquisition, all
subjects were instructed to remain with their eyes closed and to not think of anything
in particular. The described framework was applied to all the subjects and the corresponding   adjacency   matrices   were   obtained.   Using   Matlab   (v.2009b,
www.mathworks.com), the fisher r-to-z transform was applied to all subjects, to assure normality of distribution. To allow the evaluation of the networks created, a one-
sample t-test was done to calculate the connections significantly different from zero,
correcting for multiple compari-
sons with the Bonferroni procedure, which controls for the Family Wise Error Rate (FWER).
The resulting networks were
averaged across all subjects and
then further thresholded to keep
values of z greater then 0,45. The
resulting adjacency matrix can be
visualized in Figure 5. It is possible to observe that sub-cortical
regions are more interconnected
with each other then with cortical
regions. Furthermore it is also
possible to observe a stronger
tendency for regions to be connected with regions in the same
hemisphere. Finally, it can be
seen a strong tendency for cortical regions to have a stronger

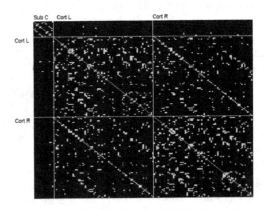

Fig. 5. Average z-connectivity matrix of the significant connections, FWE corrected and thresholded at
0.45. Sub C: Sub-Cortical areas; Cort L: Areas of the
left cortex; Cort R: Areas of the right cortex.

connection to their counterpart in the other hemisphere, as observed by the diagonal
crossing the Cort L/Cort R quadrant (Figure 5).

Using the Brain Connectivity Toolbox [17] the key network properties were calculated. When compared to equivalent random networks, the network built was found to have a much higher tendency to from clusters (ratio of 5.11) and a slightly lower

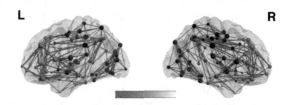

Fig. 6. Graph visualized using BrainNet Viewer; Connection color represents its z value. The node size codes the degree and the color the community of the node.

er global efficiency (ratio of 0.7), meaning that they are small-world networks, as was expected. Finally, a modularity algorithm was applied to determine communities within the network. BrainNet Viewer [18] was used to build a 3D visualization of the graph (Figure 6). It is possible to observe a tendency for communities to be composed of spatially proximal areas.

6   Conclusion

In the present work we have described the implementation of a complete framework for the construction of connectivity graphs from functional MRI data, as well as the necessary pre-processing steps. The framework uses open source software, capable of running on any Unix system. It uses shell and python scripts, in such a way that it can be easily integrated into other frameworks.

The implemented fMRI data pre-processing workflow was specifically designed for the construction of connectivity graphs and tested in a dataset of real rs-fMRI data.

Our team in the ICVS is currently applying the complete framework for the analysis of larges amount of data in projects related to the study of aging, stress, education, neurological and psychiatric disorders such as obsessive compulsive disorders.

Acknowledgements. This work has been supported by FCT - Fundação para a Ciência e Tecnologia within the Project Scope UID/CEC/00319/2013. RM is supported by a fellowship of the project FCT-ANR/NEU-OSD/0258/2012 founded by FCT/MEC (www.fct.pt) and by Fundo Europeu de Desenvolvimento Regional (FEDER). PM is supported by a fellowship of the project SwitchBox-FP7-HEALTH-2010-grant 259772-2 (www.switchbox-online.eu).

References

1. Sporns, O.: From simple graphs to the connectome: networks in neuroimaging. NeuroImage 62(2), 881–886 (2012), doi:10.1016/j.neuroimage.2011.08.085
2. Gross, J.L., Yellen, J.: Graph Theory and Its Applications. Chapman & Hall/CRC (2006)
3. Bullmore, E., Sporns, O.: Complex brain networks: graph theoretical analysis of structural and functional systems. Nature Reviews Neuroscience 10(3), 186–198 (2009), doi:10.1038/nrn2575

4. Bullmore, E.T., Bassett, D.S.: Brain graphs: graphical models of the human brain connectome. Annual Review of Clinical Psychology 7, 113–140 (2011), doi:10.1146/annurev-clinpsy-040510-143934

5. Penny, W.D., Friston, K.J., Ashburner, J.T., Kiebel, S.J., Nichols, T.E.: Statistical parametric mapping: the analysis of functional brain images: the analysis of functional brain images. Academic Press (2011)

6. Cox, R.W.: AFNI: software for analysis and visualization of functional magnetic resonance neuroimages. Computers and Biomedical Research, An International Journal 29(3), 162–173 (1996)

7. Cox, R.W.: AFNI: what a long strange trip it's been. NeuroImage 62(2), 743–747 (2012), doi:10.1016/j.neuroimage.2011.08.056

8. Smith, S.M., Jenkinson, M., Woolrich, M.W., Beckmann, C.F., Behrens, T.E.J., Johansen-Berg, H., Bannister, P.R., De Luca, M., Drobnjak, I., Flitney, D.E.: Advances in functional and structural MR image analysis and implementation as FSL. NeuroImage 23, S208–S219 (2004)

9. Song, X.W., Dong, Z.Y., Long, X.Y., Li, S.F., Zuo, X.N., Zhu, C.Z., He, Y., Yan, C.G., Zang, Y.F.: REST: a toolkit for resting-state functional magnetic resonance imaging data processing. PloS One 6(9), e25031 (2011), doi:10.1371/journal.pone.0025031

10. Marques, P., Soares, J.M., Alves, V., Sousa, N.: BrainCAT - a tool for automated and combined functional magnetic resonance imaging and diffusion tensor imaging brain connectivity analysis. Frontiers in Human Neuroscience 7, 794 (2013), doi:10.3389/fnhum.2013.00794

11. Magalhaes, R., Marques, P., Soares, J., Alves, V., Sousa, N.: The Impact of Normalization and Segmentation on Resting-State Brain Networks. Brain Connectivity (2014), doi:10.1089/brain.2014.0292

12. Fischl, B.: FreeSurfer. NeuroImage 62(2), 774–781 (2012), doi:10.1016/j.neuroimage.2012.01.021

13. Whitfield-Gabrieli, S., Nieto-Castanon, A.: Conn: a functional connectivity toolbox for correlated and anticorrelated brain networks. Brain Connectivity 2(3), 125–141 (2012), doi:10.1089/brain.2012.0073

14. Millman, K.J., Brett, M.: Analysis of functional magnetic resonance imaging in Python. Computing in Science & Engineering 9(3), 52–55 (2007)

15. Behzadi, Y., Restom, K., Liau, J., Liu, T.T.: A component based noise correction method (CompCor) for BOLD and perfusion based fMRI. NeuroImage 37(1), 90–101 (2007), doi:10.1016/j.neuroimage.2007.04.042

16. Power, J.D., Barnes, K.A., Snyder, A.Z., Schlaggar, B.L., Petersen, S.E.: Spurious but systematic correlations in functional connectivity MRI networks arise from subject motion. NeuroImage 59(3), 2142–2154 (2012), doi:10.1016/j.neuroimage.2011.10.018

17. Rubinov, M., Sporns, O.: Complex network measures of brain connectivity: uses and interpretations. NeuroImage 52(3), 1059–1069 (2010), doi:10.1016/j.neuroimage.2009.10.003

18. Xia, M., Wang, J., He, Y.: BrainNet Viewer: a network visualization tool for human brain connectomics. PloS One 8(7), e68910 (2013), doi:10.1371/journal.pone.0068910

Part II

Special Session on AI–Driven Methods for Multimodal Networks and Processes Modeling (AIMPM 2015)

Multimodal Processes Optimization Subject to Fuzzy Operation Time Constraints

R. Wójcik[1], I. Nielsen[2], G. Bocewicz[3], and Z. Banaszak[4]

[1] Institute of Computer Engineering, Control and Robotics,
Wrocław University of Technology, Wrocław, Poland
`robert.wojcik@pwr.wroc.pl`
[2] Dept. of Mechanical and Manufacturing Engineering, Aalborg University, Aalborg, Denmark
`izabela@m-tech.aau.dk`
[3] Dept. of Computer Science and Management, Koszalin University of Technology, Poland
`bocewicz@ie.tu.koszalin.pl`
[4] Dept. of Business Informatics, Warsaw University of Technology, Poland
`Z.Banaszak@wz.pw.edu.pl`

Abstract. Different material handling transport modes provide movement of work pieces between workstations along their manufacturing routes in the Multimodal Transportation Network (MTN). In this context the multimodal processes standing behind multiproduct production flow while executed in MTN can be seen as processes realized with synergic utilization of various local periodically acting processes. Such processes play a determining role in the evaluation of functioning efficiency inter alia in public transport systems, goods transport, energy and data transmission etc. The optimization of Automated Guided Vehicles (AGVs) fleet schedule subject to fuzzy operation times constraints is our main contribution. In the considered case both production rate (takt) and operations execution time are described by imprecise (fuzzy) data.

Keywords: AGVs, multimodal process, fuzzy constraints, optimization.

1   Introduction

The productivity of an AGV-served flow shop repetitively producing a set of different products, depends on both the job flow sequencing, and the material handling system required to achieve a pre-specified throughput, i.e. AGVs fleet sizing, assignment and scheduling [2, 3]. Opposite to usually studied deterministic cases where demand is known in advance and the processing times of each job on each machine is a known constant, the problem considered in this paper assumes a given transportation system encompassing the network of AGVs periodically circulating along cyclic routes while servicing work pieces load/unload operations. A takt of production flow and its profitability depend on uncertainty requirements caused by availability of employed workers. The investigated question concerns AGVs fleet match-up scheduling subject to assumed productivity constraints with reference to the schedule of the production flow imposed by uncertainty implied by workers' availability for operations handling. In that context it can be seen as an extension of the problems formulated in [2].

© Springer International Publishing Switzerland 2015   313
S. Omatu et al. (eds.), *Distributed Computing & Artificial Intelligence, 12th Int. Conference*,
Advances in Intelligent Systems and Computing 373, DOI: 10.1007/978-3-319-19638-1_36

A number of papers are concerned with the fleet assignment and maintenance planning problems [1, 3]. However, few papers have considered the integration of fleet assignment and maintenance planning as well as and inventory policy [5, 6].

As most real-life available manufacturing processes manifest a cyclic steady state, an alternative approach to AGVs fleet dispatching and scheduling can be considered as well [2]. Since the steady state of production flows has a cyclic character, hence servicing AGV-driven transportation processes (following loop-like routes) also exhibit cyclic behavior. This means, the periodicity of a Flexible Manufacturing System (FMS) [4] depends on both the periodicity of the production flow cyclic schedule and, following this schedule, the AGVs periodicity.

Besides of many models and methods have been considered to date, the mathematical programming approach [11], max-plus algebra [8], artificial neural networks [9], constraint logic programming [2] frameworks are the more frequently used. Most of these methods, however, are oriented at finding of a minimal production takt while assuming deadlock-free processes flow. In that context our main contribution is to propose a declarative framework aimed at refinement and prototyping of the cyclic steady states for concurrently executed material handling systems composed of AGVs. Since our main question is: Can the assumed AGVs fleet assignment reach its goal subject to constraints assumed on concurrent multi-product manufacturing on hand? hence the problem considered can be seen as an extension of one's formulated in [2], i.e. aimed at AGVs fleet match-up scheduling while taking into consideration the itineraries of variety of concurrently manufactured product types.

The remainder of the paper is organized as follows: Section 2 introduces the concept of multimodal processes prototyping. AGVs flow modeled in terms of Systems of Cyclic Concurrently flowing Processes (SCCP) enables to formulate a problem concerning optimization of multi-product manufacturing flow imposed by fuzzy operation time constraints. Solution to the stated problem formulated in a declarative modeling framework aimed at AGVs fleet scheduling is illustrated in Section 3 and concluding remarks are presented in Section 4.

2   Multimodal Processes Prototyping

2.1   Systems of Concurrent Cyclic Processes

The example of the Multimodal Transportation Network (MTN) is shown on Fig. 1a). This is the transportation system of the FMS layout encompassing the network of AGVs periodically circulating along cyclic routes. This kind of system can be modelled in terms of Systems of Concurrently flowing Cyclic Processes (SCCPs) shown in Fig. 1b), where four **local cyclic processes**: P_1, P_2, P_3, P_4 and their *streams*: P_1^1, P_2^1, P_3^1, P_4^1, associated with the operation of four AGVs, and two **multimodal processes** [2, 7] mP_1, mP_2 representing two products W_1 and W_2 (i.e. production routes), are considered. The processes are executed along given routes composed of six workstations ($R_1 - R_6$) and nine transportation sectors ($R_7 - R_{15}$): $R = \{R_1, R_2, \ldots, R_{15}\}$.

Specific local P and multimodal mP processes are defined in the following way:

- $P = \{P_i | i = 1, \ldots, n\}$ – the set of local processes, where each P_i is specified by the set of streams: $P_i = \left\{P_i^1, P_i^2, \ldots, P_i^k, \ldots, P_i^{ls(i)}\right\}$,
- $mP = \{mP_i | i = 1, \ldots, w\}$ – the set of multimodal processes described by sequences of some sub-sequences from local processes P, where each mP_i is specified by the set of streams: $mP_i = \left\{mP_i^1, mP_i^2, \ldots, mP_i^k, \ldots, mP_i^{lms(i)}\right\}$.

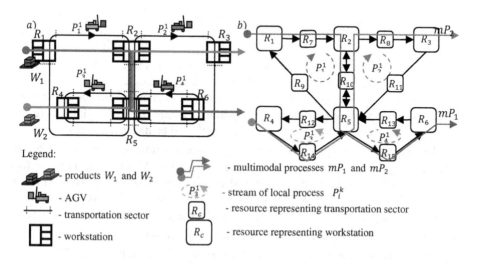

Fig. 1. Exemplary MTN a), and its SCCP representation b)

The operations of multimodal processes' streams require execution of some local processes. For example, transport operations between resources in mP_1 (product W_1 - Fig. 1) require streams of local processes P_1, P_2, P_3, respectively. This means the routes of multimodal processes are also determined by the subsequences of routes of the local processes through which they have to be processed. The resources belonging to these routes are simultaneously shared by both local and multimodal processes.

It is assumed that both kind of processes are cyclic and act concurrently, however are synchronized by a mutual exclusion protocol, guaranteeing the only one process (one product) simultaneously can be executed on a common shared resource.

In order to describe the system shown in Fig. 1b) let us introduce the notations [2]:

- $p_i^k = \left(p_{i,1}^k, p_{i,2}^k, \ldots, p_{i,j}^k, \ldots, p_{i,lr(i)}^k\right)$ specifies **the route of the k-th stream of a local process P_i^k** (the k-th stream of the i-th local process P_i), and its components define the resources used in course of process operations execution, where: $p_{i,j}^k \in R$, and $R = \{R_1, R_2, \ldots, R_c, \ldots, R_m\}$ – denotes the set of resources used for the j-th operation in the k-th stream of the i-th local process; **the j-th operation executed on the resource $p_{i,j}^k$ in the stream P_i^k** is denoted by $o_{i,j}^k$; $lr(i)$ – is the length of the cyclic process route (all streams of P_i have the same length).

- $t_i^k = (t_{i,1}^k, t_{i,2}^k, \ldots, t_{i,j}^k, \ldots, t_{i,lr(i)}^k)$ – specifies **the i-th process operation times**, where $t_{i,j}^k$ denotes the time of execution of the j-th operation $o_{i,j}^k$ in stream P_i^k.
- $x_{i,j}^k(l) \in \mathbb{N}$ – the moment of operation $o_{i,j}^k$ beginning in the l-th cycle,
- $mp_i^k = \left(mpr_{i_1}^{q_1}(a_{i_1}, b_{i_1}) mpr_{i_2}^{q_2}(a_{i_2}, b_{i_2}) \ldots mpr_{i_y}^{q_y}(a_{i_y}, b_{i_y}) \right)$ – specifies **the route of the stream mP_i^k from the multimodal process mP_i** (the k-th stream of the i-th multimodal process mP_i), where:

$$mpr_i^q(a,b) = \begin{cases} (p_{i,a}^q, p_{i,a+1}^q, \ldots, p_{i,b}^q) & a \leq b \\ (p_{i,a}^q, p_{i,a+1}^q \ldots, p_{i,lr(i)}^q, p_{i,1}^q, \ldots, p_{i,b-1}^q, p_{i,b}^q) & a > b \end{cases}$$

$a, b \in \{1, 2, \ldots, lr(i)\}$, uv – concatenation of sequences u and v. If $u = (u_1, \ldots, u_a)$, $v = (v_1, \ldots, v_b)$ and $u_a = v_1$ then $uv = (u_1, \ldots, u_a, v_2, \ldots, v_b)$.

The route mp_i^k is a sequence of sections of local streams p_i^q and determines the production routes taking into account transportation means (e.g., AGVs). In the rest of the paper, **the j-th operation of the stream mP_i^k** will be denoted by $mo_{i,j}^k$.

The operation times $mt_{i,j}^k$ of multimodal processes are: the operation times of AGV moving between the workstations (i.e. assigned to AGVs going along sectors $R_7 - R_{15}$) and the working times assigned to workstations $R_1 - R_6$. Transportation times are the same as relevant operations' in local processes. However, the working time $mt_{i,j}^k$ assigned to workstation R_g is equal to: $mt_{i,j}^k = x_{a,b}^c(l) - x_{d,e}^f(l)$, where: $x_{a,b}^c(l)$ - is the moment the process mP_i^k leaves R_g (taken by local stream P_a^c) and $x_{d,e}^f(l)$ – is the moment the process mP_i^k enters R_g (delivered by stream P_d^f).

- $mx_{i,j}^k(l) \in \mathbb{N}$ – moments of operation $mo_{i,j}^k$ beginning in the l-th cycle,
- $\Theta^l = \{\sigma_1^l, \sigma_2^l, \ldots, \sigma_c^l, \ldots, \sigma_m^l\}$ – is the set of the priority dispatching rules of local ($l = 0$) and multimodal processes ($l = 1$), where σ_c^l is the sequence components of which determine an order in which the streams of processes can be executed on the resource R_c.

Therefore the SCCP can be defined as the pair: $SC = ((R, SL), SM)$, \hfill (1)

where: $R = \{R_1, R_2, \ldots, R_k, \ldots, R_m\}$ – the set of resources, $SL = (P, U, T, \Theta^0)$ – characterizes the structure of local processes of SCCP, i.e. P – the set of local processes, U– the set of local process routes, T – the set of local process operations times, Θ^0 – the set of dispatching priority rules. $SM = (mP, m\Pi, mT, \Theta^1)$ – characterizes the structure of multimodal processes of SCCP, i.e. mP – the set of multimodal processes, M – the set of multimodal process routes, mT – the set of multimodal process operation times, Θ^1 – the set of dispatching priority rules.

The considered SC model (1) of the SCCP [2] enables one to state a search problem aimed at determining the structural parameters (i.e., local U and multimodal M routes, operations times T, mT, priority rules Θ, etc.) guaranteeing cyclic execution of processes. Since parameters guaranteeing a cyclic behavior of a SCCP under uncertain operation time constraints are sought, the following uncertainty assumptions have to be taken into account:

- Working time $mt_{i,j}$ assigned to a workstation belongs to a so called availability zone specified by fuzzy set $s_{i,j} = \mu(mt_{i,j})$, $s_{i,j} \in [0,1]$. Membership function $\mu(mt_{i,j})$ determines worker's availability for operations handling. $s_{i,j} = 0$ means that corresponding value $mt_{i,j}$ is out of worker's availability, while $s_{i,j} = 1$ – means available, well acceptable working time value $mt_{i,j}$.
- Accepted production takt mTc_i of the i-th multimodal process has to follow assigned production profitability sT_i. measured along mP_i. The takt time of the process mP_i, is the time between two consecutive moments of its subsequent streams completion: $mTc_i = mx_{i,j}^k(l) - mx_{i,j}^{k-1}(l)$ (and for one pipeline process: $mTc_i = mx_{i,j}^k(l) - mx_{i,j}^k(l-1)$). Similar to the availability zone $s_{i,j}$, the assigned production profitability zone sT_i is specified by fuzzy set $sT_i = \mu(mTc_i)$, $sT_i \in [0,1]$. Membership function $\mu(mTc_i)$ determines production profitability following mTc_i. $sT_i = 0$ means that production performed with takt mTc_i is not profitable, while $sT_i = 1$ means that takt mTc_i ensures production's profitability.

Representative workers' availability for operations handling and production profitability zones (specified by the membership functions: $\mu(mt_{i,j})$, $\mu(mTc_i)$) for the system from Fig. 1, are shown in Tab. 1 and Tab. 2. Following from these are uncertainty requirements as well as constraints guaranteeing deadlock-free processes execution contribute to a declarative model of SCCP behavior.

2.2   Constraint Programming Model

The behavioral characteristics guaranteeing deadlock-free execution of processes while following above mentioned uncertainty requirements can be specified by the following cyclic schedule:

$$X_{SC} = \big((X, \alpha), (mX, m\alpha)\big). \tag{2}$$

This kind of schedule X_{SC} is defined as a sequence of ordered pairs describing behaviors of local (X, α) and multimodal ($mX, m\alpha$) processes, where: $X = \left\{ x_{1,1}^1, \dots, x_{i,j}^k, \dots, x_{n,lr(n)}^{ls(n)} \right\}$ – is a set of moments $x_{i,j}^k$ (for $l = 0$-th cycle) of operations $o_{i,j}^k$ beginning from the stream P_i^k. By analogy, the set $mX = \left\{ mx_{1,1}^1, \dots, mx_{i,j}^k, \dots, mx_{w,lm(w)}^{lms(w)} \right\}$ consists of moments of multimodal processes operations beginning, where: w – the number of multimodal processes, and $lm(i)$ – the number of operations of i-th multimodal process.

Variables $x_{i,j}^k / mx_{i,j}^k \in \mathbb{Z}$ determine moments of operations beginning in the l-th cycle of the SCCP cyclic steady state: $x_{i,j}^k(l) = x_{i,j}^k + l \cdot \alpha$ / $mx_{i,j}^k(l) = mx_{i,j}^k + l \cdot m\alpha$.

Since values of $x_{i,j}^k / mx_{i,j}^k$ follow the system structure parameters the cyclic behavior X_{SC} is determined by SC. Moreover, the multimodal processes behavior $(mX, m\alpha)$ also depends on the local cyclic processes behavior (X, α).

The constraints determining the admissible cyclic schedule (2) are following:

- **uncertainty requirements** (Tab. 1 and Tab. 2): level of workers' availability for operations handling S and level of production profitability E (caused by X_{SC}) for the whole SCCP, are determined as a minimal value $s_{i,j}$ among the all of workers' availabilities, and a minimal production profitability sT_i among the all multimodal processes, respectively:

$$S = \min_{i=1,2; j=1,...,4}\{s_{i,j}\} \text{ and } E = \min_{i=1,2; j=1,...,4}\{sT_i\} \tag{3}$$

- **constraints describing the local processes execution:** the moment of operation $o_{i,j}^k$ beginning states for a maximum of both: the completion time of the operation $o_{i,j-1}^k$ preceding $o_{i,j}^k$, and the release time of the resource $p_{i,j}^k$ awaiting for $o_{i,j}^k$ execution:

$$moment\ of\ o_{i,j}^k\ begining = max\{(moment\ of\ p_{i,j}^k\ release + lag\ time\ \Delta t),$$
$$(moment\ of\ operation\ o_{i,j-1}^k completion)\} \tag{4}$$

- **constraints describing multimodal processes execution:** the moment of operation $mo_{i,j}^k$ beginning is equal to the nearest admissible value (determined by set $\mathcal{X}_{i,j}^k$ of values $mx_{i,j}^k$) being a maximum of both: the completion time of operation $mo_{i,j-1}^k$ preceding $mo_{i,j}$, and the release time (lag time Δtm) of the resource $mp_{i,j}^k$ awaiting for $mo_{i,j}$ execution.

$$moment\ of\ mo_{i,j}^k\ begining = \lceil max\{(moment\ of\ mp_{i,j}^k\ release + lag\ time\ \Delta tm),$$
$$(moment\ of\ operation\ mo_{i,j-1}^k completion)\}\rceil_{\mathcal{X}_{i,j}^k} \tag{5}$$

where: $\mathcal{X}_{i,j}^k = \{x_{a,b}^c(l)\mid x_{a,b}^c(l) = x_{a,b}^c + l \cdot \alpha; l \in \mathbb{Z}\}$ – the set of admissible values of $mx_{i,j}^k$ determined by $x_{a,b}^k$, where: $x_{a,b}^k$ – the moment of operation $o_{a,b}$ beginning enabling execution of the operation $mo_{i,j}^k$, $\lceil a \rceil_B = \min\{k \in B : k \geq a\}$.

Constraints (3), (4) and (5) describe the conditions guaranteeing cyclic execution of SCCP processes subject to arbitrary given uncertainty requirements assuming employers availability S and production profitability E. A number of admissible schedules X_{SC} (2), denoted by L, depends on SC (1), and especially on the assumed operations times of local processes T (including transportation and layover times of AGVs).

Table 1. Awaiting working times of operations for multimodal processes $mP_1 mP_2$

Awaiting working times for mP_1	Awaiting working times for mP_2
$S_{1,1}$ graph, $mt_{1,1}$ axis (3 4 5 6 7 8) Working time $mt_{1,1}$ of operation assigned to R_1 is "about 5".	$S_{2,1}$ graph, $mt_{2,1}$ axis (3 4 5 6 7 8) Working time $mt_{2,1}$ of operation assigned to R_4 to "about 5".
$S_{1,2}$ graph, $mt_{1,2}$ axis (1 2 3 4 5) $mt_{1,2}$ is "about 3" (on R_2)	$S_{2,2}$ graph, $mt_{2,2}$ axis (1 2 3 4 5) $mt_{2,2}$ is "about 3" (on R_5)
$S_{1,3}$ graph, $mt_{1,3}$ axis (1 2 3 4 5) $mt_{1,3}$ is "about 3" (on R_5)	$S_{2,3}$ graph, $mt_{2,3}$ axis (1 2 3 4 5) $mt_{2,3}$ is "about 3" (on R_2)
$S_{1,4}$ graph, $mt_{1,4}$ axis (4 5 6 7 8 9) $mt_{1,4}$ is "about 6" (on R_6)	$S_{2,4}$ graph, $mt_{2,4}$ axis (3 4 5 6 7 8) $mt_{2,4}$ is "about 5" (on R_3)

Table 2. Awaited profitability measured by production takt of processes mP_1, mP_2

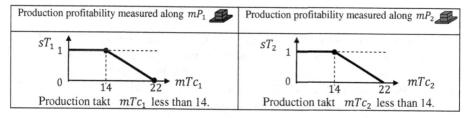

Production profitability measured along mP_1	Production profitability measured along mP_2
sT_1 graph (14, 22) mTc_1 Production takt mTc_1 less than 14.	sT_2 graph (14, 22) mTc_2 Production takt mTc_2 less than 14.

In that context different values of transportation and layover times $t_{i,j}^k$, determining different values of working time $mt_{i,j}^k$ and takts mTc_i, lead to schedules with different levels of workers' availability S and following these, levels of production profitability E (according to Tab. 1 and Tab. 2). Relevant illustration of that fact can be seen on Fig. 2 showing a number L of admissible schedules X_{SC} following (4) and (5) for SCCP from Fig. 1, while taking into account uncertainty requirements S and E (3). The results observed are obtained for operation times $t_{i,j}^k \in \{1 \ldots 10\}$. Space $\mathbb{L}$ from Fig. 2 consists 35559 (among 388979) cyclic schedules X_{SC} following constraints $E, S > 0$.

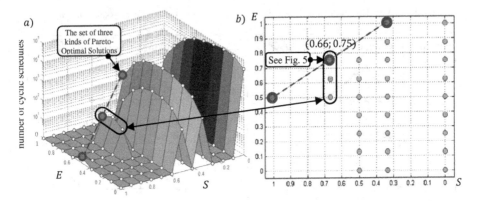

Fig. 2. Space of admissible schedules $\mathbb{L}$ including L solutions for system from Fig. 1) parameterized by S and E a), "top-down view" – elevation of $\mathbb{L}$ onto plane $S \times E$ b)

2.3   Problem Statement

Consider the SCCP specified (due to (1)) by the given sets of resources R, dispatching rules Θ^0 and Θ^1, local U, and multimodal process routes mU, as well. The response to the following question is sought: Does there exist a cyclic schedule X_{SC} which satisfies the requirements of employers availability S and production profitability E (S and E should be not less than assumed S_{TH} and E_{TH}) in the SCCP with dispatching rules Θ guaranteeing its assumed periodicity (in local and multimodal processes sense)?

The considered problem belongs to the class of reverse optimization problems:

Given are:

- the model (1) specified by the set of resources R, sets of routes U and M, sets of priority dispatching rules Θ^0 and Θ^1 while following constraints (3) - (5), i.e. resulting in the cyclic schedule X^{lp} encompassing the SCCP behavior,
- the lower limits of employers availability S and production profitability E, denoted by S_{TH} and E_{TH} respectively.

The response to the following question is sought:

What operation times T, mT, guarantee there exists the cyclic schedule X^{lp} guaranteeing maximum values of S and E : $S \rightarrow max$ and $E \rightarrow max$?

This kind of problem boils down to the following optimization constraint satisfaction problem.

$$CS_{XT} = \left(\left(\{X_{SC}, T_{SC}\}, \{D_X, D_T\} \right), C \right) \tag{6}$$

where: X_{SC}, T_{SC} – decision variables, X_{SC} – cyclic schedule (2), T_{SC} – sequence of operation times: $T_{SC} = (T, mT)$, D_X, D_T – domains determining admissible value of decision variables: D_X: $mx_{i,j}^k$, $x_{i,j}^k \in \mathbb{Z}$; D_T: $mt_{i,j}^k$, $t_{i,j}^k \in \{1 \dots 10\}$, C – the set of constraints describing execution of local and multimodal processes (3), (4), (5) as well as requirements constraints: $S \rightarrow max$ and $E \rightarrow max$.

Values of the variables X_{SC}, T_{SC}, following constraints C provide a solution to the problem CS_{XT}. That means they provide a solution to the problem (6), i.e. guaranteeing cyclic behavior of local and multimodal processes, while following: $S \rightarrow max$ and $E \rightarrow max$. Since S, E have to be optimized simultaneously, hence the solution following a trade-off between those conflicting objectives belongs to Pareto optimal ones. Well known constraints programming driven software platforms such as ILOG, OzMozart, ECL'PSE [10] can be used for solving (6).

3   Computational Experiments

Consider AGV system from Fig. 1. Given awaited profitability measured by production takts mTc_1 and mTc_2 of multimodal processes mP_1, mP_2, see Tab. 2 as well as awaiting working times of operations, see Tab. 1.

Request: Does there exist production flow specified by $mt_{i,j}$ and X_{SC} guaranteeing maximum values of S and E: $S \rightarrow max$ and $E \rightarrow max$?

This problem has been implemented and then computed in the constraint programming environment OzMozart system (Dual Core 2.67, GHz, 2.0, GB RAM). Due to Fig. 2 the search process took into account the space $\mathbb{L}$ of admissible solutions $\mathbb{L}$ containing over 3.8×10^5 items. The set of Pareto-Optimal Solutions (see Fig. 2) was obtained in less than one second.

Response: Three kinds of alternative solutions can be considered: (1;0.5), (0.66;0.75), (0.33;1) - see Fig. 2b (orange-blue points). Schedule X_{SC}, following (0.66;0.75), is shown on Fig. 3.

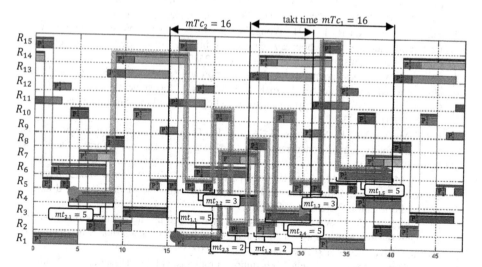

Fig. 3. Optimal schedule for $S = 0.66$ and $E = 0.75$

4   Conclusions

The proposed approach, aimed at AGVs fleet scheduling stated in terms of constraint satisfaction problem representation, provides a unified method for performance evaluation of local as well as supported by them multimodal processes. The sufficient conditions developed allow a designer to compose elementary systems in such a way as to obtain the final AGVs fleet schedule guaranteeing required quantitative and qualitative features. That leads to a method allowing one to replace the exhaustive search for the admissible control by a step-by-step design of a transportation system design following its required behavior. The domain of rescheduling problems subject to fuzzy operation time constraints and comparison this kind of approach with existing ones, determines our further research perspective.

References

1. Abara, J.: Applying integer linear programming to the fleet assignment problem. Interfaces 19, 4–20 (1989)
2. Bocewicz, G., Nielsen, I., Banaszak, Z.: Automated Guided Vehicles Fleet Match-up Scheduling with Production Flow Constraints. Engineering Applications of Artificial Intelligence 30, 49–62 (2014)
3. Hall, N.G., Sriskandarajah, C., Ganesharajah, T.: Operational Decisions in AGV-Served Flowshop Loops: Fleet Sizing and Decomposition. Annals of Operations Research 107, 189–209 (2001)
4. Krenczyk, D., Kalinowski, K., Grabowik, C.: Integration Production Planning and Scheduling Systems for Determination of Transitional Phases in Repetitive Production. In: Corchado, E., Snášel, V., Abraham, A., Woźniak, M., Graña, M., Cho, S.-B. (eds.) HAIS 2012, Part II. LNCS, vol. 7209, pp. 274–283. Springer, Heidelberg (2012)
5. Lu, S.P., Kong, X.T.R., Luo, H., Wong, W.R., Luo, K.B.: Dynamic scheduling of AGVS for tobacco automatic warehouse: A case study. In: Proc. of International Conference on Computers and Industrial Engineering, CIE, pp. 725–733 (2013)
6. Moudani, W., Mora-Camino, F.: A dynamic approach for aircraft assignment and maintenance scheduling by airlines. Journal of Air Transport Management 6, 233–237 (2000)
7. Pawlewski, P.: Multimodal approach to modeling of manufacturing processes. Procedia CIRP 17, 716–720 (2014)
8. Polak, M., Majdzik, P., Banaszak, Z., Wójcik, R.: The performance evaluation tool for automated prototyping of concurrent cyclic processes. Fundamenta Informaticae 60(1-4), 269–289 (2004)
9. Relich, M., Jakabova, M.: A Decision Support Tool for Project Portfolio Management with Imprecise Data. In: 10th International Conference on Strategic Management and its Support by Information Systems, pp. 164–172 (2013)
10. Sitek, P., Wikarek, J.: A hybrid framework for the modelling and optimisation of decision problems in sustainable supply chain management. International Journal of Production Research, 1–18 (2015), doi:10.1080/00207543.2015.1005762
11. Von Kampmeyer, T.: Cyclic scheduling problems. Ph.D. Dissertation, Mathematik/Informatik, Universität Osnabrück (2006)

The Effect of the Use of Motivators of Knowledge Sharing on the Outcome of Realized Business Strategies in Polish Manufacturing Enterprises

Justyna Patalas-Maliszewska

University of Zielona Góra
ul. Licealna 9 65-417 Zielona Góra/Poland
J.Patalas@iizp.uz.zgora.pl

Abstract. This article aims to elaborate on a model as a decision-making model for assessing the value of the outcome of a realized business strategy by using the GMDH method in a manufacturing company. The construction of the model includes values of significant motivators of knowledge sharing from the 119 Polish manufacturing companies. In particular, this study pays attention to the likely consequences and results of the use of motivation factors on the output of a realized strategy in a manufacturing company. This is followed by a discussion of the results of the empirical studies and of the supporting literature. The summary indicates potential directions for further work.

Keywords: manufacturing company, GMDH, motivators of knowledge sharing, strategy.

1 Introduction

Business strategy realization in a company can be a key to successfully obtaining a competitive advantage on the market. Mintzberg (1987) [16] defined a strategy as a plan, ploy, pattern, position and perspective. In accordance with Regner (2008) [21], this study takes the position that a strategy can be viewed and realized by an employee of a company. Machuca et al. (2011) [14] argued that formal strategy realization in a manufacturing company has a rather positive effect on operational performance. A realized strategy can act as a marker to indicate if a set of defined and desired results have been achieved. This study aims at bridging the gap of relationships between realized business strategies in manufacturing companies and motivators of knowledge sharing used in those enterprises.

Knowledge sharing among workers is a key factor of the development of an organization [5]. According to Liu and Phillips (2011) [12], encouraging employees to share knowledge across the organization can increase a firm's competitive advantages. In the literature, it is pointed out that employee knowledge sharing enhances a firm's performance [11]. According to Vera-Munoz, Ho and Chow (2006) [24], motivation is one of the most important factors that influences an employee to share their knowledge. In this study, the effect of the use of motivators of knowledge sharing in a

© Springer International Publishing Switzerland 2015
S. Omatu et al. (eds.), *Distributed Computing & Artificial Intelligence, 12th Int. Conference,*
Advances in Intelligent Systems and Computing 373, DOI: 10.1007/978-3-319-19638-1_37

manufacturing company is investigated via a simulation of a hypothetical model using the Group Method of Data Handling. This study tries to ascertain what the effects are of the use of motivation factors on the output of a realized strategy in a manufacturing company.

The remainder of this paper is organized as follows. Section 2 presents the theoretical background of the study. Section 3 describes the research model. Section 4 explains the research methodology and examines the research results. Section 5 discusses the implications of the results, provides a conclusion and highlights the limitations of the research.

2   Theoretical Background

Xu and Bernard [27] stated that knowledge in a manufacturing company is the interaction between actors and products which results in the change of product states. In this study, it is pointed out that knowledge sharing among workers in a manufacturing company should provide an improvement to the working methods which can be defined as the output of the implementation of a business strategy.

A few attempts have been made to define motivators of knowledge sharing among employees. Namjae et al. (2007) [18] identified agreeableness, conscientiousness, expertise, extrinsic and intrinsic motivation and strategy content as those factors. Kankanhalli et al. (2005) [10] argued that repositories, rewards, reciprocity, knowledge, self -efficacy, enjoyment in helping others, pro-sharing norms and the degree of usage of the repository are those factors. George and Chattopadhyay (2005) [9] stated that organizational identification is the most critical factor and Allen, Shore, & Griffeth (2003) [1] defined perceived organizational support as the indicator that most motivates workers to share their knowledge.

In this study the following the motivators of knowledge sharing among employees in a manufacturing company are defined: (1) Organizational identification (OF) [2], [3], [4], [9], [15], [22], [25], [26]: a good work atmosphere, troubleshooting, work experience, contact with customers, organizational commitment and internal communication and (2) Perceived Organizational Support (POS) [1], [6], [7], [17], [19]: training, coaching, information technology, internally organized meetings.

In accordance with Terziovski (2010) [23], this study defines the improvement of working methods as an outcome of realized business strategies in manufacturing companies.

The process of knowledge sharing in a company includes the externalization of knowledge by the knowledge transmitter and the internalization of knowledge by the receiver [13]. This study posits that knowledge (both explicit and tacit) is sent by the sender(s) to the receiver(s) by the use of different methods/tools, for example: IT systems, social software: wikis, blogs, etc.

The conceptual model shown in Fig. 1 depicts the relationships that it is examined:

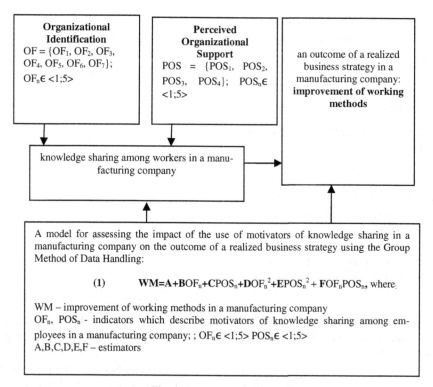

Fig. 1. A conceptual model

In order to facilitate the description of the defined relationships, the outcome of a realized business strategy should be organized in a standardized way, shown as follows:

The outcome of a realized business strategy (WM) is a set of indicators, noted as:

$$WM_1 = \begin{bmatrix} WM_1 \\ WM_2 \\ ... \\ WM_n \end{bmatrix}, \text{ where } n \in N \tag{1}$$

In this definition, WM_i is the i-th WM, 'and also included in the matrix are the indicators which describe the improvement of working methods in each of the 119 Polish manufacturing companies examined in this study. Each WM_i is associated with indicators that motivate workers to share their knowledge.

Therefore, motivators of knowledge sharing among employees in a manufacturing company base (KSM) provides a set of indicators, noted as:

$$KSM = \begin{bmatrix} GA_1 & TS_1 & WE_1 & CC_1 & OC_1 & IC_1 & T_1 & CO_1 & IT_1 & IOM_1 \\ GA_2 & TS_2 & WE_2 & CC_2 & OC_2 & IC_2 & T_2 & CO_2 & IT_2 & IOM_2 \\ \ldots & \ldots & \ldots & \ldots & \ldots & \ldots & \ldots & \ldots & \ldots & \ldots \\ GA_n & TS_n & WE_n & CC_n & OC_n & IC_n & T_n & CO_n & IT_n & IOM_n \end{bmatrix} \tag{2}$$

where $n \in N$

- GA   - A Good Work Atmosphere
- TS -   Troubleshooting
- WE – Work Experience
- CC -   Contact with Customers
- OC -   Organizational Commitment
- IC - Internal Communication
- T – Training
- CO – Coaching
- IT -   Information Technology
- IOM - Internal Organized Meetings.

In this definition, GA_i, TS_i, WE_i, CC_i, OC_i, IC_i, T_i, CO_i, IT_i, IOM_i are the i-th KSM, and they are the indicators which describe motivators of knowledge sharing in each manufacturing company.

A model for assessing the impact of the use of the motivators of knowledge sharing among employees in a manufacturing company on the outcome of a realized business strategy was built using the Group Method of Data Handling.

The multilevel GMDH allows for an optimized synthesis of a mathematical model for a given class of regression functions, and it can be used in both evaluating criteria and in quality assessment [8],[20]. Both elements of the algorithm are defined arbitrarily by the author. In this study the elements are defined as: the significance of motivators of knowledge sharing in each of the 119 Polish manufacturing companies involved in the study and the outcome of business strategy realization.

3   Research Results

The research model posits, from the aforementioned argument, that when workers in a manufacturing enterprise are motivated to share knowledge, they are able to improve the working methods of that enterprise. Before the survey was carried out, it was assumed that those companies which took part in the research would have a business strategy and be able to realize it. Additionally, it was assumed that the workers of the companies would be motivated to share their knowledge.

The surveys used for testing the research model were developed by defining scales to fit the knowledge codification context. A zero-one scale was used for an outcome of realized business strategy survey items and a five-point scale was used for motivators of knowledge sharing survey items. The data for this study were collected from 119 Polish manufacturing companies between January to September, 2014 (see Table 1):

Table 1. Profile of companies and respondents

	Items	Frequency (N=119)
Manufacturing companies	Industry	88 (74%)
	Construction	16 (13%)
	Others	15 (13%)
Department of the company in which the respondent works	Management	95 (80%)
	Sales and Marketing	24 (20%)

This study presents the possibility of defining a model as a decision-making model for assessing the value of the outcome of a realized business strategy by using the GMDH method. The construction of the model includes values of significant motivators of knowledge sharing from the 119 Polish manufacturing companies. Additionally, it includes a statement of the outcome of realized business strategies based on an empirical analysis of the 119 Polish manufacturing companies. In accordance with the data received from 119 Polish manufacturing companies, all the variations of the GMDH algorithms were investigated in the Consulting IT computer software system written by the author [20]. As a result of the implementation of the algorithm, the best possible polynomial was obtained; this was characterized by the lowest value criteria for regularity assigned to the pair object. In this way, the best polynomials are chosen; which is the one with the smallest error of modelling:

$$WM(OC,T) = 1.83 - 0.76OC - 0.83T + 2.71OC^2 + 2.86T^2 - 4.73OC \times T \qquad (3)$$

where,

- OC - the significance of the use of the motivator of knowledge sharing: organizational commitment in a manufacturing company
- T - the significance of the use of the motivator of knowledge sharing: training in a manufacturing company
- WM (OC,T) - the statement of the outcome as an improvement of working methods in a manufacturing company as a result of a realized business strategy

The estimated value of the outcome of a realized business strategy in a manufacturing company using an WM model is presented in Fig 2:

The highest values of the outcome of a realized business strategy as an improvement of working methods in a manufacturing company can be achieved if the significance of one of the motivators of knowledge sharing is very high (5 points), and the second one is not important (1 point). For example if the significance of the use of the motivator of knowledge sharing defined as organizational commitment is assessed by 5 points, and training by 1 point, it may be possible to achieve the value of the outcome of a realized business strategy as 44.16. Unfortunately, if each motivator of knowledge sharing is equally important; then the value of the outcome of the realized business strategy is decreased.

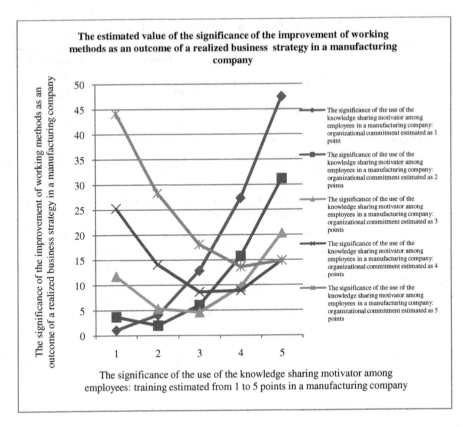

Fig. 2. The estimated value the outcome of a realized business strategy as an improvement of working methods in a manufacturing company using a WM model

4   Conclusions

The results of this study demonstrate the clear and measurable existence of a positive effect of the use of motivators of knowledge sharing in a manufacturing company on the outcome of a realized business strategy based on research results from 119 Polish manufacturing companies. Specifically, the results reveal these effects on the significance of the defined motivators of knowledge sharing.

By proposing an WM model which addresses the influence of knowledge sharing among employees using the motivators, this study contributes to a filling of the gap which exists in the literature. This study suggests that it may be a good idea to share useful knowledge because it could increase the outcome of realized business strategies in Polish manufacturing companies.

Like all studies, this one has certain limitations that further research should aim to overcome. Firstly, because the intention is to analyze Polish manufacturing companies, this study focuses on Polish industry. It would be unwise to generalize the findings too broadly to other countries. Furthermore, all the variables were measured at the same moment in time. So, it would be useful to provide such research over a longer time period and at different stages. These conclusions and limitations suggest proposals for future research directions.

References

1. Allen, D.G., Shore, L.M., Griffeth, R.W.: The role of perceived organizational support and supportive human resource practices in the turnover process. Journal of Management 29(1), 99–118 (2003)
2. Allen, N.J., Meyer, J.P.: The measurement of antecedents of affective, continuance, and normative commitment to the organization. Journal of Occupational Psychology 63, 1–18 (1990)
3. Bennet, R.: Employers' demands for personal transferable skills in graduates: A content analysis of 1000 job advertisements and an associated empirical study. Journal of Vocational Education & Training 54(4), 457–476 (2002)
4. Bodrow, W.: Toward European knowledge enterprises. Journal of Intelligent Manufacturing 18, 459–466 (2007)
5. Carmeli, A., Gelbard, R., Goldriech, R.: Linking perceived external pres-tige and collective identification to collaborative behaviors in R&D teams. Expert Systems with Applications 38(7), 8199–8207 (2011)
6. Chang, E.: Employees' overall perception of HRM effectiveness. Human Relations 58(4), 523–544 (2005)
7. Evers, W., Brouwers, A., Tomic, W.: A quasi-experimental study on management coaching effectiveness. Consulting Psychology Journal: Practice and Research 58(3), 174–182 (2006)
8. Farlow, S.J. (ed.): Self-organizing Methods in Modelling: GMDH-type Algorithms. Marcel Dekker Inc., New York (1984)
9. George, E., Chattopadhyay, P.: One foot in each camp: The dual identificationof contract workers. Administrative Science Quarterly 50(1), 68–99 (2005)
10. Kankanhalli, A., Tan, B., Wei, K.: Contributing knowledge to electronic repositories: An empirical investigation. Mis Quarterly 29(1), 113–143 (2005)
11. Liao, S.H., Fei, W.C., Chen, C.C.: Knowledge sharing, absorptive capacity, and innovation capacity: An empirical study of Taiwan's knowledge intensive industries. Journal of Information Science 33, 340–359 (2007)
12. Liu, Y., Phillips, J.S.: Examining the antecedents of knowledge sharing in facilitating team innovativeness from a multilevel perspective. International Journal of Information Management 31, 44–52 (2011)
13. Liu, M., Liu, N.: Sources of knowledge acquisition and patterns of knowledge sharing behaviors—An empirical study of Taiwanese high-tech firms. International Journal of Information Management 28(5), 423–432 (2008)
14. Machuca, J., Jiménez, C., Garrido-Vega, P., de los Ríos, J.: Do technology and manufacturing strategy links enhance operational performance? Empirical research in the auto supplier sector. International Journal of Production Economics 133, 541–550 (2011)
15. McDermott, R.: Learning across teams. Knowledge Management Review 8(3), 32–36 (1999)
16. Mintzberg, H.: The strategy concept 1: Five Ps for strategy. California Management Review 30, 11–24 (1987)

17. Mullany, L.: "Girls on tour": Politeness, small talk, and gender in managerial business meetings. Journal of Politeness Research 2, 55–77 (2006)
18. Namjae, C., Guo Zheng, L., Che-Jen, S.: An empirical study on the effect of individual factors on knowledge sharing by knowledge type. Journal of Global Business and Technology 3 (2007)
19. Park, K.H., Favrel, J.: Virtual enterprise - information system and networking solution. Computers & Industrial Engineering 37(1-2), 441–444 (1999)
20. Patalas-Maliszewska, J.: Knowledge Worker Management: Value Assessment, Methods, and Application Tools. Springer, Heidelberg (2013)
21. Regnér, P.: Strategy-as-practice and dynamic capabilities: Steps towards a dynamic view of strategy. Human Relations 61, 565–588 (2008)
22. Sharma, P., Tam, J.L.M., Kim, N.: Demystifying intercultural service encounters. Toward a comprehensive conceptual model. Journal of Service Research 12, 227–242 (2009)
23. Terziovski, M.: Innovation practice and its performance implications in small and medium enterprises (SMEs) in the manufacturing sector: a resource-based view. Strategic Management Journal 31 (2010)
24. Vera-Muñoz, S.C., Ho, J.L., Chow, C.W.: Enhancing knowledge sharing in public accounting firms. Accounting Horizons 20, 133–155 (2006)
25. Vomlelová, M., Vomlel, J.: Troubleshooting: NP-hardness and solution methods. Soft Computing 7(5), 357–368 (2003)
26. Welch, M., Jackson, P.R.: Rethinking internal communication: A stakeholder approach. Corporate Communications: An International Journal 12(2), 177–198 (2007)
27. Yang, X., Alain, B.: Quantifying the value of knowledge within the context of product development 24, 166–175 (2011)

A Declarative Approach to Decision Support in Sustainable Supply Chain Problems

Paweł Sitek

Department of Control and Management Systems, Kielce University of Technology, Poland
sitek@tu.kielce.pl

Abstract. The paper presents the concept of a declarative approach to decision support for sustainable supply chain (SSC) problems. This approach proposed here combines the strengths of mathematical programming (MP) and constraint logic programming (CLP), which leads to a significant reduction in the search time necessary to find the solution, and allows solving larger problems.

It also presents the implementation of the proposed approach in the form of the declarative decision support platform (DDSP). The DDSP allows the implementation of complete decision-making models, constraints as well as a set of questions for these models. The illustrative examples presented in the paper illustrate efficiency and possibilities of this approach.

Keywords: Decision Support, Constraint Logic Programming, Mathematical Programming, Optimization, Sustainable Supply Chain.

1   Introduction

The principal challenge of supply chain management is to maintain a regular and undisrupted flow of goods, services, information, and financial resources while minimizing costs. Along with complexity, enormous numbers of constraints relating to resources, time, production capacity, distribution capacity, transportation, etc. characterize today's supply chains. Nowadays, given the new constraints related to the availability of non-renewable resources (coal, petroleum, natural gas, etc.), enterprises are more than ever obliged to rethink their strategies to ensure the sustainability of their operations. Another group of new constraints results from novel research areas dealing with the actions related to one or more phases of the product life cycle such as product design, production planning and control for remanufacturing, product recovery, reverse logistics and carbon emissions reduction. In this context, sustainable development means the interdependence between three aspects: the economic, the environmental, and the social performance of an enterprise. An integrated approach that links supply chain decisions to the three key elements of sustainability is the ultimate goal of sustainable supply chain management (SSCM) [7]. Introduction of the new challenges and redefinition of management goals and objectives have to result in the modification and expansion of decision and optimization models. Most decision and optimization models for solving problems in supply chain management (SCM)

© Springer International Publishing Switzerland 2015

S. Omatu et al. (eds.), *Distributed Computing & Artificial Intelligence, 12th Int. Conference,*
Advances in Intelligent Systems and Computing 373, DOI: 10.1007/978-3-319-19638-1_38

[6] and in sustainable supply chain management are derived from the operations research (OR) areas, mathematical programming (MP) in particular. Due to a variety of constraint types and logical interdependencies between them, the constraint-based environment [1,11] can offer a very good framework for representing the knowledge and information needed for the decision support in SCM and SSCM.

The main contribution is proposing a declarative approach with the transformation of the problem in the area of SSCM. The implementation of the Declarative Decision Support Platform (DDSP) and its application in the context of building hybrid models with logic and symbolic constraints are presented.

The rest of the article is organized as follows. Research methodology and motivation are provided in Section 2. In Section 3 declarative is described and the implementation aspects discussed. The decision model presented as an illustrative example is provided in Section 4. Computational examples and tests of the implementation platform are presented in Sub-section 4.4. The possible extensions of the proposed approach as well as the conclusions are included in Section 5.

2   Methodology and Motivation

Based on numerous studies and our own experience, it shows that the constraint-based environment [1,10,11,12,13], offers a very good framework for representing the knowledge, information and methods needed for the decision support. The constraint logic programming (CLP) is a particularly interesting option in this context. The CLP is a form of constraint programming (CP), in which logic programming is extended to include concepts from constraint satisfaction [1]. A constraint logic program contains constraints in the body of clauses (predicates). Effective search for the solution in the CLP depends considerably on the effective constraint propagation, which makes it a key method of the constraint-based approach. Constraint propagation embeds any reasoning that consists in explicitly forbidding values or combinations of values for some variables of a problem because a given subset of its constraints cannot be satisfied otherwise [1].

Based on [1,2,4,11] and our previous work [8,13,14], we observed some advantages and disadvantages of MP and CLP environments. An integrated approach of CLP and MP can help to solve optimization and decision problems that are intractable with either of the two methods alone [3,5].

Both the MP and the finite domain CLP involve variables and constraints. However, the types of the variables and constraints that are used, and the ways the constraints are solved are different in the two approaches [3]. In both, the MP and CLP, there is a group of constraints that can be solved with ease and a group of constraints that are difficult to solve. The easily solved constraints in the MP are linear equations and inequalities over rational numbers. Integrity constraints are difficult to solve using mathematical programming methods and often the real problems of the MIP make them NP-hard. In the CP/CLP, domain constraints with integers and equations between variables are easy to solve. The system of such constraints can be solved over integer variables in polynomial time. The inequalities between variables, general linear constraints, and symbolic constraints are difficult to solve in the CP/CLP (NP-hard). This type of constraints reduces the strength of constraint propagation. The MP approach focuses mainly on the methods of solving and, to a lesser degree, on the structure of the problem. The data, however, is completely outside the model. The same model without any changes can be solved for multiple instances of data. In the CLP approach, due to its declarative nature, the methods are already in place. The data and structure of the problem are used for its modeling.

Observations above together with the knowledge of the properties of CLP and MP systems enforce the integration. The integration called hybridization consists in the combination of both systems and the transformation of the modeled problem.

The motivation underlying this research was to implement this approach as a DDSP to support managers in the modeling and optimization of decision problems in SSCM. This solution is better than using the MP or the CLP separately. What is difficult to solve in one environment can be easy to solve in the other. And the declarative character of the approach enables the managers to ask all kinds of questions at any level of the management process.

3   A Declarative Approach – Concept and Implementation

The declarative approach to modeling and optimization of decision problems in SSCM is able to bridge the gaps and eliminate the drawbacks that occur in both MP and CLP approaches. To support this concept, the Declarative Decision Support Platform (DDSP) is proposed, where:

- all types of questions can be asked: general questions1: *Is it possible ...? Is it feasible ...?* and specific questions2: *What is the minimum ...? What is the maximum ...?* (the list of example questions in the DDSP is shown in Table 1);
- knowledge related to the sustainable supply chain can be expressed in the form of linear, logical and symbolic constraints;
- complete models can be implemented, such as decision models, linear models, integer programming models, declarative and hybrid models as well as dynamically appearing constraints;
- the problem can be transformed for an improved constraint propagation;
- for general questions, only the CLP environment and domain solutions fare used;
- for wh-questions the MP environment is used additionally.

Table 1. A set of sample questions for the DDSP

Question	Description
Q1	Is timely execution of orders possible with a given production cost K_5?
Q2	Is timely execution of orders possible with a given transportation cost $K_3+K_4+K_8$?
Q3	What is the minimum environmental cost K_2 of timely execution of orders?
Q4	What is the minimum cost of execution of orders for not more than N1 distribution centers?
Q5	What is the minimum cost of execution of orders for not more than N2 recycling centers?
Q6	Is timely execution of orders possible with the specified number of means of transport d?
Q7	Is timely execution of orders possible with the specified production capacity W?
Q8	What is the maximum profit K_7 from recycling?
Q9	What is the minimum cost of timely execution of orders?

Linking various types of constraints in one environment and using the best and already proved problem optimization and transformation methods constitute the base of the DDSP architecture. The concept and architecture of this platform is presented in Figure 1. From a variety of tools for the implementation of CLP techniques in the hybrid solution platform, ECL^{i}PS^e software [15] was selected. ECL^{i}PS^e is an open-source software system for the cost-effective development and deployment of constraint programming applications. Environment for the implementation of MP in DDSP was EPLEX. EPLEX is MP-library in ECL^{i}PS^e .

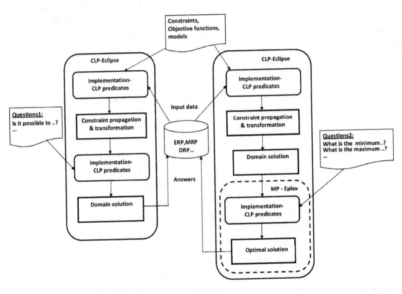

Fig. 1. Detailed scheme of the declarative decision support platform (DDSP)

4   Illustrative Example

The DDSP proposed was verified and tested for the illustrative example, which is the authors' original model of cost optimization for the supply chain with multimodal transportation and recycling [8]. The proposed model is the cost model taking into account different types of parameters, i.e., space (area/volume occupied by the product, distributor capacity and capacity of transportation unit, recycling center capacity), time (duration of delivery and service by distributor, etc.) and a transportation mode. Multimodality in this example is understood as the possibility of using different modes of transportation: railway, commercial vehicles, heavy trucks, etc. It is a simplified SSCM problem in which financial and environmental dimensions are included. The environmental dimensions refer to both transportation and recycling. The introduced environmental costs of the use of the given transportation means are constant and dependent on the transportation means type. The additional environmental costs are

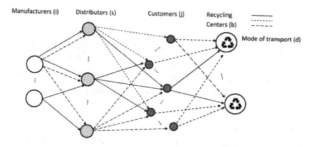

Manufacturers (i)   Distributors (s)   Customers (j)   Recycling
Centers (b)

Mode of transport (d)

Fig. 2. The simplified structure of the sustainable supply chain network

hidden in the transportation costs and depend on the route followed and on the quantity of the goods transported. The simplified structure of the sustainable supply chain network for this example has been shown at the Figure 2. The parameters, indices, decision variables for illustrative model have been presentenced in Table 2.

4.1   Objective Function

The objective function (1) defines the aggregate costs of the entire chain and consists of nine elements. The first element K_1 comprises the fixed costs associated with the operation of the distributor involved in the delivery (e.g. distribution center, warehouse, etc.). The second K_2 and eight K_8 elements correspond to environmental costs of using various means of transportation. Those costs are dependent on the number of courses of the given means of transportation, and additionally, on the environmental levy, which in turn may depend on the use of fossil fuels and carbon-dioxide emissions. The third K_3 component determines the cost of the delivery from the manufacturer to the distributor. Another component K_4 is responsible for the costs of the delivery from the distributor to the end user (the store, the individual client, etc.). The next component K_5 of the objective function determines the cost of manufacturing the product by the given manufacturer. The sixth K_6 component determines the fixed costs associated with recycling. The seventh K_7 component of the objective function represents the profit earned from the sale of products after recycling as the raw material for producers or suppliers. There is a minus sign next to it because the remaining components of the objective function are cost components, and the profit reduced the costs. The last component defines the costs of the product delivery from customers to the recycling centers. The objective function is minimized.

$$\sum_{s=1}^{E}(F_s \cdot Tc_s) + \sum_{d=1}^{L}Od_d(\sum_{i=1}^{N}\sum_{s=1}^{E}Xb_{i,s,d} + \sum_{s=1}^{E}\sum_{j=1}^{M}Yb_{j,s,d}) + \sum_{i=1}^{N}\sum_{s=1}^{E}\sum_{d=1}^{L}Koa_{i,s,d} + \sum_{s=1}^{E}\sum_{j=1}^{M}\sum_{d=1}^{L}Kog_{s,j,d} + \sum_{i=1}^{N}\sum_{k=1}^{O}(C_{ik} \cdot \sum_{s=1}^{E}\sum_{d=1}^{L}X_{i,s,k,d})$$

$$+ \sum_{b=1}^{U}(Fr_b \cdot Tb_b) - \sum_{b=1}^{U}\sum_{k=1}^{O}(Cb_{bk} \cdot \sum_{j=1}^{M}\sum_{d=1}^{L}Zr_{j,b,k,d}) + \sum_{d=1}^{L}Od_d(\sum_{b=1}^{U}\sum_{j=1}^{M}Zb_{j,b,d}) + \sum_{j=1}^{M}\sum_{b=1}^{U}\sum_{d=1}^{L}Kr_{j,b,d} \cdot Zb_{j,b,d} \qquad (1)$$

4.2   Constraints

The model was based on constraints (2) .. (23). Constraint (2) specifies that all deliveries of product k produced by the manufacturer i and delivered to all distributors s using mode of transportation d do not exceed the manufacturer's production capacity. Constraint (3) covers all customer j demands for product k $(Z_{j,k})$ through the implementation of delivery by distributors s (the values of decision variables $Y_{j,s,k,d}$). The flow balance of each distributor s corresponds to constraint (4). The possibility of delivery is dependent on the distributor's - constraint (5a) and recycling centres' - (5b) technical capabilities. Time constraint (6) ensures the terms of delivery are met. Constraints (7) and (8) guarantee deliveries with available transportation taken into account. Constraints from (9) to (11) set values of decision variables based on binary variables Tc_s, $Xa_{i,s,d}$ $Ya_{s,j,d}$ $Za_{j,b,d}$ Tb_b. The remaining constraints up to (23) arise from the nature of the model (MILP-Mixed Integer Linear Programming). Formalization of the constraints and a detailed description of the model are presented in [8].

Table 2. Summary indices, parameters and decision variables

Symbol	Description		
Indices			
N	number of manufacturers/factories	k	product type k=1..O
M	number of customers	j	delivery point/customer/city j=1..M
E	number of distributors	i	manufacturer/factory i=1..N
O	number of product types	s	distributor /distribution center s=1..E
L	number of mode of transportation	d	mode of transportation d=1..L
U	number of recycling centers	b	recycling center b=1..U
Input parameters			
F_s	the fixed cost of distributor/distribution center s		
Od_d	the environmental cost of using mode of transportation d		
$C_{i,k}$	the cost of product k at factory i		
Fr_b	the fixed cost of recycling center b		
$Cb_{b,k}$	the value of product k at recycling center b		
$Kr_{j,b,d}$	the fixed cost of delivery from customer j to recycling center b using mode of transportation d		
Decision variables			
Tc_s	if distributor s participates in deliveries, then Tc_s=1, otherwise Tc_s=0		
$Xb_{i,s,d}$	the number of courses from manufacturer i to distributor s using mode of transportation d		
$Yb_{s,j,d}$	the number of courses from distributor s to customer j using mode of transportation d		
$X_{i,s,k,d}$	delivery quantity of product k from manufacturer i to distributor s using mode of transportation d		
Tb_b	if recycling center b participates in deliveries, then Tb_b=1, otherwise Tb_b=0		
$Zb_{j,b,d}$	the number of courses from customer j to recycling center b using mode of transportation d		
$Zr_{j,b,k,d}$	delivery quantity of product k from customer j to recycling center b using mode of transportation d		
Values calculated			
$Koa_{i,s,d}$	the total cost of delivery from manufacturer i to distributor s using mode of transportation d		
$Kog_{s,j,d}$	the total cost of delivery from distributor s to customer j using mode of transportation d		

4.3   Model Transformation

The transformation is an important and inseparable part of the DDSP (see Figure 1). Due to the nature of the decision and optimization problems in SCM, SSCM (adding up decision variables and constraints involving a lot of variables), the constraint propagation efficiency decreases dramatically. The idea was to transform the problem by changing its representation without changing the very problem. At the stage of transformation, the structure of, and maximum knowledge about the problem have to be used, including the knowledge about the orders, technical capacity of the producers, distributors and recycling centers. All permissible routes were first generated based on the fixed data and a set of orders, then the specific values of parameters i, s, k, d, b were assigned to each of the routes. In this way, only decision variables X and Zr (deliveries) had to be specified. This transformation fundamentally improved the efficiency of the constraint propagation and reduced the number of backtracks and decision variables. This is due to the simple fact that it should be set-values for the two decision variables instead of seven.

4.4   Numerical Experiments

All the experiments relate to the supply chain with five manufacturers (i=1..5), four distributors (s=1..4), ten customers (j=1..10), three recycling centers (b=1..3), four modes of transportation (d=1..4), fifteen types of products (k=1..15), and five sets of orders (P1(10), P2(20), P3(40), P4(60), P5(120), -(n)-the number of orders in set P).

Computational experiments consisted in asking questions (Table 1) for the model (Section 4.1,4.2), [8] implemented in the DDSP. Question Q1 to Q8 were asked many times, for the set of different parameters for example P4(60). In order to compare the results and effectiveness of the DDSP, the model (Section 4.1, 4.2) was also implemented in mathematical programming environment for question Q9.

Table 3. Results of numerical experiments

| P(No) | Q9 | | | | | | | | Q1 | |
| | Declarative Approach-DDSP | | | | Mathematical Programming | | | | | |
	Fc	Time	V	C	Fc	Time	V	C	K₅	Answ
P5(120)	79068.70	178	5706(3847)	1029	83456.30*	900**	10293(7927)	39300	5270	Yes
P4(60)	44473.00	145	4266(2407)	1029	46303.00*	900**	10293(7927)	24900	5200	No
P3(40)	28954.00	123	3616(1767)	1029	29664.00*	900**	10293(7927)	18500	Q3	Q8
P2(20)	14289.30	57	2976(1127)	1029	14434.70*	900**	10293(7927)	12100	K₂	K₇
P1(10)	7297.67	48	2646(607)	1019	7348.67*	900**	10293(7927)	6900	39810	3434,07

| Q2 | | Q4 | | Q5 | | Q6 | | Q7 | |
K₃+K₄+K₈	Answ	N1	Fc	N2	Fc	d	Answ	W	Answ
42526	Yes	1	NFSF	1	NFSF	50	No	50	No
42000	No	2	44473.00	2	45008.00	70	No	75	No
41000	No	3	44473.00	3	44473.00	80	Yes	100	Yes

Fc   the value of the objective function   d the number of means of transport
Time time of finding solution (in seconds)   W production capacity
**   the calculation was stopped after 900 s   * the feasible value of the objective function after the time T
V(V)/ C   the number of decision variables (integer decision variables)/constraints

5   Conclusion

This paper provides a robust and effective declarative approach to modeling and decision support of SSCM problems, implemented with the DDSP which incorporates two environments (i) mathematical programming (Eplex) and (ii) constraint logic programming (ECL^{i}PS^e). The model presented as an example is transformed using the DDSP, which results in the new representation of the problem. For the specific questions that have to be optimized, the application of this approach leads to a substantial reduction in (i) the number of decision variables (up to four times), (ii) the number of constraints (up to seven times) (iii) computing time. The proposed solving method with DDSP is recommended for decision-making problems whose structure is similar to that of the presented models (Section 4). This structure is characterized by (i) constraints and the objective function, in which decision variables are added up, and (ii) logical constraints that are difficult to implement in mathematical programming-based models. The DDSP allows providing decision support (questions Q8, Q9) for three-fold larger order sets than those in the approach proposed in [8]. The classical approach does not provide answers to question Q9 within acceptable time (Table 3).

In the versions to follow, implementation is planned of other supply chain layers such as remanufacturing, reverse logistic, etc. and agile supply chain [9].

References

1. Apt, K., Wallace, M.: Constraint Logic Programming using Eclipse. Cambridge University Press, Cambridge (2006)
2. Bocewicz, G., Nielsen, I., Banaszak, Z.: Iterative multimodal processes scheduling. Annual Reviews in Control 38(1), 113–132 (2014)
3. Bockmayr, A., Kasper, T.: A Framework for Combining CP and IP, Branch-and-Infer, Constraint and Integer Programming: Toward a Unified Methodology. Operations Research/Computer Science Interfaces 27, 59–87 (2014)
4. Dang, Q.-V., Nielsen, I.E., Bocewicz, G.: A genetic algorithm-based heuristic for part-feeding mobile robot scheduling problem. In: Rodríguez, J.M.C., Pérez, J.B., Golinska, P., Giroux, S., Corchuelo, R. (eds.) Trends in PAAMS. AISC, vol. 157, pp. 85–92. Springer, Heidelberg (2012)
5. Milano, M., Wallace, M.: Integrating Operations Research in Constraint Programming. Annals of Operations Research 175(1), 37–76 (2010)
6. Mula, J., Peidro, D., Diaz-Madronero, M., Vicens, E.: Mathematical programming models for supply chain production and transportation planning. European Journal of Operational Research 204, 377–390 (2010)
7. Seuring, S., Müller, M.: From a Literature Review to a Conceptual Framework for Sustainable Supply Chain Management. Journal of Cleaner Production 16, 1699–1710 (2008)
8. Sitek, P., Wikarek, J.: A hybrid framework for the modelling and optimisation of decision problems in sustainable supply chain management. International Journal of Production Research, 1–18 (2015), doi:10.1080/00207543.2015.1005762
9. Grzybowska, K., Kovács, G.: Developing Agile Supply Chains – System Model, Algorithms, Applications. In: Jezic, G., Kusek, M., Nguyen, N.-T., Howlett, R.J., Jain, L.C. (eds.) KES-AMSTA 2012. LNCS, vol. 7327, pp. 576–585. Springer, Heidelberg (2012)
10. Wikarek, J.: Implementation aspects of hybrid solution framework. In: Szewczyk, R., Zieliński, C., Kaliczyńska, M. (eds.) Recent Advances in Automation, Robotics and Measuring Techniques. AISC, vol. 267, pp. 317–328. Springer, Heidelberg (2014)
11. Rossi, F., Van Beek, P., Walsh, T.: Handbook of Constraint Programming. Foundations of Artificial Intelligence. Elsevier Science Inc., New York (2006)

12. Wikarek, J.: Implementation Aspects of Hybrid Solution Framework. In: Szewczyk, R., Zieliński, C., Kaliczyńska, M. (eds.) Recent Advances in Automation, Robotics and Measuring Techniques. AISC, vol. 267, pp. 317–328. Springer, Heidelberg (2014)
13. Sitek, P., Wikarek, J.: A Hybrid Approach to the Optimization of Multiechelon Systems. Mathematical Problems in Engineering 2014, Article ID 925675 (2014), doi:10.1155/2014/925675
14. Sitek, P., Wikarek, J.: Hybrid Solution Framework for Supply Chain Problems. In: Omatu, S., Bersini, H., Corchado Rodríguez, J.M., González, S.R., Pawlewski, P., Bucciarelli, E. (eds.) Distributed Computing and Artificial Intelligence 11th International Conference. AISC, vol. 290, pp. 11–18. Springer, Heidelberg (2014)
15. Eclipse 2014, Eclipse - The Eclipse Foundation open source community website (2014), http://www.eclipse.org (accessed August 12)

A Knowledge-Based Approach to Product Concept Screening

Marcin Relich[1], Antoni Świć[2], and Arkadiusz Gola[3]

[1] Faculty of Economics and Management, University of Zielona Gora, Zielona Gora, Poland
m.relich@wez.uz.zgora.pl
[2] Faculty of Mechanical Engineering, Lublin University of Technology, Lublin, Poland
a.swic@pollub.pl
[3] Faculty of Management, Lublin University of Technology, Lublin, Poland
a.gola@pollub.pl

Abstract. This paper is concerned with developing a knowledge-based approach for selecting portfolio of product concepts for development. The critical success factors for new product development are identified on the basis of information acquired from an enterprise system, including the fields of sales and marketing, project management, and production. The model of new product screening consists of enterprise functional domains and business information systems. The model has been described in terms of a constraint satisfaction problem (CSP) that contains a set of decision variables, their domains, and the constraints. Knowledge base is specified according to CSP framework and it reflects the company's resources and relationships identified. The illustrative example presents the use of fuzzy neural network to estimating the success of new products and constraint programming to product concept screening in the context of the different search strategies.

Keywords: project management, new product development, concept selection, decision support system, constraint satisfaction problem, constraint programming.

1 Introduction

New product development (NPD) is the critical process that aims to maintain company's competitive position and continue business success. This process consists of the stages such as idea generation, evaluation and screening of product concepts (ideas), development of the selected concepts, testing and commercialization of new products [1]. In order to increase a chance of successful product, the project manager should select the most promising set of concepts according to company's constraints. One of the characteristics of many industrial companies concerns the management of several simultaneously developed new products using the same resources. The complexity of the NDP process results in developing a task-oriented tool to support the decision-makers, especially in the field of evaluation and screening of product concepts.

© Springer International Publishing Switzerland 2015

S. Omatu et al. (eds.), *Distributed Computing & Artificial Intelligence, 12th Int. Conference,*
Advances in Intelligent Systems and Computing 373, DOI: 10.1007/978-3-319-19638-1_39

As the success of the NPD projects is closely connected with the success of the entire company, the NPD projects play an important role in an organisation's growth and development [2]. To survive and succeed in the dynamic business environment, companies usually focus on several areas to improve their new product development, such as identifying customer needs for continuous new product development, improving product quality, and accelerating the process of commercialization [3].

The concept selection aims to identify the most promising products for development, and consequently, to reduce the potential expenses on the unsuccessful NPD projects. From this point of view, the concept selection is the crucial stage of the NPD process. Selection of new products for further development usually bases on metrics of the product success, and it should also take into account the company's resources. The success of a product is estimated in this study on the basis of the previous developed products which specification can be retrieved from an enterprise system. This system includes project management software, enterprise resource planning (ERP) system, customer relationship management (CRM) system, and computer aided design (CAD) system.

Knowledge creation and management through the new product development and management processes is of significant interest in the context of recent technology and infrastructure changes. Knowledge management is understood as the process of identification, sharing, and applying knowledge [4]. The advancement of information technology helps today's organisations in business management processes and collecting data that is potential source of information [5]. A main task faced by NPD projects is how to acquire knowledge and sustain success rate among the products. This paper aims to present the use of fuzzy neural network to estimating the success of new products and constraint programming to seeking the most promising set of potential products. The proposed approach takes into account data of the previous projects that is stored into an enterprise system. The next section presents the model of new product screening that is specified in term of a constraint satisfaction problem that in turn can be considered as a knowledge base enabling the design of a decision support system.

2   Model of New Product Screening

The development of new product projects depends on the external and internal factors. The external factors include customer demand, changes in regulations, technology or financial limitations. In turn, internal factors concern the resources and processes that appear in the different fields of an organization. Specifications of the previous products, customer requirements, design parameters, and product portfolios are registered and stored in an enterprise system that mainly includes ERP, CRM or CAD. The selected attributes of an enterprise system, resources, and the identified relationships are stored in knowledge base that aims to facilitate the project manager to evaluate the success of a new product, and finally, select a set of the most promising products for further development. Figure 1 illustrates new product screening in the context of the NPD process and enterprise system.

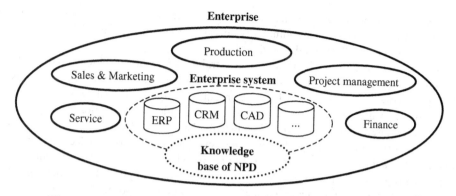

Fig. 1. Reference model of knowledge acquisition for new product screening

The proposed model consists of variables that describe the product success (output variable), and that are suspected of significant impact on this success (input variables). Taking into account the product lifetime and return on product development expense, as the output variable and the measure of the product success, the average net profit from a product per month is chosen. The model contains a set of decision variables, their domains, and the constraints that can be referred to the performance indicators and company's resources. The decision problem concerning the selection of the most promising set of products for further development has been described in terms of a constraint satisfaction problem (CSP). The model description encompasses the limitations of a company, parameters of new products that are considered for development, and a set of routine queries (the instances of decision problems) that are formulated in the framework of CSP. The structure of the constraint satisfaction problem may be described as follows [6]:

$$CSP = ((V, D), C)$$

where: $V = \{v_1, v_2, ..., v_n\}$ – a finite set of n variables, $D = \{d_1, d_2, ..., d_n\}$ – a finite set of n discrete domains of variables, and $C = \{c_1, c_2, ..., c_k\}$ – a finite set of k constraints limiting and linking variables.

Consider a set of new products for development $P = \{P_1, ..., P_i, ..., P_I\}$, where P_i consists of J activities. The following variables have been chosen: duration of new product development (PD_i) and its cost (PC_i), number of project team members (PTM_i), percentage of existing parts used in a new product (PEP_i), production cost of the product (PPC_i), duration of marketing campaign of the product (PMD_i) and its cost (PMC_i), number of customer requirements for a new product (PCR_i), percentage of customer requirements translated into technical specification (PTS_i), and net profit of the product (PNP_i). The company's limitations include the total number of R&D employees (project team members) $C_{1,t}$ and financial means $C_{2,t}$ in the t-th time unit. Variables are limited and linked through constraints, e.g. $\sum PC_i + \sum PMC_i \leq C_2$. The problem solution can be stated as seeking the answers to the following two types of questions: what products should be selected to the product portfolio by a fixed amount of resources to ensure the maximal total net profit from the products, and what resources in which quantities are minimally necessary to complete the product portfolio by the desired net profit from the products?

The model description in terms of constraint satisfaction problem enables the design of a decision support system taking into account the available specifications, routine queries, and expert knowledge. Consequently, the model integrates technical parameters, available resources, expert experience, identified relationships (rules), and user requirements in the form of knowledge base. Knowledge base is a platform for query formulation and for obtaining answers, and it comprises of facts and rules that are relevant to the system's properties and the relations between its different parts [7]. The proposed method of developing a decision support system for evaluating the net profit from a new product and selecting product concept portfolio is presented in the next section.

3 Decision Support System for Product Concept Screening

As the amount of available data in companies becomes greater and greater, companies have become aware of an opportunity to derive valuable information from their databases, which can be further used to improve their business [8]. The process of identifying novel and potentially useful patterns in data is known as knowledge discovery, and it consists of the stages such as data selection, data preprocessing, data transformation, data mining, and evaluation of the identified patterns [9]. Figure 2 illustrates the framework of the knowledge discovery in the context of developing a decision support system for product concept screening.

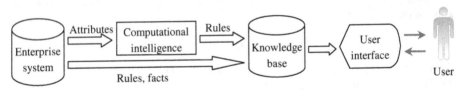

Fig. 2. Framework of decision support system for product concept screening

The selected attributes from an enterprise system are led to a fuzzy neural network (FNN) as input and output variables. The identified relationships are stored in the form of rules in knowledge base that also includes rules and facts acquired by an expert from an enterprise system (e.g. financial, temporal, personal constraints). Knowledge base is used to estimate the success of product concepts, to select the most promising concepts for further development, and to identify amount of resources that ensure the desired value of total net profit for the selected product concepts. Knowledge base is specified in term of CSP that in turn can be implemented in constraint programming environment providing the solution in the effective way. The user interface presents the project manager a set of products that are selected according to user's preferences, and the evaluation of these products in the context of their estimated profits. Moreover, the project manager can check which amount of resources is needed to obtain the desire total net profit from the selected products.

The use of computational intelligence techniques (FNN) aims to identify the hidden relationships between the success of a product and the variables that are suspected of impact on the project success. FNN is able to identify these nonlinear and complex relationships (if there are any), and consequently, develop the knowledge base. The fuzzy neural network has the advantages of both neural networks (e.g. learning abilities, optimization abilities and connectionist structures) and fuzzy systems (e.g. if-then reasoning, simplicity of incorporating expert knowledge). FNN can learn from training samples, and after learning it is possible to obtain fuzzy if-then rules that can be further used to perform nonlinear predictive modelling, simulation, and forecasting. Taking into account good forecasting properties and the possibility of obtaining if-then rules that can be saved in knowledge base, among computational intelligence techniques, fuzzy neural network has been chosen in this study.

The constraint programming (CP) is an emergent software technology for a declarative constraints satisfaction problem (CSP) description and can be considered as a pertinent framework for the development of decision support system software [10]. In the case of extensive search space, the processing time of calculations can be significantly reduced with the use of constraints programming techniques [11-12].

CP consists of a set of techniques for solving a CSP that is specified as a set of constraints on a set of variables. CP approach tackles with CSPs with the use of paradigms such as propagate-and-search or propagate-and-distribute. CP has embedded ways to solve constraints satisfaction problems with greatly reduction of the amount of search needed [13]. This is sufficient to solve many practical problems such as supply chain problem [14-16] or scheduling problem [11], [17-18]. The next section presents an example of using the proposed decision support system to product concept screening.

4   Illustrative Example

The success of a new product is estimated on the basis of information about the previous NPD projects. With the use of fuzzy neural networks, there are sought the relationships between the input variables and net profit from a product (PNP). The following input variables have been chosen from an enterprise database: number of activities in the NPD project (J), duration of the NPD project (PD), cost of the NPD project (PC), number of project team members (PTM), percentage of existing parts used in a new product (PEP), unit cost of production for the product (PPC), duration of marketing campaign of the product (PMD), cost of marketing campaign of the product (PMC), number of customer requirements for a new product (PCR), and percentage of customer requirements translated into technical specification (PTS).

The relationships between the input variables and net profit from a product has been sought with the use of the adaptive neuro-fuzzy inference system (ANFIS). In order to eliminate the overtraining of ANFIS and increase the estimation quality, the data set concerning the past NPD projects has been divided into learning and testing sets. In studies, the ANFIS has been trained according to subtractive clustering method that was implemented in the Matlab® software. The identified relationships can be described as if-then rules and used to estimating net profit from a new product. Figure 3 presents the membership functions for 11 rules that are used to estimating net profit for P_1.

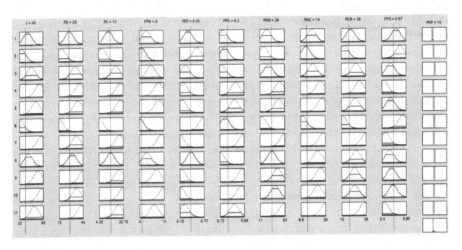

Fig. 3. Estimation of net profit for product P_1

The estimated net profits of the products are further used to seeking the NPD project portfolio that ensures the maximal total net profit from all products, and selecting the NPD project portfolios that ensure the desired level of total net profit according to the user's preferences. The R&D department generated 28 concepts for new products that are evaluated according to net profit, cost of development, and the number of people in a NPD project. In the NDP projects can participate 19 employees (C_1) and the total budget for NDP projects (C_2), including the R&D and marketing campaign budget, equals 225,000 €. The number of possible solutions, i.e. combinations of new product portfolio, is very large. This imposes the use of techniques that enable the reduction of the amount of search needed, such as constraint programming.

The considered problem has been implemented in the Oz Mozart programming environment that includes CP paradigms, and tested on an AMD Turion(tm) II Dual-Core M600 2.40GHz, RAM 2 GB platform. The number of the admissible solutions equals 5,660 instances. The set of the NPD projects that ensures the maximal total net profit (78,000 € per month) consists of product P_3, P_6, P_{20}, and P_{21}. Let us assume that the project manager wants to check what amount of resources is needed to obtain the total net profit equals 90,000 € per month. In order to compare the time consumption, among the distribution strategies, naïve and first-fail distribution has been chosen. The naïve distribution strategy chooses the first not yet determined variable from a list, i.e. it simply checks variables in their order. In turn, the first fail distribution checks next a variable with smallest domain. The selection of a suitable distribution strategy is crucial for the performance of the search process. Table 1 presents the results of finding solution for the different distribution strategies.

Table 1. Comparison of distribution strategies

Case	Product portfolio	Net profit	Distribution strategy	Time [sec]	Number of solutions	Number of nodes
$C_1 \cdot 22$; $C_2 \cdot 225,000$	$P_3, P_{20}, P_{21}, P_{23}$	90,000	Naïve	0.78	1	8877
			First-fail	0.72		
$C_1 \cdot 25$; $C_2 \cdot 225,000$	$P_{19}, P_{21}, P_{25}, P_{27}$	92,000	Naïve	1.54	577	18041
			First-fail	1.50		
$C_1 \cdot 25$; $C_2 \cdot 235,000$	$P_{10}, P_{20}, P_{21}, P_{23}$	98,000	Naïve	6.84	2509	22963
			First-fail	5.22		
$C_1 \cdot 25$; $C_2 \cdot 245,000$	$P_{10}, P_{15}, P_{16}, P_{20}$	109,000	Naïve	31.4	23241	40772
			First-fail	25.4		

The presented comparison consists of four cases for the additional resources (employees and budget increase) to illustrate the impact of the different distribution strategies on time of searching solution, the number of solutions and nodes. The product portfolio presented in Table 1 concerns the optimal solution, in which the selection criterion is portfolio with the maximal total net profit. The results indicate that the first-fail distribution strategy outperforms the naïve strategy, especially in the case of the large number of admissible solutions. The proper choice of distribution strategy can shorten time taken in searching solution and facilitate development of an interactive decision support system.

5   Conclusions

The characteristics of the presented approach includes the use of an enterprise system database to knowledge creation, fuzzy neural network to seek the relationships and their description in the form of if-then rules, and framework of constraint satisfaction problem to specify a knowledge base. This knowledge base includes the rules identified by fuzzy neural network or/and an expert, facts (including company's resources), and it allows the project managers to obtain an answer to the routine questions such as what is the most promising set of products for development? The use of constraint programming to implementing the constraint satisfaction problem allows developing a decision support system in a pertinent framework.

The proposed approach has several advantages such as the problem description in terms of CSP that can be solved effectively in constraint programming environment, the low effort of data retrieval from an enterprise system, the possibility of what-if analysis, and the identification of resource amounts that are needed to obtain the desired value of net profit from new products. Moreover, the identified relationships are used in the decision support system to help the managers in conducting simulation of the NPD projects, selecting the most promising product portfolio, and reducing the risk of unsuccessful product development. On the other hand, the application of the proposed approach encounters some difficulties, for instance, by collecting enough amount of data of the past similar NPD projects and ambiguous principles to build structure of fuzzy neural network. Nevertheless, the presented approach seems to have the promising properties for acquiring information from an enterprise system and improving the decision making process in the context of product concept screening.

References

1. Sun, H., Wing, W.: Critical Success Factors for New Product Development in the Hong Kong Toy Industry. Technovation 25, 293–303 (2005)
2. Spalek, S.: Does Investment in Project Management Pay Off? Industrial Management & Data Systems 114(5), 832–856 (2014)
3. Chan, S.L., Ip, W.H.: A Dynamic Decision Support System to Predict the Value of Customer for New Product Development. Decision Support Systems 52, 178–188 (2011)
4. Doskocil, R., Smolikova, L.: Knowledge Management as a Support of Project Management. In: International Scientific Conference on Knowledge for Market Use, pp. 40–48 (2012)
5. Relich, M.: Using ERP Database for Knowledge Acquisition: A Project Management Perspective. In: International Scientific Conference on Knowledge for Market Use, pp. 263–269 (2013)
6. Rossi, F., van Beek, P., Walsh, T.: Handbook of Constraint Programming. Elsevier Science (2006)
7. Bocewicz, G.: Robustness of Multimodal Transportation Networks. Eksploatacja i Niezawodnosc–Maintenance and Reliability 16(2), 259–269 (2014)
8. Li, T., Ruan, D.: An Extended Process Model of Knowledge Discovery in Database. Journal of Enterprise Information Management 20(2), 169–177 (2007)
9. Woolliscroft, P., Relich, M., Caganova, D., Cambal, M., Sujanova, J., Makraiova, J.: The Implication of Tacit Knowledge Utilisation Within Project Management Risk Assessment. In: 10th International Conference of Intellectual Capital, Knowledge Management and Organisational Learning (ICICKM 2013), Washington, DC, pp. 645–652 (2013)
10. Do, N.A.D., Nielsen, I.E., Chen, G., Nielsen, P.: A Simulation-Based Genetic Algorithm Approach for Reducing Emissions from Import Container Pick-up Operation at Container Terminal. Annals of Operations Research (2014) (article in press)
11. Banaszak, Z., Zaremba, M., Muszynski, W.: Constraint Programming for Project-Driven Manufacturing. International Journal of Production Economics 120, 463–475 (2009)
12. Sitek, P., Wikarek, J.: A Hybrid Approach to Supply Chain Modeling and Optimization. In: Federated Conference on Computer Science and Information Systems, pp. 1223–1230 (2013)
13. Van Roy, P., Haridi, S.: Concepts, Techniques and Models of Computer Programming. Massachusetts Institute of Technology (2004)
14. Grzybowska, K., Kovács, G.: Logistics Process Modelling in Supply Chain – Algorithm of Coordination in the Supply Chain – Contracting. In: de la Puerta, J.G., et al. (eds.) International Joint Conference SOCO'14-CISIS'14-ICEUTE'14. AISC, vol. 299, pp. 311–320. Springer, Heidelberg (2014)
15. Grzybowska, K., Awasthi, A., Hussain, M.: Modeling Enablers for Sustainable Logistics Collaboration Integrating Canadian and Polish Perspectives. In: The Federated Conference on Computer Science and Information Systems, pp. 1311–1319 (2014)
16. Sitek, P., Wikarek, J.: Hybrid Solution Framework for Supply Chain Problems. In: Omatu, S., Bersini, H., Corchado Rodríguez, J.M., González, S.R., Pawlewski, P., Bucciarelli, E. (eds.) Distributed Computing and Artificial Intelligence 11th International Conference. AISC, vol. 290, pp. 11–18. Springer, Heidelberg (2014)
17. Baptiste, P., Le Pape, C., Nuijten, W.: Constraint-Based Scheduling: Applying Constraint Programming to Scheduling Problems. Kluwer Academic Publishers, Norwell, Massachusetts (2001)
18. Relich, M.: Identifying Relationships Between Eco-innovation and Product Success. In: Golinska, P., Kawa, A. (eds.) Technology Management for Sustainable Production and Logistics, pp. 173–192. Springer, Heidelberg (2015)

Scheduling Part-Feeding Tasks for a Single Robot with Feeding Quantity Consideration

Izabela Nielsen, Ngoc Anh Dung Do[*], and Peter Nielsen

Department of Mechanical and Manufacturing Engineering, Aalborg University, Denmark
{izabela,ngoc,peter}@m-tech.aau.dk

Abstract. Scheduling part-feeding tasks for a single robot have been considered in manufacturing. Maximum the feeding quantity can save the traveling time of robot but it can make an infeasible problem due to the conflict of carrying quantity of robot and the window time for feeding. This study considers feeding quantity as a decision variable which is not investigated in literature. The problem is formulated as a nonlinear program, and hence, a heuristic algorithm is developed to solve the problem. The conducted numerical experiment shows the good performance of the proposed algorithm which can be implemented to solve large scale problem.

Keywords: robot scheduling, part-feeding task scheduling.

1 Introduction

Automatic vehicles have now been widely used in production system to improve the smooth flow of materials, and hence, improve the productivity. Moreover, changing from traditional production policy where high-leveled customization is produced with high production cost to mass customization with lower production cost and the same quality of product and on-time delivery to satisfy diversified demands will bring success to the company (Dang et al. 2013). The need of using automatic vehicles (e.g. AGV and mobile robots) which are required in a highly flexible production system such as mass customization system is more and more increasing in real world. An automatic vehicle is able to move unfinished parts from warehouses and feed unfinished parts to machines following the flow of production lines. Some special mobile robots also assist the machines during the production process. The problem to schedule the mobile robots to serve the production process is necessarily considered and more complicated when the mobile robots are assigned many different tasks.

In this paper, the scheduling problem for a mobile robot is considered. The single mobile robot has the task to feed the unfinished parts for a set of machines. Each machine has a minimum and maximum level of unfinished parts at its buffer. The mobile robot can only feed unfinished parts to a machine when the number of unfinished parts at the buffer is lower than the minimum value. It is not allowed that the machine

---

[*] Corresponding author.

© Springer International Publishing Switzerland 2015
S. Omatu et al. (eds.), *Distributed Computing & Artificial Intelligence, 12th Int. Conference,*
Advances in Intelligent Systems and Computing 373, DOI: 10.1007/978-3-319-19638-1_40

is idle because of no unfinished part at that machine's buffer. Therefore, the mobile robot must be scheduled so that there are no lack of unfinished parts at all machines.

The energy consumed by the mobile robot is proportional to the moving distance. In a planning horizon, the robot can feed a machine many times. The number of feeds is inversely proportional to the feeding quantity (i.e. the quantity of unfinished parts per feed). The problem is more complicated when the system has more than one machine. Starting from warehouse, the robot can feed many machines before it goes back to the warehouse to load unfinished parts for feeding other machines. If the mobile robot carries too many unfinished parts to feed a machine, it has no room to carry other unfinished parts for other machines due to the limit of carrying capacity. Consequently, the mobile robot can move longer distance because it has to go back to the warehouse more frequently. It is noted that the length of the cycle to feed a machine is proportional to the feeding quantity. Furthermore, different cycle length leads to different feeding schedule of the mobile robot. Therefore, the idea for the case of single machine cannot be applied to the case of multiple machines. The problem to find the optimal feeding quantity for each machine and the feeding schedule for the mobile robot is the novelty of this paper.

2   Literature Review

The problem of scheduling part-feeding tasks of mobile robot has been modelled similarly to Traveling Salesman Problem (TSP). Aschuer et al. (1993) used cutting plane approach to the sequential ordering problem. Edan et al. (1991) developed a near minimum harvesting task planning problem by solving the TSP to find a near optimum path between N fruit locations. Dang et al. (2011) proposed an MIP model to obtain the optimal feeding sequence of a mobile robot in a manufacturing cell. Some studies have been done by Relich and Muszyński (2014), Pawlewski (2014), and Do et al. (2014) for planning and scheduling in manufacturing and services. There have been some papers solving the robot task-scheduling using heuristic algorithm. Han et al. (1987) developed the nearest neighbor rule in which robot travels to nearest pick up point from its current position. Suarez and Rosell (2005) used dispatching rules to determine the sequence of tasks. Other papers implemented metaheuristic to solve the combinatorial optimization problem, especially for robot task scheduling problem. Sitek and Wikarek (2014 and 2015) developed hybrid approach to solve optimization problem for supply chain management. Hurink and Knust (2002) proposed a tabu search algorithm to solve the scheduling problem for a single robot in a job-shop environment. Maimon et al. (2000) presented a neutral network approach for robot task-scheduling problem. Bocewicz et al. (2013) and Bocewicz (2014) considered the task scheduling problem for scheduling and rescheduling production process based on constraint satisfaction programming.

The problem of scheduling single robot with time windows has been investigated. The differences from researches in literature are the limit of carrying capacity and the smooth flow material requirements. The robots have to go back to ware house to pick up unfinished parts for its new route due to the limit of carrying capacity. Moreover, the robots have to be scheduled to assure that the machines have enough unfinished parts to work continuously. Another issue is that the robot cannot feed many unfinished parts to a machine at a time point due to the limit of machine's buffer. This makes the robot task scheduling more practical and more complicated. Dang et al. (2013) developed GA to solve the robot task scheduling with considering (s,Q) policy for part feeding. However, they assumed that the robot fills unfinished parts up to the maximum capacity of buffer. This may lead to a situation that the robot can fail to feed unfinished parts to the next machine on time due to limit of carrying capacity.

3   Problem Description and Mathematical Model

3.1   Problem Description

A system including a mobile robot and several machines is considered. Each machine can process on one type of unfinished part with constant processing time meaning that the demand rate on unfinished parts is also constant. Each machine has a buffer with finite capacity. The mobile robot, which has a finite carrying capacity, collects different types of unfinished parts at a warehouse and distributes these unfinished parts to the machines. The machine consumes unfinished parts for its process. Each machine has its own maximum and minimum levels of unfinished parts at the buffer and these levels are pre-determined. When the number of unfinished parts at the buffer is less than the minimum level, the mobile robot has to feed new unfinished parts before the buffer is empty. Feeding quantity has to assure that the level of unfinished parts at the buffer does not exceed maximum level when the machine is fed.

The feeding quantity ≤ maximum level – minimum level

The feeding period starts when the level of unfinished parts at the buffer reaches the minimum level and ends when the buffer empties. This period can be determined if the feeding quantity is known in advanced (Figure 1). The length of feeding period and the length of feeding cycle are calculated as follows:

Length of feeding period = processing time per unfinished part * minimum level
Length of feeding cycle = processing time per unfinished part * feeding quantity

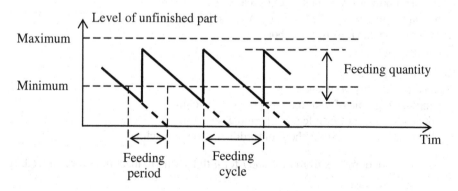

Fig. 1. Level of unfinished part at the buffer of a machine

A route of the mobile robot is started at the ware house, and ended at the warehouse as well. Based on the feeding quantity and the carrying capacity, the mobile robot can visit one or more machines on a route. The sequence of tasks on a route of the mobile robot is described as follows. First, unfinished parts are loaded to the mobile robot. After that, the mobile robot moves to each machine assigned to that route and unloads unfinished parts to the buffer of that machine. Finally, the mobile robot backs to the warehouse and start the sequence of tasks on the next route. The mobile robot moves with constant speed on all routes. The decision making in this problem are the feeding quantity at each machine and the feeding sequence to minimize the total travel distance of the mobile robot.

3.2   Mathematical Formulation

Assumption:

- The system includes an automatic mobile robot and multiple machines
- The robot has a constant moving speed
- The duration for unloading unfinished parts is proportional to the feeding quantity.
- All unfinished parts are located at the warehouse
- Each machine can process a type of unfinished part with constant processing time
- All distances from warehouse/machines to warehouse/machines are known
- The robot must feed the unfinished part before the buffer of a machine is empty.
- At the beginning, the number of unfinished parts is at the maximum level

Notation
M: set of machines, $M = \{1,2,..., I\ \}$
N: set of location, $N = M \cup \{0\}$, 0 represents warehouse

Parameters
p_i: processing time on an unfinished part at machine i
c_i: unfinished part unloading rate (unit of time/ part)
t_{ij}: the travel time from machine i (or warehouse) to machine j (or warehouse), $i, j \in N$
u_i: the maximum level at the buffer of machine i
v_i: the minimum level at the buffer of machine i
K: the carrying capacity of the robot
T: planning horizon

Decision variables
Q_i: feeding quantity at machine i
S_{ih} : the starting time of h–th feeding period at machine i
E_{ih}: the ending time of h-th feeding period at machine i
F_{ih}: the time that robot starts the h-th feeding period at machine i

$$X^{ih}_{jl} = \begin{cases} 1 & \text{if robot moves from machine } i \text{ for } h\text{-th feeding to machine } j \text{ for } l\text{-th feeding} \\ 0 & \text{otherwise} \end{cases}$$

Objective function:

$$\text{Min} \sum_{i \in N} \sum_{1 \le h \le H_i} \sum_{j \in N} \sum_{1 \le l \le H_j} t_{ij} X_{jl}^{ih}$$

Subject to

$$\sum_{i \in N} \sum_{1 \le h \le H_i} X_{jl}^{ih} = 1 \qquad\qquad \forall j \in N, l \in \left[1, H_j\right] \qquad\qquad (1)$$

$$\sum_{j \in N} \sum_{1 \le l \le H_j} X_{jl}^{ih} = 1 \qquad\qquad \forall i \in N, h \in \left[1, H_i\right] \qquad\qquad (2)$$

$$\sum_{j \in N} \sum_{1 \le l \le H_j} X_{jl}^{ih} = \sum_{m \in N} \sum_{1 \le n \le H_m} X_{ih}^{mn} \qquad \forall i \in M, h \in \left[1, H_i\right] \qquad (3)$$

$$\sum_{i \in M} X_{i1}^{01} = 1 \qquad\qquad\qquad\qquad\qquad\qquad\qquad\qquad (4)$$

$$S_{ih} \le F_{ih} \le E_{ih} \qquad\qquad \forall i \in M, h \in \left[1, H_i\right] \qquad\qquad (5)$$

$$F_{jl} \ge F_{ih} + c_i Q_i + t_{ij} X_{jl}^{ih} - L\left(1 - X_{jl}^{ih}\right) \qquad \forall j \in N, l \in \left[1, H_j\right], i \in M, h \in \left[1, H_i\right] \qquad (6)$$

$$F_{jl} \ge F_{0h} + t_{ij} X_{jl}^{0h} - L\left(1 - X_{jl}^{0h}\right) \qquad \forall j \in N, l \in \left[1, H_j\right], h \in \left[1, H_i\right] \qquad (7)$$

$$\sum_{\substack{i \in M, l \in [1, H_i], \\ F_{0h} \le F_{il} \le F_{0,h+1}}} Q_i \le K \qquad\qquad \forall h \in \left[1, H_i\right] \qquad\qquad (8)$$

$$F_{ih} \le F_{i,h+1} \qquad\qquad \forall i \in N, h \in \left[1, H_i - 1\right] \qquad (9)$$

$$Q_i \le u_i - v_i \qquad\qquad \forall i \in M \qquad\qquad\qquad\qquad (10)$$

$$Q_i, S_{ih}, E_{ih}, F_{ih} \ge 0, X_{jl}^{ih} \in \{0,1\} \qquad \forall i, j, h, l$$

The objective is to minimize the traveling distance of the mobile robot. Constraints (1) – (4) are to assure a feasible route for the robot in which the robot will start at the ware house. Constraint (5) requires the robot has to feed a machine on time so that the machine can work continuously. Constraints (6) and (7) show the sequence of task. The carrying capacity of robot is shown in constraint (8). Constraint (9) assures that the next visit at a machine should happen after the current visit. Constraint (10) is used to guarantee that the machine is not over feeding.

4   Solution Algorithm

The formulated problem is a nonlinear programming due to constraint (8). Therefore, an algorithm should be developed to solve the problem. When Q_i is fixed, the problem becomes similarly to a task sequencing problem of robot. S_{ih} and E_{ih} are fixed and they have a role of release time and due date. Then, the sequence to feed machines can be listed in the order of ending time of each machine. It is noted that the robot should feed as many machines as possible on one route because it will save the travel distance. Suppose that two machines of which starting time and ending time are closed are served on two routes, they should be merged in to one route. In order to merge in one route, the robot should have enough capacity to carry unfinished parts to feed those two machines. Merging two routes is done when there is no routes can be possibly merged. The value of Q_i should be as large as possible because it will reduce the number of times to feed a machine. Last but not least is the time horizon. The systems will repeat after a period of time because there is a feeding cycle for each machine. If we call T_i as the feeding cycle of machine i ($T_i = Q_i p_i$) and T^S as the cycle of the system, then T^S is the minimum common multiply of T_i. This implies that we can do the robot task scheduling in a period of time with the duration $\min(T^S, T)$. This may reduce the computation time of the problem.

The procedure to solve the problem is shown as follows:

Step 1. Set $Q_i = u_i - v_i$ and calculate S_{ih} and E_{ih}.

Step 2. Sort E_{ih} in non-descending order. If $E_{ih} = E_{jl}$, i is listed before j if $S_{ih} \leq S_{jl}$; otherwise, j is listed before i. The list proposes the sequence of feeding.

Step 3. Consider two machines i and j which are consecutively listed and i before j

- If $F_{ih} + c_i Q_i + t_{i0} + t_{0j} > E_{jl}$, check the condition $Q_i + Q_j \leq K$
  o If the condition is satisfied, i and j are assigned to the same route
  o Otherwise, decrease Q_i and go back to Step 2
- If $S_{ih} > S_{jl}$:
  o Machine j should be feed before machine i when robot can arrive i before the ending time of i.
  o Otherwise, machine i should be feed before machine j with the condition that robot can arrive j before the ending time of j.

Step 4. Assign route where each route served one machine
Step 5. Merging two routes if $Q_i + Q_j \leq K$

It is noted that the procedure is not guaranteed to find optimal solution. GA or constraint programming can be applied for both determining value of Q_i and robot routes to find the better solution (optimal solution may be optioned). This will be considered for further research.

5   Numerical Experiment

The numerical experiment is conducted with a system with a mobile robot which feeds 4 machines. The maximum and minimum level at the buffer of each machine, the unloading rate, the processing time on an unfinished part at a machine, and the traveling time from warehouse and machines are given in Tables 1 and 2. The planning horizon is about 45 minutes. The carrying capacity will be 1500 parts

Table 1. Minimum level, maximum level, unloading rate and processing time at each machine

Machine	1	2	3	4
Maximum level (part)	250	1500	1500	250
Minimum level (part)	125	400	400	125
Unloading rate (second/part)	0.168	0.04	0.04	0.168
Processing time (second/part)	4.5	1.5	1.5	1.5

Table 2. Traveling time of robots among warehouse and machine (in seconds)

	Warehouse	Machine 1	Machine 2	Machine 3	Machine 4
Warehouse	0	34	37	34	40
Machine 1	39	0	17	34	50
Machine 2	35	17	0	35	49
Machine 3	34	33	35	0	47
Machine 4	36	47	48	46	0

The result from the algorithm is present in Table 3 with the computation time less than 1 second. The result shows that the mobile robot can served many machines on one route (routes 1,2, and 4). However, because the feeding time of machine 2 and 3 are close and it is required big feeding quantity for both machines, there is only machine 2 is served on route 2.

Table 3. The result of numerical experiment

	Machine/ warehouse	Arrive at (seconds)	Feeding at (seconds)	Feeding quantity (parts)
Route 1	4	40	187.5; 375; 562.5; 750	125
0-4-1-4-1-4-0	1	818	818	125
	4	889	937.5; 1125; 1312.5	125
	1	1380.5	1380.5	125
	4	1451.5	1500; 1687.5; 1875	125
	0	1932		
Route 2	2	1969	1969	1100
0 – 2 – 0	0	2048		
Route 3	3	2082	2082	1100
0-3-1-4-0	1	2159	2159	125
	4	2230	2230	125
	0	2287		
Route 4	4	2327	2327; 2437.5	125
0-4-1-4-0	1	2505.5	2505.5	125
	4	2576.5	2625	125
	0	2682		

6   Conclusion

In this study, the feeding task scheduling for a mobile robot is considered. The problem is more practical when feeding quantity is taken in account. The problem is formulated as a nonlinear programming, and hence, a heuristic algorithm is developed. The numerical result shows that the proposed algorithm can be applied to solve the practical problem with large scale in short time. The proposed methodology can be applied in a mass production line with a high variety of products. The problem can be extended in two folds. On one hand, more practical factors can be taken in to account.

One of practical issues, the different feeding quantities on different visits at a machine, can be further investigated. Robot battery charging is another issue which is very promising in further research. On the other hand, solution algorithm can be considered. GA and constraint programming, can be applied on both determining the feeding quantity at a machine and the feeding sequencing to improve the quality of the solution.

References

1. Ascheuer, N., Escudero, L.F., Grötschel, M., Stoer, M.: A cutting plane approach to the sequential ordering problem (with applications to job scheduling in manufacturing). SIAM Journal on Optimization 3(1), 25–42 (1993)
2. Bocewicz, G.: Robustness of Multimodal Transportation Networks. Eksploatacja i Niezawodnosc–Maintenance and Reliability 16(2), 259–269 (2014)
3. Bocewicz, G., Wójcik, R., Banaszak, Z., Pawlewski, P.: Multimodal Processes Rescheduling: Cyclic Steady States Space Approach. Mathematical Problems in Engineering 2013, Article ID 407096, 24 pages (2013), doi:10.1155/2013/407096
4. Dang, Q.V., Nielsen, I., Steger-Jensen, K.: Scheduling a single mobile robot for feeding tasks in a manufacturing cell. In: Proc. of Int. Conf. Adv. Prod. Manag. Syst., Norway (2011)
5. Dang, Q.V., Nielsen, I., Steger-Jensen, K., Madsen, O.: Scheduling a single mobile robot for part-feeding tasks of production lines. Journal of Intelligent Manufacturing, 1–17 (2013)
6. Do, N.A.D., Nielsen, I.E., Chen, G., Nielsen, P.: A simulation-based genetic algorithm approach for reducing emissions from import container pick-up operation at container terminal. Annals of Operations Research (2014) (article in press)
7. Edan, Y., Flash, T., Peiper, U.M., Shmulevich, I., Sarig, Y.: Near-minimum-time task planning for fruit-picking robots. IEEE Transactions on Robotics and Automation 7(1), 48–56 (1991)
8. Han, M.H., McGinnis, L.F., Shieh, J.S., White, J.A.: On sequencing retrievals in an automated storage/retrieval system. IIE Transactions 19(1), 56–66 (1987)
9. Hurink, J., Knust, S.: A tabu search algorithm for scheduling a single robot in a job-shop environment. Discrete Applied Mathematics 119(1), 181–203 (2002)
10. Maimon, O., Braha, D., Seth, V.: A neural network approach for a robot task sequencing problem. Artificial Intelligence in Engineering 14(2), 175–189 (2000)
11. Pawlewski, P.: Multimodal Approach to Modeling of Manufacturing Processes. Procedia CIRP 17, 716–720 (2014); Variety Management in Manufacturing — Proceedings of the 47th CIRP Conference on Manufacturing Systems (2014)
12. Relich, M., Muszyński, W.: The use of intelligent systems for planning and scheduling of product development projects. Procedia Computer Science 35, 1586–1595 (2014)
13. Sitek, P., Wikarek, J.: Hybrid Solution Framework for Supply Chain Problems. In: Omatu, S., Bersini, H., Corchado Rodríguez, J.M., González, S.R., Pawlewski, P., Bucciarelli, E. (eds.) Distributed Computing and Artificial Intelligence 11th International Conference. AISC, vol. 290, pp. 11–18. Springer, Heidelberg (2014)
14. Sitek, P., Wikarek, J.: A hybrid framework for the modelling and optimisation of decision problems in sustainable supply chain management. International Journal of Production Research, 1–18 (2015), doi:10.1080/00207543.2015.1005762
15. Suárez, R., Rosell, J.: Feeding sequence selection in a manufacturing cell with four parallel machines. Robotics and Computer-Integrated Manufacturing 21(3), 185–195 (2005)

Approving with Application of an Electronic Bulletin Board, as a Mechanism of Coordination of Actions in Complex Systems

Katarzyna Grzybowska and Patrycja Hoffa

Poznan University of Technology, Faculty of Engineering Management, Chair of Production Engineering and Logistics, Strzelecka 11, 60-965 Poznan, Poland
katarzyna.grzybowska@put.poznan.pl,
patrycja.hoffa@doctorate.put.poznan.pl

Abstract. The article presents works, which refer to the coordination of actions in complex systems. The coordination of actions aims at integrating and synchronizing various elements of the complex system and their compliant operation. The publication presents one of the mechanisms of coordinating actions, applied by the complex systems. What is more, it embraces a demonstration of its variance. It also presents results of experiments with proper commentaries.

Keywords: coordination, coordination mechanisms, supply chain, complex systems, multimodal networks, simulation.

1   Introduction

Supply chains (SC) are a system with "multiple actors". A supply chain can create very complex interrelation networks at every stage [7]. But the supply chain management is a decision process that not only integrates all of its participants, but also helps to coordinate the basic flows: products/services, information and funds [11]. Supply Chain Management is an integrating function with primary responsibility for linking major business functions and business processes within and across companies into a cohesive and high-performing business model [3]. SC are complex systems, dynamic, dispersed and open. Those elements together with other factors (e.g. multiple subjects, independence of cooperating enterprises) determine difficulties in the field of management, or more broadly, of coordination of commonly take up and independently realized actions. The discussed systems are affected, as a whole, by a lack of internal rationality, unverified information and insufficient knowledge. The problem is also posed by uncertainty and a lack of precision [8, 9], indispensable in the realized projects and complex undertakings. Hence, these systems frequently constitute a row of poorly interconnected internal actions (inside the supply chain) [5], as well as those taken outside the enterprise, but remaining within the system. Therefore, there is the necessity for an activity, whose aim is to coordinate the realized common actions. Such an activity is on application of the so called mechanisms of coordination of actions, whose application may cause that the complex systems, including isolated and independent enterprises, will operate much more effectively. Multiechelon systems have been introduced in different areas: for example supply chains [10].

© Springer International Publishing Switzerland 2015   357
S. Omatu et al. (eds.), *Distributed Computing & Artificial Intelligence, 12th Int. Conference,*
Advances in Intelligent Systems and Computing 373, DOI: 10.1007/978-3-319-19638-1_41

The coordination of actions means combined operation of the subjects, which: (1) is directed towards achievement of set, common, mutually compliant targets, (2) encompasses systematizing, ordering and approving process and various components of the system, (3) takes place in the agreed time and (4) exerts some influence on behavior of the cooperating entities. The coordination of actions consists in approving them [6]. The mechanism of coordination of actions is needed to maintain connectivity of a dynamic multimodal network.

The publication aims at presenting a selected mechanism of coordination of actions, with regards to potential variants, as well as the presentation of results of the simulation experiment, with application of the FlexSim program. In the study one hypothesis was assumed: coordinate activities under the responsibility of one cell (coordinating body) will accelerate the process using the concept of reconciliation, called: Electronic Bulletin Board. The structure of the work is as follows: section 2 presents the mechanism of coor-dination of actions called "Approving with application of an Electronic Bulletin Board". This part of the paper describes the mechanism, its reference models according to the modified IDEF0 methodology, together with an ICOM cube. The subsequent part of the article discussed the simulation experiment performed through the FlexSim program. Section 4 focuses on and comments results of the experiment. The paper is finished with a summary.

2   Reference Model of the Selected Mechanism of Coordination of Actions

The coordination mechanism has a form of interactions among differentiated (in terms of form, targets, intentions, the manner of organization, etc.) and independent entities. [2] Adequate selection of coordination mechanisms allows not only the efficient cooperation of organization, but it also generates benefits arising from the synergy effect, through reduction of costs and a higher level of competitiveness. Furthermore, it supports realization of the assumed targets connected with costs reduction, elevation of the clients' satisfactions, an increase of flexibility, etc. However, if the supply chain is more complex (network), the actions coordination in such a structure becomes harder and more demanding.

The mechanism of the coordination of actions, with the application of the electronic bulletin board, poses a modified form of classic coordination - contracting [1]. Coordination with application of the electronic bulletin board is used, when the request has a well defined structure of sub-requests or sub-tasks. Hence, the request may be structured into its simpler sub-tasks.

2.1   The Electronic Bulletin Board – Serial Coordination of Actions

The applied mechanism of coordination of actions, called "Approving with application of an Electronic Bulletin Board", encompasses serial presence of two roles - the commissioner (receiver) and the contractor (supplier). It is a task role, assumed consciously, regarding the performed actions, and resulting from the ascribed task. The same enterprise may (regarding the business processes that take place), play a role of both a supplier and a receiver. This results from the complexities of the actions realized within the supply chain. The commissioner (of the first degree), decomposes the primary requests into sub-requests. They also allocate those sub-requests to the verified contracting parties that they cooperate with. What is more, they use their own (most often closed) database of subcontractors and a so called Electronic Bulletin Board [2]. The role of the contractor is complementary towards the role of the commissioner. They perform sub-requests directly or commission the task to another entity. They change their role into the commissioner of a lower rank (second degree). They use their own (closed) database of subcontractors (Fig. 1). While analyzing this variant, we may determine a so called distance - between one cell to the others. Each cell communicates directly with the closest cell.

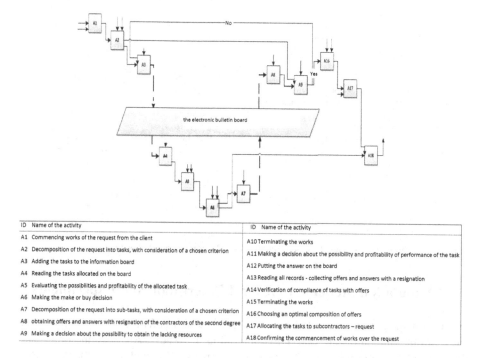

ID	Name of the activity	ID	Name of the activity
A1	Commencing works of the request from the client	A10	Terminating the works
A2	Decomposition of the request into tasks, with consideration of a chosen criterion	A11	Making a decision about the possibility and profitability of performance of the task
A3	Adding the tasks to the information board	A12	Putting the answer on the board
A4	Reading the tasks allocated on the board	A13	Reading all records - collecting offers and answers with a resignation
A5	Evaluating the possibilities and profitability of the allocated task	A14	Verification of compliance of tasks with offers
A6	Making the make or buy decision	A15	Terminating the works
A7	Decomposition of the request into sub-tasks, with consideration of a chosen criterion	A16	Choosing an optimal composition of offers
A8	obtaining offers and answers with resignation of the contractors of the second degree	A17	Allocating the tasks to subcontractors – request
A9	Making a decision about the possibility to obtain the lacking resources	A18	Confirming the commencement of works over the request

Fig. 1. Reference model: The electronic bulletin board - serial coordination of actions, own study

2.2   The Electronic Bulletin Board – Concentrated Coordination of Actions

The second variant of the presented mechanism of coordination of tasks called "Approving with application of an Electronic Bulletin Board", embraces coordination and supervision of all works, even those that remain on the lowest level of complexity, through the main commissioner. If the subcontractor for a certain sub-request is not found, the scope of works of this sub-request becomes decomposed into the sub-requests of a lower level by the main commissioner. While analyzing this variant, we may determine a so called smaller distance - be-tween one cell to the others. Each cell established direct communication with a one, main cell, being a leader or an integrator (Fig. 2).

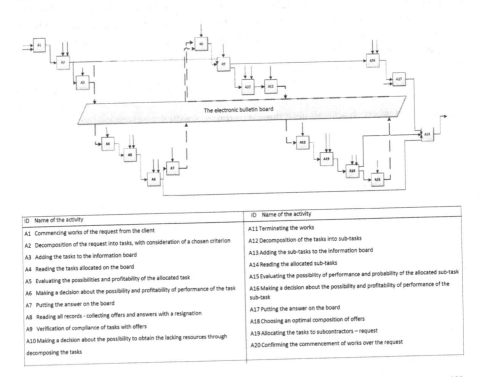

ID	Name of the activity	ID	Name of the activity
A1	Commencing works of the request from the client	A11	Terminating the works
A2	Decomposition of the request into tasks, with consideration of a chosen criterion	A12	Decomposition of the tasks into sub-tasks
A3	Adding the tasks to the information board	A13	Adding the sub-tasks to the information board
A4	Reading the tasks allocated on the board	A14	Reading the allocated sub-tasks
A5	Evaluating the possibilities and profitability of the allocated task	A15	Evaluating the possibility of performance and probability of the allocated sub-task
A6	Making a decision about the possibility and profitability of performance of the task	A16	Making a decision about the possibility and profitability of performance of the sub-task
A7	Putting the answer on the board	A17	Putting the answer on the board
A8	Reading all records - collecting offers and answers with a resignation	A18	Choosing an optimal composition of offers
A9	Verification of compliance of tasks with offers	A19	Allocating the tasks to subcontractors – request
A10	Making a decision about the possibility to obtain the lacking resources through decomposing the tasks	A20	Confirming the commencement of works over the request

Fig. 2. Reference model: The electronic bulletin board - concentrated coordination of actions [2]

3   Simulation Model – A General Description and Experiments

Simulation models were created by using the Flexsim software. It is a simulation software, in which tools are both powerful and user-friendly. Analysis and optimization any systems and processes in any industry are possible with using FlexSim. Besides, it is possible to 3D visualization and customize our modeled systems by added layout in CAD, import our own shapes etc. [4].

The simulation models were created for both presented mechanism of coordination in complex systems. Thanks to these simulation models, duration of all activities in analyzed mechanism is determined. Time depends on the assumed size of the order. In addition, it is possible to observe the impact of each decision on the analyzed process.

3.1 Simulation Model – General Information

In order to discuss experiments and results of these experiments, it is necessary to present some description of models and assumptions which was made. Some assumptions were made for both variants. For the purposes of this article the assumptions which have the most impact on the result will be presented.

Firstly, three sizes of the order were identified: 1-small, 2-medium, 3-large. Depending on the size, the time of each activity is different, time for variant 2 are presented (in variant 1 they are similar) (Tab. 1). Moreover, it determines for how many tasks and subtasks one order will be devided. Depending on the size of the order, it is divided into a predetermined number of tasks, and the one task is divided into the number of subtasks. Small order (1) is divided into 5-10 task, medium (2) to 11-30 tasks and the large (3) to 31-80 tasks. Number of subtasks for them is as follows: (1) – 2 subtasks, (2) – 3-4, and (3) – 5-10 subtasks. To map activities in the model in the best way, authors decided to describe time of the processes as a random value (the activities carried out by people never last forever as much). These times are drawn using the normal distribution: normal (mean, deviation, stream).

Table 1. Time of every process depending on order size – variant 2. Source: own study

Process	Small		Medium		Large	
	Mean [min]	Deviation [min]	Mean [min]	Deviation [min]	Mean [min]	Deviation [min]
A1	24	4,8	30	6	36	7,2
A2	288	57,6	360	72	432	86,4
A3	48	9,6	60	12	72	14,4
A4	4,8	0,96	6	1,2	7,2	1,4
A5, A6	96	19,2	120	24	144	28,8
A7	12	2,4	15	3	18	3,6
A8, A9	24	4,8	30	6	36	7,2
A10, A11	48	9,6	60	12	72	14,4
A12	4,8	0,96	6	1,2	7,2	1,4
A13, A14	96	19,2	120	24	144	28,8
A15	12	2,4	15	3	18	3,6
A16	96	19,2	120	24	144	28,8
A17	48	9,6	60	12	72	14,4
A18	4,8	0,96	6	1,2	7,2	1,4

The last important information is the fact that in models are decision-making points. Flow in model depends on decision taken at these points. These decisions relate to the profitability of the job. In first variant of coordination mechanisms is one decision point (A5), in second variant two decision points have been distinguished (A5 and A13). Share of positive decisions (shared in %) in the highlighted areas is defined. This element is a basic variable in model, ie. in the different experiments value of these variables will be change, and it will have a huge impact on the experiments results.

3.2   Simulation Experiment

For each presented variant of coordination mechanism, several experiments have been made. In these experiments, the values of selected variables have been changed. The observed element was assumed time of realized each processes. The main goal was to show the time of whole process depending on the adopted values (in simulation it is determine as performance measure). Decision variables in presented experiments are: size of the order and % share of positive decisions in the distinguished points decision.

In order to carry out experiments, authors have used a built-in experimenter tool. As we can read at Flexsim website [4]: "The experimenter allows you to quickly change multiple sets of variables and see the results. Create scenarios by choosing which model parameters to vary. Try a handful of scenarios, or hundreds! Save time by running many replications of multiple scenarios automatically – set it up and walk away."

It is worth to add, that in connection with the occurrence of normal distributions in the model, each of the experiments was carried out ten times (10 replications). The aim was to show differences in the analyzed order flow time, which depends of the duration of each activities (which are random values).

Table 2. Definition of simulation experiments – variant 1 of coordination. Source: own study

No Experiment	Order Size	% positive decision in A5	% positive decision in A13 (only for variant 2.X)
1.1 & 2.1	1	80	80
1.2 & 2.2	1	100	100
1.3 & 2.3	1	50	50
1.4 & 2.4	2	80	80
1.5 & 2.5	3	80	80

For both presented variants of coordination, five experiments were carried out (for each 10 replication were made). The variables in each experiment are: the size of the order and % share of positive decisions in selected point (variant 1 in A5, variant 2 in A5 and A13). Defined value in experiment 1.1 and 2.1 (Tab. 2) are basic value. Other values are defined for the purposes of the experiment, in order to show the effect of each variable on the observed main aspect, which is the flow time.

4   Results of Simulation Experiments

The purpose of the experiments was to determine the flow time for the order, ie. time to realized all the processes in mechanism of coordination. By using the built-in experimenter tool in Flexsim software, referring to the flow time, two types of results were obtained: (1) the flow time for each scenario and each replication (in minutes) and (2) the determination of the interval of values of the flow time for the defined confidence interval as 90%. Analyzing the results obtained for the variant 1 (Tab. 3) it can be noted the correct behavior - with increasing size of the order, the time of realized all processes is increased (scenarios 1.1, 1.4, 1.5). In the case with large order, the analized time is over 17 times longer since the case in small order. So huge prolongation is a result from the greater number of tasks and subtasks (after splitting the order). It can be noted that with increasing % of positive decision in A5, time of all process is shorter, in case of decrease % of this decision the time is longer (scenarios 1.1/1.2/1.3).

Analyzing the results obtained in the variant 2 (Tab. 4) it can be noted the similar dependences as in variant 1. Besides, flow time in every scenarios are shorter then in variant 1. It is result of change the structure of mechanism - in variant 1 some activities make only contractor (longer time of single activity), in variant 2 - contractor and principal. The second reason is different loops in each variants.

Table 3. Variant 1 – flow time – results of experiments. Source: own work

	Rep 1	Rep 2	Rep 3	Rep 4	Rep 5
Scenario 1.1	2703,28	3034,17	3159,13	2660,38	2650,60
Scenario 1.2	2276,23	1625,18	2827,07	2660,38	2140,24
Scenario 1.3	4012,64	3320,97	4301,73	5812,22	2650,60
Scenario 1.4	8111,11	7100,24	12117,93	10905,05	9430,48
Scenario 1.5	46379,47	37277,95	46750	50460,36	46147,43
	Rep 6	**Rep 7**	**Rep 8**	**Rep 9**	**Rep 10**
Scenario 1.1	1592,56	1752,03	1625,78	1590,16	3453,19
Scenario 1.2	1678,52	1752,03	1625,78	1590,16	2373,48
Scenario 1.3	2673,54	3106,85	2097,70	1949,01	4706,32
Scenario 1.4	5895,21	5117,98	8266,50	3488,06	8387,40
Scenario 1.5	28253,15	43327,44	39577,37	24493,56	47181,37

Table 4. Variant 2 – flow time – results of experiments. Source: own work

	Rep 1	Rep 2	Rep 3	Rep 4	Rep 5
Scenario 2.1	1510,03	1687,13	1992,13	2023,87	2132,58
Scenario 2.2	1510,03	1188,83	1697,81	1755,61	1415,70
Scenario 2.3	2838,12	2266,50	2885,90	3392,08	2269,38
Scenario 2.4	4537,37	3184,19	5974,66	6573,08	5019,41
Scenario 2.5	18586,62	12442,84	22877,46	26071,75	17934,77
	Rep 6	**Rep 7**	**Rep 8**	**Rep 9**	**Rep 10**
Scenario 2.1	1520,33	1340,66	1888,75	1802,65	2102,25
Scenario 2.2	1098,43	1340,66	1294,26	1217,64	1754,20
Scenario 2.3	2788,05	1643,62	2566,75	1983,03	3504,94
Scenario 2.4	3640,56	3540,99	4294,12	3156,73	5636,93
Scenario 2.5	10127,79	22746,00	20807,75	16543,89	21753,81

5   Conclusion

Thanks to the modeling of individual variants of coordination it is possible to observe the changes which are taking place in the whole process, as a result of changes in a single aspect. By using a simulation model and a built-in experimenter tool, it is easy to obtained a series of results. Based on these results, for the example, the range of values of a flow time of the 90% confidence interval can be defined.

The experiment confirmed the hypothesis. Serial coordinating the activities is a solution, which requires a long time. It requires the exchange of information between the various links in the supply chain, which is not efficient it this case. Decentralised coordination of actions in complex systems also leads to local optimization. In contrast, coordination concentrated in a single decision-making medium allows for effective (in relation to time) reaction. Designated distance between elements of the supply chain is also smaller. So it seems, that communication between them can be more effective. Centralized coordination leads to global optimization in created complex system.

Acknowledgements. Presented research are carried out under the LOGOS project (Model of coordination of virtual supply chains meeting the requirements of corporate social responsibility) under grant agreement number PBS1/B9/17/2013.

References

1. Grzybowska, K., Kovács, G.: Logistics Process Modelling in Supply Chain – Algorithm of Coordination in the Supply Chain – Contracting. In: de la Puerta, J.G., et al. (eds.) International Joint Conference SOCO'14-CISIS'14-ICEUTE'14. AISC, vol. 299, pp. 311–320. Springer, Heidelberg (2014)
2. Grzybowska, K.: Reference models of selected action coordination mechanisms in the supply chain. LogForum 11(2), 151–158 (2015), doi:10.17270/J.LOG.2015.2.3
3. Hoffa, P., Pawlewski, P., Nielsen, I.: Multimodal Processes Approach to Supply Chain Modeling. In: Omatu, S., Bersini, H., Corchado Rodríguez, J.M., González, S.R., Pawlewski, P., Bucciarelli, E. (eds.) Distributed Computing and Artificial Intelligence 11th International Conference. AISC, vol. 290, pp. 29–37. Springer, Heidelberg (2014)
4. https://www.flexsim.com (date of access: January 23, 2015)
5. Jasiulewicz-Kaczmarek, M.: Participatory Ergonomics as a Method of Quality Improvement in Maintenance. In: Karsh, B.-T. (ed.) EHAWC 2009. LNCS, vol. 5624, pp. 153–161. Springer, Heidelberg (2009)
6. Kotarbiński, T.: Traktat o dobrej robocie, Zakład Narodowy im. Ossolińskich Wydawnictwo Polskiej Akademii Nauk, Wrocław-Warszawa-Kraków-Gdańsk-Łódź (1982)
7. Kramarz, M.: The nature and types of network relations in distribution of metallurgical products. LogForum 4, 57–66 (2010)
8. Relich, M.: Knowledge Acquisition for New Product Development with the Use of an ERP Database. In: The Federated Conference on Computer Science and Information Systems, Krakow, Poland, pp. 1285–1290 (2013)

9. Relich, M., Muszynski, W.: The use of intelligent systems for planning and scheduling of product development projects. Procedia Computer Science 35, 1586–1595 (2014)

10. Sitek, P., Wikarek, J.: A Hybrid Approach to the Optimization of Multiechelon Systems. Mathematical Problems in Engineering 2015, Article ID 925675 (2015), doi:10.1155/2015/925675

11. Sitek, P., Wikarek, J.: Hybrid Solution Framework for Supply Chain Problems. In: Omatu, S., Bersini, H., Corchado Rodríguez, J.M., González, S.R., Pawlewski, P., Bucciarelli, E. (eds.) Distributed Computing and Artificial Intelligence 11th International Conference. AISC, vol. 290, pp. 11–18. Springer, Heidelberg (2014)

Part III
Special Session on Multi-agents Macroeconomics (MAM 2015)

Bilateral Netting and Contagion Dynamics in Financial Networks

Edoardo Gaffeo and Lucio Gobbi

Department of Economic and Management, University of Trento, Italy
{edoardo.gaffeo,lucio.gobbi}@unitn.it

Abstract. The bilateral netting of mutual obligations is an institutional arrangement employed in payment systems to reduce settlement risks. In this paper we explore its advantages and pitfalls when applied to an interbank lending market, in which banks extend credit to – and borrow from – other banks. Bilateral netting considerably reduce the potential for default cascades over an interbank network whenever the source of contagion is a negative shock to the assets of a randomly chosen bank. When the shock hits the liability side of the balance sheet, the mitigating effect of a bilateral netting agreement depends critically on the topological characteristics of the interbank network, however.

Keywords: Financial networks, Contagion dynamics, Bilateral netting.

1 Introduction

Driven by the unfolding of the recent global financial crises, the last few years have witnessed an explosion of studies combining computational techniques and network analysis to explore the issue of systemic risk due to contagious failures in interbank markets.[1] This work has allowed to enrich the path-breaking analytical approach developed by Allen and Gale (2000), showing that there appears to be a non-monotonic (inverted U-shaped or M-shaped) relation between the density in an interbank market - expressed by the proportion of the bilateral lending/borrowing relationships actually present out of all possible combinations - and the number of defaults as an idiosyncratic shock hits the system.

In this paper we compare a real gross time settlement interbank market with one in which counterparties are allowed to sign bilateral netting agreements, here defined as contractual arrangements according to which every individual bilateral combination of participants settles its net credit-debit position on a bilateral basis. In doing so, we try to clarify two issues that are still controversial.

First, there are discordant views about the effect of this kind of agreements on the unfolding of systemic risk associated to default cascades in financial networks. Upper and Worms (2004) show that the bilateral netting of otherwise gross interbank

---

[1] Chinazzi and Fagiolo (2013) provide a comprehensive survey.

© Springer International Publishing Switzerland 2015
S. Omatu et al. (eds.), *Distributed Computing & Artificial Intelligence, 12th Int. Conference,*
Advances in Intelligent Systems and Computing 373, DOI: 10.1007/978-3-319-19638-1_42

exposures decreases the aggregate losses suffered by the system after a contagion occurs from 75% to 10% of total assets, a result confirmed by Degryse and Nguyen (2007). Elsinger et al. (2006), on the contrary, suggest that the final effect of a bilateral netting scheme is ex-ante indeterminate, although their simulation show that from a practical viewpoint it is almost irrelevant. Thus, it appears that the output of existing simulation models based on bilateral netted interbank markets might be affected by network dependence, even if it is not clear what the connections between network topology and financial stability are. To address this issue, we compare the resilience of a network hit by idiosyncratic shocks on bank external assets with and without netting, under different configurations. We show that netting can affect systemic risk considerably, reducing the number of defaults in the best scenario up until the 85%. Since this result does not change for a wide range of network parameters and configurations, we claim that netting agreements represent an institutional arrangement capable to consistently reduce the probability of default cascades and equity losses.

Second, we explore the effect of netting on financial crisis generated by liquidity shortages. While credit losses are transmitted from debtors to creditors, liquidity shortages (like for instance un unexpected withdrawal of deposits) run from creditors to borrowers. It follows that a deficit (i.e., net borrower) bank can suffer a liquidity shortage even if it is not directly hit by the shock, and that it can shield against a liquidity shock in two ways: holding more liquid assets or increasing the number of interbank linkages. By applying bilateral netting settlements to a wide array of network topologies in a system hit by an idiosyncratic run on deposits at one institution, we find that that the netting of bilateral exposures scores some drawbacks if compared to the standard way of conducting payments. Hence our analysis, while trying to clarify the trade-off embedded in this type of payment agreement, illustrates that a one-size-fits-all institutional arrangement fails to provide a clear-cut policy answer to liquidity shocks.

2   Building the Banking Network

Following a standard approach in the literature, we set-up a computational platform aimed at modeling the functioning of a financial network by moving from two key drivers. The first takes the form of a probability p that two intermediaries are linked together by a credit-debit obligation, while the second is a simple network size parameter M. Our starting point is a classic Erdös-Renyi random network, in which every bank has the same attachment probability. The interbank market can therefore be represented by the $M \times M$ adjacency matrix X, in which each element x_{ij} represents a loan bank i extends to bank j. The principal diagonal of X is zero (a bank does not lend to itself) and the other elements can take values 0 or 1 with probability p.

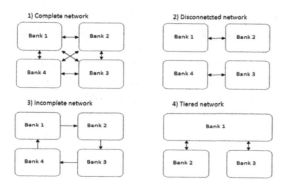

Fig. 1. Typical topologies of financial networks

In Figure 1 we represent four typical network topologies as applied to a financial market. While Panels *1)* to *3)* represent *regular* networks characterized by a degenerate degree distribution (each node has the same number of edges pointing from and towards it) and a decreasing connectivity, the network in Panel *4)* is characterized by the presence of a *hub*, that is a node having a higher number of links if compared to the others. Note that as we move from a deterministic to a probabilistic setting, the first three types of topologies are attainable by means of suitable restrictions of an Erdös-Renyi scheme, whereas the latter requires a connection structure which is not compatible with the former ones. Therefore, after having investigated purely random structures with a varying degree of connectivity we change the attachment procedure in order to be able to replicate also a money center network.

Once the network has been shaped, each bank is endowed with a balance sheet which has to be consistent with double-entry book keeping. On the liabilities side, a generic bank i is endowed with a given amount of interbank borrowing (b_i), deposits (d_i) and equity (e_i), so that total liabilities are given by $TL_i = b_i + d_i + e_i$. Likewise, total assets are defined by $TA_i = a_i + l_i$, where $l = \Sigma l_{ij}$ represents interbank lending and a external assets (like loans to firms and households). Clearly, $TL_i = TA_i$ must be true.

The procedure that assigns a balance sheet to each bank consists of three steps. As in Nier et al. (2007), the first step is to fix the amount of total assets available in the network as a whole, as well as the percentage of external assets (ω). Having fixed the aggregate volumes of total and external assets (A, EA), we can obtain by difference the aggregate volume of interbank assets, $I = A - EA$. The second step consists in dividing interbank assets by the total number of links, in order to obtain the size of every loan in the interbank market. Now, the value of interbank assets\liabilities for each bank is computed multiplying the number of creditors\borrowers by the number of links. In this way we obtain an adjacency matrix whose entries are l_{ij} and b_{ij}. The third step of the procedure deals with the determination of the external assets, deposits and equity at an individual level. In particular, we derive the amount of equity for each bank by multiplying the bank's total assets by a capitalization parameter γ, that can immediately be interpreted as a leverage ratio. Having found values for equity and interbank borrowing, we finally derive deposits by subtracting the interbank borrowing and the equity from total assets. In this benchmark treatment, interbank obligations are regulated on a gross settlement basis.

Given the total amount of assets and four parameters (M, p, ω, γ), we are able to fully describe the financial network. While M and ω will be held constant, in simulations we will explore the dynamics of contagion for different values of p and γ.

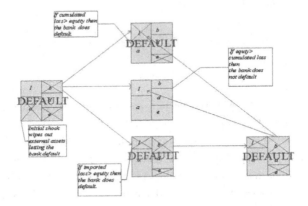

Fig. 2. Graphical representation of the default algorithm

The network is then buffeted by an idiosyncratic shock that wipes out all the external assets of a randomly chosen bank, forcing it to default. The part of the shock not offset by its equity is spread evenly among creditor banks. That happens because we are implicitly assuming that deposits are senior to interbank liabilities, for instance due to a public deposit insurance. Loss spreading ignites a contagion according to a sequential default scheme (Furfine, 2003). A sketch of how the default algorithm works is in Figure 2. The Matlab code is available from the authors upon request.

3   Default Cascades with Bilateral Netting

In order to see whether bilateral netting can affect the magnitude of contagion crisis, we introduce an additional arrangement regarding how credit-debt interbank obligations are settled. In this new treatment, we allow gross interbank credits and debts to be offset whenever two banks have a bilateral relationship, i.e. whenever bank i is at the same time debtor and creditor of bank j. The performances – defined in terms of the deepness of contagion ensuing from an idiosyncratic shock – obtained under the two treatments are then compared for identical parameterizations of the system.

We fix the total amount of assets at 100,000 monetary units, the percentage ω of interbank assets at 20% of total assets, and the number of operating banks at $M = 25$. Then we let the attachment probability p vary from 0 to 1, and the leverage ratio γ from 1% to 3%.

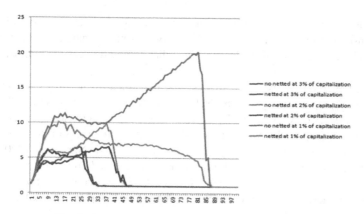

Fig. 3. Contagion profiles for purely random networks

As we focus on purely random networks (Figure 3), we stress two findings. First, contagion profiles for different values of leverage - defined by the total number of failures after the contagion ended for different values of the connectivity parameter p - are invariably M-shaped. Second, bilateral netting agreements change this result significantly. Even if the profile remains M-shaped, the number of defaults drops and, for low levels of capitalization, the global maximum of contagion profile for the netted market is approximately lower than the local minimum of the no netted one.

This happens because the netting procedure cuts the channels through which a contagion spreads out. In particular, we find that for a disconnected network the contagion profile does not change too much if netting is applied, due to the fact that - given a low attachment probability - there are not many cases in which credits and debits can be offset. The number of defaults decreases consistently when the probability of connection is set in the 0.2-0.4 interval, however, where the number of failures decreases on average by 30%. The advantage of allowing bilateral netting agreements is maximized for very low capitalized and high interconnected systems.

Table 1. Comparison in terms of equity losses for purely random networks. Figures represent savings in total equity when using bilateral netting agreements vs. gross settlement

Level of capitalization	Degree of interconnectedness (ranges of variation for the p parameter)				
	$0-20$	$20-40$	$40-60$	$60-80$	$80-90$
1%	11%	24%	44%	67%	85%
2%	6%	28%	50%	70%	85%
3%	10%	30%	50%	70%	85%

Table 1 shows that the savings in terms of disrupted equity are rather explicit and uniform. The decrease in equity losses one registers when moving from a gross settlement scheme to bilateral netting agreements varies from 6% for a low connected network, up to 85% for complete ones. Even in the case where no default occurs at all, netting agreements allow substantial equity savings and act as a kind of sanitary belt against the propagation of shocks. The only drawback regards the fact that depositors can be charged part of the losses, given that netting reduces the number of creditors in a seniority class lower to that of deposits.

In fact, the empirical evidence suggest that real financial networks are not purely random but characterized by a high level of tiering (Craig and Von Peters, 2014). In a tiered network, two types of banks operate: *i*) money-centers, who act as liquidity poolers; *ii*) standard banks, who interact with money-centers to satisfy their liquidity needs. In order to shape a topology of this kind, we follow Nier et al. (2007) in assigning different attachment probabilities to two pre-defined groups of bank. We isolate a subsample of $n = 21$ banks, each with a probability of connection $p_n = 0.20$, and we let the parameter p_h for the remaining $h = 4$ hubs to vary between 0.20 and 1.

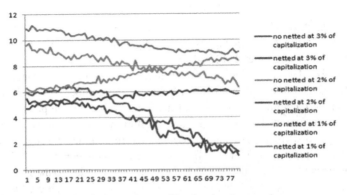

Fig. 4. Contagion profiles for tiered networks

The number of defaults decreases sensibly as we move from lowly to highly tiered networks (Figure 4). That happens because netting agreements isolate money centers from other banks, eliminating in this way many channels of contagion. Notice that this implies that netting saves equity losses as well. This fact appears clearly as we look at Table 2, where we report figures for savings in equity losses when comparing the gross settlement benchmark with the alternative netted treatment. The decrease in equity losses is more pronounced for well capitalized networks and it shows an improvement ranging from 17% to more than 30% for every level of equity tested.

Table 2. Equity losses for tiered networks. Figures represent savings in total equity when using bilateral netting agreements vs. gross settlement

Level of capitalization	Degree of interconnectedness (ranges of variation for the p parameter)			
	20 – 40	40 – 60	60 – 80	80 – 100
1%	17%	19%	24%	31%
2%	19%	22%	24%	29%
3%	20%	25%	27%	31%

We finally investigate the network resilience with money centers in structural surplus or deficit of liquidity. Moving from the former algorithm, to each money center we add (subtract) an additional stock of interbank assets proportional to its number of links, obtaining four types of networks with a variable number of surplus-deficit money centers: *i*) 1 deficit/4 surplus; *ii*) 2 deficit/3 surplus; *iii*) 3 deficit/2 surplus; *iv*) 4 deficit/1 surplus. Table 3 summarizes simulation results achieved using the following parameterization: $M = 25$ banks; $h = 5$, $I = 0.3$, $\gamma = 0.02$, $p_h = (0.9, 1)$ and $p_n = 0.4$, $\omega = 0.2$. Figures represent the percentage difference in the aggregate number of defaults due to contagion for the two types of payment agreements.

Table 3. Defaults in networks with hubs in structural deficit/surplus. Figures represent savings in the number of defaults when using bilateral netting agreements vs. gross settlement

	Shock on deficit bank	Shock on surplus bank	Shock on normal bank
1 deficit - 4 surplus	54%	31%	7%
2 deficit - 3 surplus	43%	38%	7%
3 deficit - 2 surplus	39%	48%	8%
4 deficit - 1 surplus	31%	53%	8%

When a shock forces to default a normal bank, the benefit of netting agreements is limited to a few percentage points. As we apply a shock to large banks, however, the benefit of netting depends on the relative number of hubs in structural deficit/surplus and the liquidity position of the bank originally affected by the run.

4   Liquidity Shocks

In order to address the issue of liquidity shocks, we build a wide array of network topologies and we hit each network by a shock that wipes out a part of the deposits of each bank. Each network is characterized by surplus and deficit banks. An intermediary is considered a surplus one if the amount of its interbank assets is higher than the amount of its interbank liabilities.

As in Lee (2013), we assume that each bank facing a liquidity shock calls in its interbank credits on other banks and uses its buffer of liquidity. Furthermore, we assume that banks cannot short-sell their liquid assets and they can dispose of their illiquid assets only if they do not hold any other kind of liquid assets. Lastly, banks are not allowed to issue stocks or to make new borrowings in the interbank market. The aggregate level of interbank assets after netting is fixed at $\alpha = 0.1$; the size of the shock that buffets each bank is $\gamma = 0.025$, the liquidity reserve of each bank is $\delta = 0.025$, the amount of equity $e = 0$, and the external assets z of each bank is set to respect the double-entry book keeping consistency.

Our simulations show that netting can increase the risk of a systemic liquidity shortage if compared to a standard way of conducting payments. We observe that for complete and disconnected networks the amount of liquidity shortage increases by 33%. Circular networks are not affected by netted agreements, and for that reason the aggregate liquidity shortage remains the same. For what concern core-periphery structures, we register an increase in the aggregate liquidity need, amounting to 22% for the surplus money center and to more than 16% for the deficit one, respectively.

5   Conclusion

We show that netting agreements can be viewed as a tool able to diminish systemic risk in a modern financial network. This is the most important value added of the paper, because it changes the idea that increasing interconnectedness from low to intermediate values necessarily reduces financial stability.

References

1. Allen, F., Gale, D.: Financial contagion. Journal of Political Economy 108, 1–33 (2000)
2. Chinazzi, M., Fagiolo, G.: Systemic risk, contagion and financial networks: A survey, LEM Papers Series 2013/08, Sant'Anna School of Advanced Studies, Pisa (2013)
3. Craig, B., Von Peters, G.: Interbank tiering and money center banks. Journal of Financial Intermediation 23, 322–347 (2014)
4. Degryse, H., Nguyen, G.: Interbank exposures: an empirical examination of systemic risk in the Belgian banking system. International Journal of Central Banking 3, 123–171 (2007)
5. Elsinger, H., Lehar, A., Summer, M.: Using market information for banking system risk assessment. International Journal of Central Banking 2, 137–165 (2006)
6. Furfine, C.: Interbank exposures: Quantifying the risk of contagion. Journal of Money, Credit and Banking 35, 111–128 (2003)
7. Lee, S.: Systemic liquidity shortages and interbank network structures. Journal of Financial Stability 9, 1–12 (2013)
8. Nier, E., Jang, J., Yorulmazer, T., Alentorn, A.: Network models and financial stability. Journal of Economic Dynamics and Control 31, 2033–2060 (2007)
9. Upper, C., Worms, A.: Estimating bilateral exposures in the German interbank market: is there a danger of contagion? European Economic Review 48, 827–849 (2004)

Using Social Media Advertising to Increase the Awareness, Promotion and Diffusion of Public and Private Entities

Eva Lahuerta Otero and Rebeca Cordero Gutiérrez

University of Salamanca, Department of Business Administration,
Campus Miguel de Unamuno, 37007, Salamanca, Spain
{eva.lahuerta,rebecacg}@usal.es

Abstract. The purpose of this paper is to examine the effectiveness of the on-line advertising techniques through the social network Facebook, as a broadcasting and promotion tool to disseminate information of companies. By means of two different case studies, we present the findings on two different teaching centers belonging to the public and the private sector. Results show that these techniques are able to get a large number of impressions and clicks, which impacts (in terms of awareness and recognition) go beyond the initial targeted group. Although the results are based on a single country and therefore they cannot be used to make generalizations, the conclusions of the study show the cost-effective effect relationship of these innovative techniques when promoting businesses among the online community. Consequently, these methods represent and important supporting point on the organizations' marketing strategies when addressing the challenges of this new, digital society.

Keywords: Social networks, social media analytics, promotion, case study, advertising, Facebook.

1 Introduction

Our society is experiencing an unstoppable development of the information and communication technologies (ICTs). With the rapid digitalization of media, advertising landscape has changed dramatically.

Spain accounted for 28.9 million Internet users in 2013 [10], 84% of which consulted the social media very often [9]. Furthermore, almost 56% of the Spanish Internet users buy on the Internet, which means 15.2 million shoppers spending €816 on average per year [10].

Consequently, companies invested 874.4 million Euros in digital advertising in Spain in 2013 [6]. Online and social advertising accounted for the 21% of the total share, consolidated its second position only behind a traditional media such as television.

Even if online advertising is becoming more important day by day, it has not re- place the traditional advertising yet, as both techniques are conceived as complementary to

© Springer International Publishing Switzerland 2015

S. Omatu et al. (eds.), *Distributed Computing & Artificial Intelligence, 12th Int. Conference,*
Advances in Intelligent Systems and Computing 373, DOI: 10.1007/978-3-319-19638-1_43

each other [1]. It is not yet well understood among the research community how social advertising works and so there is a need to study the use of the Internet as an advertising tool more in deep.

The Web 2.0 is understood as a new generation of Internet services based on the creation of communities, the network concept and the use of the new ICTs for information dissemination. It includes a set of social media so that users can co-create, organize, edit, share, comment and evaluate content by forming social networks that interact with each other.

This research will focus on horizontal social networks, which are those formed by users with no particular interest or objective on a particular topic [9]. Individuals participating in these types of networks are looking for entertainment and communication, so they become a powerful tool to gain and build customer loyalty from a business perspective [4].

Businesses can no longer afford to ignore the benefits of using social media. In the past, customers consulted with a limited number of family members of friends before making a purchase. The frequent use of social media increases that number to thousands of opinions from all over the world on a single click. This means online advertising is no longer a luxury; it is an affordable necessity.

Recent studies found that social media marketing techniques can generate more business exposure, increased traffic and improved search engine rankings [16]. Even if the media landscape has dramatically changed, and investments in digital advertising increase every year, is not yet well understood how firms can use this forms of publicity and their impacts on organizational goals. Today's consumers expect companies to have an online presence in social media and it is not just an-other channel for disseminating corporate information. It represents a tremendous opportunity to engage and create social bonds with customers and followers, with helps to improve corporate reputation.

To help filling this literature gap, this study presents two different empirical case studies to test the effectiveness of the online advertising techniques in horizontal social networks.

2   Theoretical Framework

2.1   Social Network Sites and Advertising

In the past, advertisers had total control over where the message was place and when customers would be exposed to it. Thanks to the improvements on the Internet and the ICTs, advertising has added to the traditional media advertisements (press, radio, television, magazines...) marketing strategies in order to increase the coverage of the targeted population. Advertisers now control the initial placement of information but they are unable to control how information is disseminated across customers' social networks.

Companies and organizations are aware of the power of advertising in the online world, as it is capable of adapt the content according to the rapid technological changes and, besides it implies a direct contact with the prospective customers.

Companies have now a unique opportunity to connect with their customers by offering unique, adapted and personalized promotions using the social network sites. This also means that online advertising has suffered from severe criticisms, as some users consider it to be deceitful, mean, abusive and annoying. Bearing in mind the fact that these social networks may not be the ideal place to show annoying and intrusive advertisements, the platforms have redesigned their strategies so they can generate customer trust by using the power of social networks and the effects they have in the contact list of the users. If publicists progressively incorporate social elements to advertisement, it is easier for users to accept, cooperate and transmit the message. Moreover, peer and social pressure can turn users' attitudes, values or behaviors in regard of a company, brand or service so they conform to the group norms. This will ultimately increase potential customers [1].

According to the relevant literature, users engage in social media to share interests and opinions. In addition, social network sites facilitate education and information [14]. These motivations are in agreement with media uses and gratifications theory [7]. This theory states that media users are willing to expose themselves to certain kind of media based on their needs and gratification-seeking objectives. Users will use media to satisfy their hedonic or utilitarian needs. Initially formulated to explain television viewership, this theory has been applied to studies of Internet usage, online advertising, mobile advertising and even the usage of social networking sites.

Following this theory, users become fans of a Facebook page, so they get the latest updates of the organization on their walls, and they can keep up with the latest promotions, latest arrivals, contests or relevant information. When users see this type of content they can indicate they like it using the "like button" and they can also share it. This makes information spread so it can reach other users that belonging to the fans' network. In this manner, a single user transmits broadcasts and expands the information generated in origin by a company. This generates word-of-mouth, one of the most powerful tools described in the literature when making promotions.

Social media can therefore be used as a powerful advertising platform to create and reinforce bonds between firms and customers. This allows for the establishment of more dependable relationships, compared to other types of media.

However, teaching centers face difficulties when building as social presence as they are marketing intangible goods, so they need to deliver a consistent message across channels to reach the proper audience [17].

Authors such [5], point out that this social benefits can also ex- tend to SMEs as brand awareness increases exponentially at a minimum cost. Furthermore, as Facebook allows for a complete personalization and segmentation of the targeted population in the campaign, unnecessary costs can be avoided. Besides, compared to traditional advertising, digital advertising provides with a set of unique characteristics: is measurable, more cost-effective, has a strong segmentation power and is faster and more flexible [11].

Companies using social media proclaim they are receiving measurable business benefits including increased sales and market size, improved customer satisfaction and relationships, improved employee relationships, better and faster technical support, reduced marketing expenses and improved search engine rankings (e.g. [8][16]). Thus, numerous firms have successfully used this technique to socially promote their brands.

Following this line, the present research analyzes the effectiveness of the Facebook advertising campaign by means of two different case studies related education and teaching both in the public and in the private sector.

3   Methodology and Analysis

Facebook is the main global social network accounting for the 23.39% of the world-wide traffic, and it is the main social network in Spain [15]. With 13.5 billion users and 864 million daily active users, this networks gets 1.8 million likes per minute and shares 41,000 publications every second [13]. Users mainly join this social network to look for friends, to get social support, to entertain themselves, to look for information and also due to its convenience. Consequently, we may say that users spend time on Facebook in search for social capital [12].

This research uses the case study method to analyze the results obtained by two different businesses (a teaching and training company and a research institute from the University of Salamanca) using Facebook ads platform to promote their activities.

— Case 1: This private firm dedicates its commercial activity to teaching and training courses. Besides, they regularly conduct a set of activities to promote the company and to improve customer engagement. The company has a website showing up-dated information on training courses, events and activities. Its presence on the social networks started in July 2013, creating Facebook and Twitter public profiles.

Table 1. Results of the advertising campaigns in both case studies

	Case 1		Case 2	
Objective	Website clicks	Page post engagement	Website clicks	Page post engagement
Reach	8,409	33,599	416,979	27,933
Frequency	1.641575	9.785232	4.263423	1.47
Impressions	13,804	328,774	1,777,758	40,998
CPM (Euros)	1.448855	0.811621	0.30224	2.09
Clicks	467	3,080	8,376	507
Total unique clicks	332	1,794	6,604	426
Click-through rate (CTR)	3.383077	0.936814	0.471155	1.237
CPC (Euros)	0.042827	0.086636	0.064149	0.17
Actions	290	1,425	6,789	333
Total spent (Euros)	20	266.84	537.31	85.66

1. Case 2: The campaign belongs to a research group from the University of Sala-manca dedicated to teaching courses, degrees, masters and seminars in the new technologies field. This research group has an updated website where all in-formation about congresses, conferences, projects and activities they carried out can be found. Besides, all information on the degree's content is available. Its pres-ence on the social network sites started in 2012 when they opened profiles in hori-zontal social networks such as Twitter, Facebook, LindedIn and Google +

As observed in table 1, the results of the advertising campaigns are positive and have had a great impact. Both firms increased significantly the number of website clicks and the reach of their publications. The most important metrics and KPIs (key performance indicators) of the campaigns are described below:

- Reach: the number of people the ad was served to. The number of people reached of case 2 ads is clearly higher than for case 1 on website clicks. In both cases, we find a positive result but for case 2 with only 670 fans we got 417.000 people reached whereas on case 1, with 2370 fans we only reached 8,409 people. This means there is a chance to increase the contacts with other Facebook members in 4 persons for every fan in case 1, and of 662 persons for each fan for case 2. Regarding the interaction results (page post engagement), the case 1 gets better results, and this is probably due to the type of content they publish.
- Impressions. The number of times the ad was served. We need to bear in mind that a high impression rate is necessary as firms are never alone when advertising at a users' wall. This is the reason why in order to get a higher reach levels we see a 10 times frequency (on average) per every individual that sees the ad (in the case study 1).
- Cost per 1,000 impressions (CMP). The average cost the firm pays to have 1,000 impressions on its ad. This cost will depend on the number of ads with similar segmentation criteria competing to appear on a users' wall. In this case, we can see that when competing with similar ads, the case study number 2 the CPM is more expensive if the objective is to get post page engagement. However, in case 1 the cost is four times higher when getting website clicks. This may be due to the fact that in case two ads have more general content and also stronger competence in this social network. The degree of engagement of the fans with the firm is also an important factor.
- Clicks. The number of total clicks the ad gets. We can observe that in both cases the number of clicks is remarkable, especially for case 2 with 8,376 clicks.
- Unique clicks. The total number of unique people who have clicked on the ad. This is a very important metric as it allows us to know if our advertisements get several clicks from the same user. This means our ad is catching users' attention so they repeat the action to get informed on the proposed offer. This fact increases the chances that an individual will enroll on some of the teaching courses or seminars advertised. We can also observed that there is a significant different between clicks and unique clicks in both case studies.

- Click-Through Rate (CTR). The number of clicks the ad gets, divided by the number of impressions. This metric indicates the effectiveness of the publicity. Case 1 is more efficient when getting website clicks, whereas case 2 gets better results in page post-engagement actions. These results are consistent with the rest of the KPI explained above.
- Cost per click (CPC). The average cost per click for these ads, calculated as the amount spent divided by the number of clicks received. As shown in table 1, the cost for clicks when the objective is to get page post engagement is more expensive as it requires a bigger effort so users take an action after having seen the ad. We can also see that CPC for case 1 is cheaper than for case 2, probably because of the fierce competence on Facebook for similar ads. Nevertheless, the relative cost of Facebook ads campaigns is lower compared to other type of similar platforms where the CPC can be 15 or more times higher (e.g. Google Adwords).
- Actions: the number of actions taken on the page, page app or event after the ad was served to someone, even if they did not click on it. Actions include page likes, comments, shares, app installs, conversions, event responses and more. Results show that the case two got the most actions when getting website clicks. Raisons may be the won content of the ad, and also the great amount of information an individual gets by visiting the landing page. This creates incentives to users to take complex (or even multiple) actions on the same ad. Case 1 results are different, as the landing page offers very low information content, which diminishes its attractiveness for users and reduces their predisposition to take a specific action.
- Money spent: total money spent on a particular campaign. The cost of the different ad sets was below the established budget for both case studies.

Once analyzed the core KPIs, we see how case 1 gets better performance when the objective of the campaign is to get page post engagement, whereas case 2 shows superior performance on website clicks. This is because the teaching center on case 1 shows a closer relationship with their customers (current and prospective students) which favors customer engagement. However, the quality and characteristics of the website of the business in case two, as their international projection of this research group facilitates better results on website clicks. Pages with greater diffusion usually get better results, as they get higher scores on the Edgerank, the algorithm Facebook uses to decide which contents will be shown to users. A better position on this algorithm improves the diffusion of contents of a business, and also their related advertisements.

4   Results and Discussion

In the light of the results obtained by the two case studies, this research shows the efficacy of advertising in horizontal social networks to promote teaching-related content. The success of both campaigns was bigger than expected, thanks to the improvement of the attention and awareness from Facebook users to the organizations.

As our results show, a well-structured and oriented campaign may have a significant reach. This makes these social platforms a valuable tool to sell and promote businesses. However, we need to highlight that in order to reach its potential, advertising campaigns need to act on Facebook pages with a sufficient number of users. This also means that Facebook pages content must be updated, relevant and carefully designed so businesses can create an own social brand among users [3].

It should be noted that neither SME nor research centers dependent on public universities have a large advertising budget. Our study reveals that the use of paid advertising in horizontal social networks can become an effective communication solution without wasting resources away. The effort in terms of time, resources and money is relatively small compared to the good results obtained. This is the reason why we can affirm that horizontal social networks are a good communication platform for organizations involved in teaching, which can increase the ratio of enrolled students in the different activities promoted such as courses, research seminars, degrees or conferences.

Consequently, teaching centers need to get social as they get several benefits [2]:

- Reach students with a single social platform.
- Build an active community that interacts and co-creates content.
- Keep audiences informed an updated on the latest news, courses and seminars available.
- Monitor reputation of the teaching centers and retrieve feedback for improvement actions.
- Measure campaign success to control the performance of the advertisements.

Thanks to the big development of the advertising tools in the social networks (especially in Facebook and Twitter), there is greater flexibility to launch diverse campaigns adapted to a segmented targeted population. These characteristics allows for further promotion and diffusion possibilities among potential users, which would be difficult to reach using traditional media.

In this line, the new 2.0 advertising techniques, more precisely the use of the Facebook Ads tool, allows for the creation of fully personalized and segmented publications, adapted to the target businesses need in every one of the ad sets. Accordingly, social network sites become a powerful and original information and promotion tool that channels and understands the needs of their users' community (which differ depending on the role of the user: follower, student, teacher, prospective student, staff...) and meets them in a personalized and specific manner.

These advertising techniques are cost-efficient and innovative, and so they become an excellent attraction tool for prospective students. It is necessary to insist once again on the intensive role of technology in the millennial generation. Consequently, we strongly recommend that businesses of all kinds, aware of this new technological change, adapt their communication strategies so they include active social profiles in the most relevant horizontal social network sites: Facebook, Twitter, LinkedIn, Google + or even Instagram.

References

1. Barreto, A.M.: Do users look at banner ads on Facebook? Journal of Research in Interactive Marketing 7(2), 119–139 (2012)
2. Bernardo, E.: Why universities need to get social (2013), http://www.wired.com/2013/06/why-universities-need-to-get-social/ (retrieved)
3. Chan, C.: Using online advertising to increase the impact of a library Facebook page. Library Management 32(4/5), 351–370 (2011)
4. Cordero-Gutiérrez, R., Santos-Requejo, L.: La confianza y la actitud hacia la red social como determinantes de aceptar herramientas de marketing. Diferencias según intensidad de uso de la red. DOCFRADIS, Colección de documentos de trabajo Cátedra Fundación Ramón Areces de Distribución Comercia l, DOC 03/2014 (2014)
5. Harris, L., Rae, A.: Social networks: the future of marketing for small business. The Journal of Business Strategy 30(5), 24–31 (2009)
6. IAB Spain. La publicidad digital aumenta su peso en el reparto de la inversión publicitaria (2014), http://www.iabspain.net/noticias/la-publicidad-digital-aumenta-su-peso-en-el-reparto-de-la-inversion-publicitaria/ (retrieved)
7. Katz, E., Foulkes, D.: On the Use of the Mass Media As "Escape": Clarification of a Concept. The Public Opinion Quarterly 26(3), 377–388 (1962)
8. Lahuerta-Otero, E., Muñoz-Gallego, P.A., Pratt, R.: Click-and-mortar SMEs: Attracting customers to your website? Business Horizons 57(6), 729–736 (2014)
9. ONTSI. Las redes sociales en Internet (2011), http://www.ontsi.red.es/ontsi/es/estudios-informes/estudio-sobre-el-conocimiento-y-uso-de-las-redes-sociales-en-espa%C3%B1 (retrieved)
10. ONTSI. Perfil sociodemográfico del internauta, análisis de datos INE 2013 (2013), http://www.ontsi.red.es/ontsi/sites/default/files/perfil_sociodemografico_de_los_internautas_2013_0.pdf (retrieved)
11. ORSI. Marketing digital para PYMES. Observatorio Regional de la Sociedad de la Información (2012), http://www.orsi.jcyl.es (retrieved)
12. Pikas, B., Sorrentino, G.: The effectiveness of online advertising: consumer's perception of ads on Facebook, Twitter and YouTube. Journal of Applied Business and Economics 16(4), 70–81 (2014)
13. Qmee. What happens online in 60 seconds (2013), http://blog.qmee.com/qmee-online-in-60-seconds/ (accessed November 22, 2014) (retrieved)
14. Safko, L., Brake, D.K.: The Social Media Bible. John Wiley & Sons, Hoboken (2009)
15. Shareaholic. Facebook and Pinterest are the King and Queen of Social Referralls (2014), https://blog.shareaholic.com/social-media-traffic-trends-04-2014/ (accessed January 1, 2014) (retrieved)
16. Stelzner, M.A.: 2011 Social Media Marketing Industry Report: How Marketers Are Using Social Media to Grow Their Businesses (2011), http://www.socialmediaexaminer.com/social-mediamarketing-industry-report-2011/ (accessed January 1, 2014) (retrieved)
17. Washenko, A.: 2 universities tell us how they're using social media (2013), http://sproutsocial.com/insights/universities-social-media/ (retrieved)

Understanding the Support of Savings to Income: A Multivariate Adaptive Regression Splines Analysis

Iacopo Odoardi[1] and Fabrizio Muratore[2]

[1] Department of Philosophical, Pedagogical and Economic-Quantitative Sciences,
University of Chieti-Pescara, Viale Pindaro 42, 65127 Pescara, Italy
[2] Department of Economics and Social Sciences,
Marche Polytechnic University, Piazzale Martelli 8, 60121 Ancona, Italy
iacopo.odoardi@unich.it, fabrizio.muratore@univpm.it

Abstract. The understanding of the complex socio-economic phenomena requires a deep insight of the dynamics through several correlated variables. Our aim is to demonstrate how some relevant macroeconomic variables could affect the evolution of the economic development path. This research arises in a broader analysis on the role of the wealth of households held in specific forms unexploited and potentially useful if properly integrated in productive cycles. At the basis of these assumptions, however, there are complex dynamics on the formation and composition of wealth, which does not include business capital and savings in the bank channel. We want to demonstrate the utility of aggregate savings for the composition and the very existence of wealth, which can be stimulated through policy instruments. We need to ask how to prove empirically that, in modern economies, encouraging savings and the accumulation of private wealth represent, under certain conditions, a not fully considered support to economic development. In this phase we test if statistical techniques inspired by Artificial Intelligence can be better exploited respect to classic approaches, given the quality of the data available. We use multivariate adaptive regression splines model, in a comparison with a multivariate model, to examine the relationship of aggregate savings, and then other related variables, on GDP in the US for the period 1970-2012.

Keywords: artificial intelligence, data mining, multivariate adaptive regression splines model, aggregate savings, economic growth and development.

1 Introduction

Data management and statistical applications are integral of socio-economic analyses to recognize causes and effects of relevant connected social and economic phenomena. Sometimes the usual econometric analysis cannot capture the complexity of social events characterized by many variables, endogenous and exogenous to the investigated system. It is interesting to examine databases that include detailed data for numerous territories and for long time series, that can explain the succession of social policies and cyclical events. An opportunity to go beyond the explanatory limits of

© Springer International Publishing Switzerland 2015
S. Omatu et al. (eds.), *Distributed Computing & Artificial Intelligence, 12th Int. Conference,*
Advances in Intelligent Systems and Computing 373, DOI: 10.1007/978-3-319-19638-1_44

econometric analysis of time series and multivariate dependence is including analysis that move towards Artificial Intelligence (AI), such as data mining techniques. Data mining is in fact a complex process useful to analyse data from different perspectives with the aim to achieve valuable information. Data mining is the process of finding correlations or patterns among dozens of fields in large relational databases.

A better understanding of the social and economic processes can result from applications of newer techniques that combines Artificial Intelligence, statistical analysis, computer science and the efficient database management [9]. The possible application of these techniques to a large number of cases, in our case of countries, could serve to discover and demonstrate empirical regularities and unexpected relationships, that are not fully detectable with traditional econometric methods. As our paper is a first step to involve in economic analyses that part of private wealth that is an unexploited resource in the economic system, the data mining techniques can extract information not readily observable, and test them on large quantitative data, otherwise unmanageable [16]. Wealth can be formed by several types of resources: material, financial, intangible (such as the the ability to generate future income), or even embedded in personal characteristics. Only the first two can be objectively quantified, transferred to other economic agents and exploitable in production cycles. In our case, policy implications are an obvious result from these findings. The study of a large number of social and economic variables for several countries, representing different visions of political organization and markets, allows to apply both descriptive and forecasting techniques of data mining. The first type is useful to clearly represent what really happens in the analysed context, and possibly justify clusters of cases with similar characteristics, to draw generalized conclusion, even more useful if we look at subgroups of countries ordered, for example, by income. For our study, these techniques are useful if we do not consider a specific dependent variable, but we can simultaneously investigate different forms of wealth held by consumer households. The second type needs an independent variable, interesting in the economic point of view if we consider the aim of encouraging the accumulation of private wealth or, more generally, if we try to find out the effect of these "potential resources" on national income. In this work, we propose the comparison of two main class of empirical analysis, such as multivariate adaptive regression splines model with a standard multivariate model. We would like to move towards the full implementation of AI techniques to the study of the relationship of relevant socioeconomic variables, in this paper observing the close relationship between the values of aggregate savings of consumer households and the values of national income. The analysis of the savings is for us the first phase of the study in the broad analysis of household wealth as a resource of the system. Usually, scholars consider the household savings directed to businesses through the banking channel. Banks are in fact those who can most efficiently shift resources from areas or sectors to surplus to those in deficit. We want to move beyond the preestablished roles of consumers who save and entrepreneurs who invest. This is a consideration of economic agents in defined classes and standardized functions representative of immovable social classes, as already known to the classical economists (also in the field of distribution of resources see, among others, [12, 11], for a full explanation of income inequality and growth see [2]). But we want to develop this vision from a multiplicity of points of view. We consider households as economic agents who make their decisions in total independence, which have a limited rationality and can modify their choices (*e.g.* consumption or investment in human capital of the offspring) based on the current conditions of the system in which they operate (*e.g.* availability and incentives to save), as well as on expectations and experiences. Then we want to not consider a behavior assumed standardized for classes of individuals [4, 5]. Furthermore, we consider that households accumulate wealth in many forms, and that some wealth remain unproductive. That is wealth that is not managed by the banks and not part of the capital of firms. This happens because in reality individuals and businesses can have more forms of income [15]. At the same time, businesses can "accept" many forms of useful capital [4]. To analyze what we said, given the difficulty of finding data on the shape and the distribution of wealth (see in particular [7]), we use a variable indicative of the phenomenon, that is the aggregate savings, analyzing its relationship with the Gross Domestic Product (GDP). We consider that if we can demonstrate the utility in general terms of specific forms of private wealth, which of course are influenced by the rates of savings, we can observe a contribution to the economic development of the countries that are "equipped", at least in the long term. For economic development we refer to the general qualitative and quantitative progress in life and economic conditions. Of course, the necessary steps are complex, since the incentives to save towards forms of "wealth useful", up to advocate for the efficient use of these resources in production cycles. We consider World Bank data (retrieved on December 2014) on GDP and households savings and we compare techniques of data mining analysis (see definition and explanation in the next Chapter) with well-known econometric methods as multivariate analysis, to see which technique better explain the relationship described above. After the above comparison we develop another comparison by comparing simultaneously other macroeconomic variables on consumption and fixed capital.

2   A Multivariate Adaptive Regression Splines Model Application on GDP

The understanding of social phenomena can benefit from AI techniques, which can for example simulate the real behavior of economic agents, as in our case could be applied to the behaviors that lead families to accumulate wealth through saving, and pass wealth to the next generation. At the base of the existence and transmission of wealth and there is a chance to save shares of income (possibly to borrow, but taking into account the imperfections of the credit market, as in [14, 3]). It is necessary to demonstrate what technique further illustrates the dynamics of the phenomena described in Chapter 1. We use data mining techniques because we have no complications due to the origin of the data, and we want to understand the relationships between multiple variables also creating clusters of countries with similar characteristics and observing regressions with the target variable, the GDP. We compare a complex dataset using two principal models, regarding a period from 1970 to 2012 for the US, with the following variables: GDP (as dependent variable) GCF (gross capital formation), GFCF (gross fixed capital formation), HFCE (household final consumption expenditure), GS (gross savings) as predictors[1]. The models that we consider in this analysis are: multivariate and multivariate adaptive regression splines. The multivariate model concern to advanced techniques for examining relationships among multiple variables [1]. We analyze a multiple regression referred to a regression analysis, that examines the effects of several independent variables (predictors) on the value of a dependent variable, or outcome. Regression calculates a coefficient for each independent variable, as well as its statistical significance, to estimate the effect of each predictor on the dependent variable, with other predictors held constant [10]. In this case we do not claim in fact to show that the considered variables can explain the complexity of GDP formation. The multivariate equation considering our case study is given as:

$$y = f(x) = \beta_0 + \sum_{i=1}^{n} x_i \beta_i + \varepsilon_i \tag{1}$$

where β_0 and β_i are parameters of the model, y is the GDP (dependent), x are predictors GCF, GFCF, HFCE, GS and ε_i is the ith independent identically distributed normal error. The multivariate adaptive regression splines is an implementation of techniques for solving regression-type problems [8] with the main purpose to predict the values of a dependent variable from a set of predictor variables considering a non-parametric regression technique[2]. Specifically, this model uses two-sided truncated functions of the form $\pm(x-t)_+$ as basis functions[3] for linear or nonlinear expansion, which approximates the relationships between the response and predictor variables. The basis functions together with the model parameters (estimated via least square estimation) are combined to produce the predictions given the inputs. The aim is to adjust the coefficient values to best fit the data; considering a recursive partitioning regression we obtain a geometrical procedure better than a classic approach. The multivariate splines algorithm builds models from two sided truncated functions of the predictors (x) of the form:

$$(x-t)_+ = \begin{cases} x-t & x > t \\ 0 & otherwise \end{cases}$$

---

[1]   All variables are considered as annual percentage change.
[2]   A nonparametric regression procedure makes no assumption about the underlying functional relationship between the dependent and independent variables.
[3]   Basis functions of predictor variables (x) play an important role in the estimation of Multivariate adaptive regression splines. We consider basis functions like $(t-x)_+$ and $(x-t)_+$ where parameter t is the knot of the basis functions (defining the "pieces" of the piecewise linear regression); these knots (parameters) are determined from the data.

Therefore, the multivariate adaptive regression splines model [13] for a dependent variable y, and M terms , considering our case study can be summarized in the following equation:

$$y = f(x) = \beta_0 + \sum_{i=1}^{n} \beta_i H_{ki}\left(x_{v(k,i)}\right) \tag{2}$$

where β_0 and β_i are parameters of the model, y is the GDP (dependent), x are predictors GCF, GFCF, HFCE, GS and function H is defined as:

$$H_{ki}\left(x_{v(k,i)}\right) = \prod_{k=1}^{K} h_{ki} \tag{3}$$

where $x_{v(k,i)}$ is the predictor in the kth of the ith product[4].

We compare the two models (classic multivariate and multivariate adaptive regression splines) calculating the mean squared error that is equal to:

$$MSE = \sum_{i=1}^{n} \frac{\left(x_i - \hat{x}_i\right)^2}{n}$$

Where x_i is the observed value of the GDP, $\hat{x}_i$ is the estimated value of the GDP in the two models.

3   Results

The analysis was done first of all considering only the variable gross savings (GS) as a predictor compared to the Gross Domestic Product (GDP); after considering all the independent indicators (GFC, GCFC, HFCE, GS) compared to the GDP: in both cases we apply the two models presented above (multivariate model and multivariate adaptive regression splines one). We report results of the two models considered in the two analysis in the following table[5]:

Moreover, we obtain the following results for *MSE*: *MV* = 0,21 *MAR Splines* = 0,15. From the two results we find a lower value in a multivariate adaptive regression splines that confirm a better estimation model than the multivariate linear model.

---

[4]   We examine a forward stepwise regression searching the space of basis functions, for each variable and for all possible knots, and add those which maximize a certain measure of goodness of fit (minimize prediction error); so we achieve five basic functions added to the model according to a pre-determined maximum considerably larger than the optimal.

[5]   The models were applied with the software STATISTICA.

Table 1. Comparison about multivariate regression and multivariate adaptive regression splines models

Dep = GDP	MV	MAR Splines	MV	MAR Splines
Intercept	2,347*** (0,235)	3,24	0,794*** (0,190)	2,34
GCF			0,104*** (0,031)	1,735 (t3)
GFCF			0,075 (0,047)	
HFCE			0,452*** (0,076)	2,256 (t1)
GS	0,219*** (0,031)	7,759 (t2)	0,023 (0,018)	7,759 (t5)
r	0,74	0,65	0,98	0,97
R square	0,55	0,63	0,95	0,96

Note: *** = high significant (p-value < 0,001); MV = Multivariate Model; MAR Splines = Multivariate adaptive regression splines model. In MAR Splines model we report basis functions of type $\max(0, independent - knot)$, otherwise $\max(0, knot - independent)$.

Source: our own elaborations on World Bank data (2014).

The first analysis considers the comparison between the predictor gross savings (GS) and the dependent gross domestic product (GDP). The aim is not to demonstrate the formation of the GDP, which is dependent on many variables also exogenous to the system, but to observe how the variables of interest influence it. The first model gets a high significance of the predictor gross savings on the gross domestic product: in fact there is a probability of error almost nothing (less than 0,001). In the multivariate adaptive regression splines model the gross savings get the optimal value with minimal error to the second knot. In both cases, the two models well represent the phenomenon under study. In the second step of the analysis we consider all indicators as predictor. In the multivariate model with the classic approach, we observe that only the variable HFCE is significant with a probability of error less than 0,001. This demonstrates the role of domestic aggregate demand as stimulus to the national economy.

In the multivariate adaptive regression splines model instead there are three indicators which influence the GDP: HFCE (the most influential because it is at the first knot), GCF (significant in our model, after other two steps, at the third knot) and GS (fifth knot) obtaining the optimal values and a minimum expected error. This analysis approximates better the phenomenon under study in order that many predictors influence positively the GDP: in this case is captured as the annual percentage change in GDP is conditioned by household savings.

Graphically, we report in the next graph the comparison between the two last mentioned models.

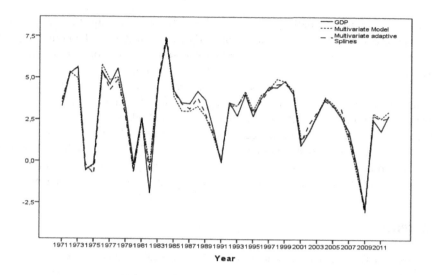

Fig. 1. Observed and estimated values of US data (by multivariate model and multivariate adaptive regression splines model, 1970-2012)

Source: our own elaborations on World Bank data (2014).

From Figure 1 we note that the changes in the expected values are very similar to the observed ones: in some cases the multivariate adaptive regression splines model report expected values more similar to the value observed, it means that there is a better approximation of this model to the reality of the phenomenon under study. The residual values, between observed and predicted values in the two models, in many cases are lower in the multivariate adaptive regression splines model.

4   Concluding Remarks and Future Research

In this work we have compared the opportunity to explain the relationship among some relevant macroeconomic variables and the GDP for the period 1970-2012 in the US, in the case of a multiple regression using multivariate and multivariate adaptive regression splines models. This analysis is for us a first step towards the use of techniques related to Artificial Intelligence to study the complex relationship between the household wealth, as specific unexploited resources, and the possible support to national income. We use the aggregate savings not directly as a proxy variable of the opportunity to accumulate wealth, and of course we are aware that the household savings is in part already redirected to businesses through the banking channel. We have considered initially only the role of the aggregate savings, and then also that of consumption and fixed capital. These relationships are widely known in the scientific literature (*e.g.* the role of aggregate demand), so our goal was the comparison of the explanatory models. The private saving represent the principal source for the creation of household wealth, in particular real estate and financial wealth, that are the two forms more potentially "exploitable" if properly managed. Given the complexity of our demand for research, which potentially affects many socio-economic variables in time series for many countries, we implement a technique involved in Artificial Intelligence combining statistical analysis and computer science. We exploit the ability of data mining techniques, in particular multivariate adaptive regression splines models, to manage and process large datasets with the aim to find relationships among the considered variables that the standard econometric techniques do not contemplate. Our further aim is to demonstrate the effectiveness of specific forms of wealth (such as real estate and financial ones) in production cycles, starting from the accumulation of specific types of "useful" wealth, from private savings through incentive mechanisms.

References

1. Anderson, T.W.: An introduction to multivariate statistical analysis, 2nd edn. Wiley, New York (1984)
2. Barro, R.J.: Inequality and growth in a panel of countries. Journal of Economic Growth 5, 5–32 (2000)
3. Bhattacharya, J.: Credit market imperfections, income distribution, and capital accumulation. Economic Theory 11, 171–200 (1998)
4. Christensen, L.R., Jorgenson, D.W.: The measurement of U.S. real capital input, 1929-1967. Review of Income and Wealth 15(4), 293–320 (1969)
5. Colander, D.: The death of neoclassical economics. Journal of the History of Economic Thought 22(2), 127–143 (2000a)
6. Colander, D.: Introduction. In: Colander, D. (ed.) Complexity and the history of economic thought. Routledge, London (2000b)
7. Davies, J.B., Sandström, S., Shorrocks, A., Wolff, E.N.: The World distribution of household wealth. UNU-WIDER, Discussion paper no. 2008/03 (2008)
8. Friedman, J.H.: Multivariate Adaptive Regression Splines. The Annals of Statistics 19(1), 1–67 (1991)
9. Han, J., Kamber, M.: Data Mining: concepts and techniques. Morgan Kaufmann Publishers, San Francisco (2006)
10. Johnson, R.A., Wichern, D.W.: Applied Multivariate Statistical Analysis, 6th edn. Prentice Hall (2007)
11. Kaldor, N.: Alternative Theories of Distribution. The Review of Economic Studies 23(2), 83–100 (1955)
12. Lewis, W.A.: Economic Development with Unlimited Supplies of Labor. Manchester School of Economic and Social Studies 22, 139–191 (1954)
13. Leathwick, J.R., Elith, J., Hastie, T.: Comparative performance of generalized additive models and multivariate adaptive regression splines for statistical modelling of species distributions. Ecological Modelling 199(2), 188–196 (2006)
14. Ma, C.H., Smith, B.D.: Credit market imperfections and economic development: theory and evidence. Journal of Development Economics 48(2), 351–387 (1996)
15. Pasinetti, L.: Rate of profit and income distribution in relation to the rate of economic growth. Review of Economic Studies 29, 267–279 (1962)
16. Witten, I.H., Frank, E.: Data Mining: practical machine learning tools and techniques. Morgan Kaufmann Publishers, San Francisco (2005)

Verification of Data Aware Business Process Models: A Methodological Survey of Research Results and Challenges

Raffaele Dell'Aversana

School of Advanced Studies G. D'Annunzio
University of Chieti-Pescara, Italy
r.dellaversana@gmail.com

Abstract. Business Process Management is a discipline that gives a systematic approach to the development of more efficient and effective organizations, enabling quick adaptation to the changes of the business environment. For this reason modeling languages such as BPMN [1] have a wide adoption in modern organizations. Such modeling languages are used for the design and reengineering of Business Processes and have the advantage of having a representation that is not only easy to understand by all the stakeholders but also machine processable.

However any executable model of a business process could contain potential problems (just like any computer software), so there are several research branches focused on formal verification of the models. The verification is not limited to check the correctness of the model but also to verify properties of the model, such as conformance to business rules.

This short paper presents a survey of recent approaches to verification of data-aware business process models and identifies a range of open research challenges.

Keywords: business process re-engineering, business process modeling, formal verification, static verification, logic programming.

1 Introduction

The adoption of Business Process Management solutions into industrial and business realities is constantly growing, due to its ability to reduce costs, enhance efficiency and productivity and also giving to organizations the opportunity to improve governance, helping executives to measure and manage company resources. The re-engineering of business processes (BPs) is an activity focused on the analysis of existing business processes and workflow within an organization and then on changing them with the main purposes of improving activities and reducing inefficiencies. The verification of properties and correctness of the BPs is a challenging activity that require specific tools and methodologies; moreover, when reengineering a BP, it would be desirable to be able to guarantee that the modified BP preserves the properties of the original process.

© Springer International Publishing Switzerland 2015   393
S. Omatu et al. (eds.), *Distributed Computing & Artificial Intelligence, 12th Int. Conference*,
Advances in Intelligent Systems and Computing 373, DOI: 10.1007/978-3-319-19638-1_45

The research on this topic starts with a fundamental paper [2] where workflow management is seen as an application domain for Petri nets; the author identify a special class of Petri nets that is suitable for the representation, validation and verification of the control-flow part of business processes using standard Petri net based techniques.

Another interesting formalism is YAWL [3] (Yet Another Workflow Language), initially designed with a focus only on the control flow and then extended to support also the data perspective of business processes. After a rigorous analysis on the existing workflow languages and management systems the authors identified a set of workflow patterns and, starting from Petri nets, they developed YAWL to have sufficient expressive power to support also complex workflow patterns, preserving the possibility to do validation and verification of the control flow.

However it must be noted that, in the enterprise context, the most widely adopted modeling language is BPMN[1], standardized by the OMG[1]. BPMN, like YAWL, can be used to develop conceptual models and also to produce a machine-executable representation of the BPs, enabling the possibility of automating process simulation, analysis and verification.

The standard approaches to business process modeling are focused on the representation of the process as a workflow graph that define all operations (*tasks*), their interactions and their planned order of execution, mostly abstracting away the underlying *data layer*. Due to this kind of representation, also the standard veryfication approach is mainly focused on the verification of the properties of the BP control flow. This kind of approach give the possibility to verify some properties of the control flows (for example termination or *soundness* [4]).

However in a real system there are also data models and consequently the need to verify properties of the BP taking into account the data part. For example, in an order process for an e-commerce company, the verification of the workflow without the data part could check properties such as "if the order completes with success also the payment must have been successful", while including data (and the possibility of reasoning on data) one can verify more complex properties that take into account the data layer, such as "for every completed order the total payment must be equal to the total cost of all the ordered items". Adding the data model to the verification of the BP it is quite challenging, and require new approaches, because the introduction of data give rise to infinite-state systems.

2   Verification of Data Aware Business Process Models

In order to provide an integrated approach to BP verification, including both workflow and data modeling, several new methodologies have been proposed. All these methodologies define an abstraction of the data layer and identify techniques to perform verification and reasoning on the BPs including the data models.

---

[1] Object Management Group: an international, open membership, not-for-profit computer industry standards consortium. http://www.omg.org/

One class of methodologies is based on the *artifact-centric* business process models (see [5], [6]). This modeling approach introduce the concept of "business artifact", that is a key object relevant to the business. Each artifact has a life cycle and expose several services that can be invoked on it (services can be also viewed as tasks that can be executed); moreover each artifact has an internal state (the data part) that can change over time. Substantially the artifact centric approach is focused on describing how business data is updated throughout the business process.

In [7] the authors introduce an automatic verification approach based on the artifact-centric model, with services that can access an underlying database and call the artifacts, updating their state. The methodology proposed is oriented to verify statically that all runs of an artifact system satisfy certain correctness properties, expressed in a first-order extension of linear-time temporal logic. The authors also identify an expressive class of artifact systems for which the verification is decidable, introducing some limitations on data. In [8] the authors extend the previous approach, adding the possibility to have some data dependencies and arithmetic on data while preserving the decidability of the verification.

In [9] the authors present a survey of the database oriented approach to data-aware BP analysis carried out in the database theory community and in [10] is presented a verification technique based on μ-calculus variants over processes described in terms of atomic actions that evolve a database; the approach has an expressive power similar to artifact centric models. While the verification is undecidable in general, they identify several notable cases where decidability is achieved. The approach has a strong theoretic foundation but the authors do not propose a practical implementation.

Another class of methodologies is based on Logic Programming (LP). In [11] the authors propose an approach based on an extension of answer set programming (ASP)[12] with temporal logic and constraints; the authors analyze YAWL workflows using bounded model checking techniques, extended with constraint solving to deal with conditions on numeric data, alllowing the verification of temporal properties of BPs and verification of compliance to business rules. This framework impose several restrictions on the workflow structure and finite domains for variables, resulting in a language with a reduced expressive power but decidable verifiable with static analysis techniques.

In [13] the authors extend an approach based on declarative workflow specifications, to include data-aware constraints. Their framework is formalized using Event Calculus (EC) and implemented using EC reasoners based on logic programming. Their approach is intended to be used for a-posteriori analysis of event and process logs, to check their compliance with the constraints but also to monitor running processes against the workflow constraints.

Another interesting and quite different approach is the framework presented in [14], that is based on Constraint Logic Programming (CLP)[15] for representing and reasoning about business processes from both the workflow and the data perspective. The framework adopts the BPMN language and notation and define a formal semantic for process model, representing explicitly data object interactions and manipulation. They define a rule-based process representation of a relevant fragment of BPMN and model the behavioral semantics of the process as a state transition system, similar to the Fluent Calculus, giving the possibility of performing several interesting verification and reasoning services using standard CLP inference engines. Data values are modeled in a symbolic way based on arithmetic constraints, giving the possibility of expressing data manipulations and database interactions of each process activity in terms of preconditions and effects. This approach does not impose any restrictions to the workflow or to the data, so the verification task is, in general, undecidable. The authors however show several examples of analysis that can be performed in the areas of simulations and testing, by keeping bounded the process runs.

3   Research Direction and Challenges

The current stream of research deals with an integrated data and workflow model for the BP, for analysis and verification purposes. Several authors are focused on more theoretical questions, such as finding decidability and complexity results, and is a research field that needs to be explored, where the main challenges are the identification of new classes of decidable and tractable data-aware BP models. Looking for future search directions in this area, new approaches are needed to deal with verification of data-aware BP models, where the decidable frameworks should be extended to more expressive workflows. Moreover there is the need to implement practical tools to exploit those theoretical results in real world BPs.

On the other side there are more pragmatic approaches that allows the implementation of reasoning and verification techniques, taking advantages of existing engines (e.g. CLP, designed to deal with constraint solving). This line of research should explore new optimization techniques and approaches to reduce the state space and enhance the performances of the reasoning services. Another possibility is to investigate the class of Business Processes that leads to a symbolic finite state space with the purpose of defining decidability and complexity of the verification tasks.

In general the frameworks for the verification of data-aware business process models should explore, in a wider way, the automated reasoning techniques for the analysis and verification, such as simulation, testing, verification of properties on workflow and data, also producing real world tool for BPM languages.

Finally, it would be interesting to study the preservation of existing properties after change in the workflow or in the data constraints (e.g. for reengineering purposes), and conversely exploring the possibility of finding automatically a set of changes in data constraint or process worklfow that can make a property to be satisfied.

References

1. OMG: Business Process Model and Notation (BPMN), Version 2.0 (January 2011)
2. van der Aalst, W.M.P.: Verification of workflow nets. In: Azéma, P., Balbo, G. (eds.) ICATPN 1997. LNCS, vol. 1248, pp. 407–426. Springer, Heidelberg (1997)
3. van der Aalst, W.M.P., ter Hofstede, A.H.M.: Yawl: Yet another workflow language. Inf. Syst. 30(4), 245–275 (2005)
4. van der Aalst, W.M.P.: The application of petri nets to workflow management. Journal of Circuits, Systems, and Computers 8(1), 21–66 (1998)
5. Nigam, A., Caswell, N.S.: Business artifacts: An approach to operational specification. IBM Syst. J. 42(3), 428–445 (2003)
6. Hull, R.: Artifact-centric business process models: Brief survey of research results and challenges. In: Meersman, R., Tari, Z. (eds.) OTM 2008, Part II. LNCS, vol. 5332, pp. 1152–1163. Springer, Heidelberg (2008)
7. Deutsch, A., Hull, R., Patrizi, F., Vianu, V.: Automatic verification of data-centric business processes. In: Proceedings of the 12th International Conference on Database Theory, ICDT 2009, pp. 252–267. ACM, New York (2009)
8. Damaggio, E., Deutsch, A., Vianu, V.: Artifact systems with data dependencies and arithmetic. ACM Trans. Database Syst. 37(3), 22 (2012)
9. Calvanese, D., De Giacomo, G., Montali, M.: Foundations of data-aware process analysis: A database theory perspective. In: Proceedings of the 32nd Symposium on Principles of Database Systems, PODS 2013, pp. 1–12. ACM, New York (2013)
10. Bagheri Hariri, B., Calvanese, D., De Giacomo, G., Deutsch, A., Montali, M.: Verification of relational data-centric dynamic systems with external services. In: Proceedings of the 32nd Symposium on Principles of Database Systems, PODS 2013, pp. 163–174. ACM, New York (2013)
11. Giordano, L., Martelli, A., Spiotta, M., Dupré, D.T.: Business process verification with constraint temporal answer set programming. TPLP 13(4-5), 641–655 (2013)
12. Gelfond, M.: Answer sets. In: van Harmelen, F., van Harmelen, F., Lifschitz, V., Porter, B. (eds.) Handbook of Knowledge Representation. Elsevier Science, San Diego (2007)
13. Montali, M., Chesani, F., Mello, P., Maggi, F.M.: Towards data-aware constraints in declare. In: Proceedings of the 28th Annual ACM Symposium on Applied Computing, SAC 2013, pp. 1391–1396. ACM, New York (2013)
14. Proietti, M., Smith, F.: Reasoning on data-aware business processes with constraint logic. In: Proceedings of the 4th International Symposium on Data-driven Process Discovery and Analysis (SIMPDA 2014). CEUR-WS, vol. 1293 (2014)
15. Jaffar, J., Maher, M.J.: Constraint logic programming: A survey. Journal of Logic Programming 19, 503–581 (1994)

Decision Making under Uncertainty: A Methodological Note

Assia Liberatore

School of Advanced Studied, University of Chieti-Pescara via dei Vestini 31 – Chieti, Italy
assia.liberatore@unich.it

Abstract. Under uncertainty, the economic agent is far from taking rational decisions, incurring in heuristics and cognitive errors such as reversal of preferences, focal points and so on. Many economists and psychologists have shown that the behavior of the decision maker in the real market is also driven by loss aversion and trust. The purpose of this study is to describe the experimental evidences that diverge from the standard economic decision theory, through a methodological review of the most significant laboratory techniques and field analysis performed in behavioral and experimental economics since the last quarter of the Twentieth Century.

Keywords: behavioral economics, experimental economics, bounded rationality, loss aversion.

1 Introduction

The decision making under conditions of uncertainty is an interdisciplinary topic that has taken hold in the second half of the Twentieth Century and involves the choices that the agent or groups of agents have to face in view of two or more alternatives. The number and breadth of the characteristics of the agent's choice options which may in part remain implicit, depend on the degree of knowledge or experience possessed by the agent in the sector in which he is to decide. However, in terms of risk, the alternatives of choice are not always measured by the objective probability of their outcome: the decision maker often used *heuristics*[1], such as availability, representativeness and anchor [16, 40], as mental shortcuts that violate logical principles, leading even to mistakes in a number of situations. Similarly to Sugden [40], this study proposes itself as a small collection of laboratory and field experiments on decision-making in experimental economics, focusing on the economic choices made in risky conditions. However Sugden's vision is complemented by boundary methodologies described by Camerer *et al.* [7]. According to the assumptions of bounded rationality, the vision here used comes from economics, cognitive psychology and neuroscience, creating hybrid disciplines such as cognitive economics and neuroeconomics. Many

---

[1] When people make predictions based on what they can imagine more easily, or what is more similar or where they can anchor with a minimal cognitive effort.

contributions are also drawn from behavioral economics that, through experiments regarding problems of choice, show tendencies of human behavior contrary to what is theorized by economic standards [45]. After recalling the main economic theories of decision making (section 2), this paper focuses on some empirical evidences (section 3), with mention of the newer methods (section 4) which may be the subject of future studies.

2   Decision Making and Bounded Rationality: Theoretical Basis

The rationality of choice was questioned by many economists [1, 35, 44]
"Human behavior in general and [...] also in the market place, is not under the constant and detailed guidance of careful and accurate hedonic calculations, but is the product of an unstable and unrational complex of reflex action, impulses, instincts, habits, customs, fashion and hysteria" - **Viner J. [44] p. 373.**

According to the theory of expected utility, the utility of the possible results are weighted depending on their probability, Von Neumann and Morgenstern [45] also stated that the agent is able to calculate the option that maximizes the gains or minimizes the losses through transitivity, independence, dominance and invariance axioms. Nevertheless, the Allais paradox [1] identified some regularity in the behavior of choice that does not obey the Bernoulli principle. Twenty-five years later, the psychologists Kahneman and Tversky [16] gave new light to the seminal work of the French economist, confirming experimentally the discrepancy between the predictions of the traditional theory of decision making and the actual behavior of economic agents, which are not equipped with Olympic rationality and unlimited computing power that the traditional economic yearned. The *common ratio effect* [40] occurs in experiments with two tasks where each requires a choice between a "riskier" lottery (R1 in one task, R2 in the other) and a "safer" lottery (S1 or S2). In contrast to expected utility theory, the choice of the first problem, risky (R1) or careful (S1) that is, is not replicated in the second problem by the majority of subjects. People tend to be less risk adverse when the probabilities of winning are smaller.

"The capacity of the human mind for formulating and solving complex problems is very small compared with the size of the problems whose solution is required for objectively rational behavior in the real world - or even for a reason - able approximation to such objective rationality" - **Simon H. [34] pp. 198-202.**

Simon not just coined the concept of bounded rationality but extended it to the business world, studying the principle of profit maximization by the entrepreneur in psychological terms[2]. An empirical analysis of managerial decisions [10] identified two behavior trends: the choices are consistent repeated under conditions as are learned, however the decision meets a high degree of uncertainty when the decision maker faces problem solving. An example is the *success trap*, proposed by Levinthal and March [21]: the tendency of organizations to focus on success may induce them

---

[2] *"Thus economists who are zealous in insisting that economic actors maximize turn around and become satisficers when the evaluation of their own theories is concerned. They believe that businessmen maximize, but they know that economic theorists satisfice."* - Simon H. [35] p. 345.

to emulate procedures and actions that have been associated with past successes, not orienting its activities to research and innovation. When economists began accepting the anomalies of the actual behavior towards the dogmas of the standard economic, psychological contributions suggested new theories[3]. Behavioral economics arises from the combination of the psychological realism and applicability to economic research both in the laboratory and in the field. It is a discipline that has placed experimentation as its foundations, through field studies and, more recently, the use of case studies and brain imaging techniques, which will be discussed in the fourth section. The economy has long been considered a hypothetical-deductive science, constrained to a core set of assumptions about the rationality, but recently the theoretical propositions are verified by empirical designs. Under the vision of "blame theory"[38], the experimental strategy consists in the following stages: after selecting a pre-existing theory from the core of the economics, set up an environment suitable analysis that reproduces the theoretical abstractions but at the same time allows agents freedom of action; finally identify anomalies that will be an inspiration to postulate theories and alternative behavioral models[4].

Means by which economists experimental leading theories on the field are defined exhibits[5]. The topics that concern how economic agents decide under uncertainty are numerous. The following section will introduce some experimental designs using game theory's technique and field experiments.

3   Experimental Empirical Applications

The decision's research can be divided into two categories: the first gives attention to probability judgment and examines the mechanisms heuristics and cognitive errors, such as endowment effect and sunk costs; the other focuses on the choice, highlighting the effects of trust or focal point, as well as the reversal of preferences.

In real life the best choice may not always be done; because of that the decision maker, in order to arrive at a satisfactory solution[6], uses strategies such as *focus* or *elimination by aspects*, typical of the advertising marketing of niche products [30]. Gamblers in Las Vegas, when sit around the card table, do not analyze their complete financial portfolio, on the contrary they are only focused on chips right in front of them and the possibility of winning (or losing) that specific amount of money. Kahneman and Tversky concluded that people's thinking about alternative outcomes is based on gains or losses, not on states of wealth. The following subsections discuss some empirical evidence on economic decision theory: preference reversals (3.1), fairness (3.2), focal points (3.3) and endowment effect (3.4).

---

[3] In the '60s cognitive psychology assimilated the brain not to a machine Stimulus - Response, typical of behavioral vision, but rather to an information processing device, allowing study new topics, including the process of decision-making [7].

[4] The final stage comes from Camerer *et al.* [7]

[5] Replicable experimental designs that reliably produces interesting results [40]

[6] *"The real economic actor is in fact a satisficer, a person who accepts "good enough" alternatives, not because less is preferred to more but because there is no choice"* - Simon, H. [36] p. 29

3.1   Preference Reversals

Behavioral economists define the human decision's process with ill-defined preferences as "constructing preferences" [27, 37]. Preference reversal takes place between joint and separate evaluations of pairs of goods [13] even in application to marketing [14]. People will often price or otherwise evaluate an item A higher than another item B when the two are evaluated independently, but evaluate B more highly than A when the two items are compared and priced at the same time. Lichtenstein and Slovic [22, 23] reported the first experimental evidence about the discrepancy between preferences revealed in the choices and preferences revealed in monetary valuations. The phenomenon of *preference reversal* was first tested in a sample of students with small or no payoff, then in gamblers of Las Vegas casino with a high payoff, finding similar results.

3.2   Fairness

"In a typical trust game, one player has a pot of money, again typically around $10, from which he can choose to keep some amount for himself, and to invest the remaining amount X, between $0 and $10, and their investment is tripled. A trustee then takes the amount 3X, keeps as much as she wants, and returns Y. In standard theory terms, the investor-trustee contract is incomplete and the investor should fear trustee moral hazard. Self-interested trustees will keep everything (Y = 0) and self-interested investors who anticipate this will invest nothing (X = 0). In fact, in most experiments investors invest about half and trustees pay back a little less than the investment. Y varies positively with X, as if trustees feel an obligation to repay trust" - **Camerer** *et al.* **[7] p. 31.**

Berg *et al.* [3] employed the trust game in order to check the hypothesis that individuals act according to their own personal interest. They took over in economic agents a motivation to trust others and to repay other agents' trust in them, thus supporting that trust is a economic primitive. Since most proposers offer large shares (around 40%) to their co-players, it can be said that they are driven by *taste of fairness,* supported by Zak, Matzner and Kurzban that, exploring the role of hormones in many points of trust games, found an increase in oxytocin.

Kahneman, Knetsch and Thaler [17] analyzed consumer perceptions of fairness through phone surveys. Rising prices of shovel snow after a snowstorm reduces consumer surplus and is considered unfair. In contrast subjects believe that it is right to raise prices when the cost of production factors of a company increases, because otherwise it would reduce the profit of the company, compared to reference. It is interesting to investigate whether the prevalence of regularity varies depending on the demographic characteristics of the subject pool. Buchan et al. [4] applied the trust game to students of four countries (China, Japan, Korea and USA), creating trust indicators. As a consequence trust becomes what in psychology is called construct, a method that measures and represents a real phenomenon.

3.3   Focal Points

The research of the focal points is evident in situations where the decision of an agent is conditioned by the decision of other makers, trying to answer the question "what would I do if I were him?". A famous example is the New York city problem[7], proposed by the father of game theory rules [32]. *"When individuals have a common interest in coordinating their expectations of one another, they are often remarkably successful at doing so, even in the absence of any means of communication. A point of convergence for expectations is a focal point"* - Sugden R., Zamarrón I. [39] p. 610.

Schelling studied the behavior of the agents using a pragmatic approach, where seemingly irrelevant details turn out to be crucial. The traditional game theory is based on the preferences of the agents and formal axioms in order to form a set of possible actions that most satisfy them, but in the case of coordination games is in trouble because it is unable to find the focal points. Schelling rejects the idea of a logical a priori preferences and approaches to rationality empirically in the sense that *"a principle of decision is rational for an agent just to the extent that, by using it, the agent tends to be successful in achieving her ends"*- Sugden R., Zamarrón I. [39] p.619.

3.4   Endowment Effect

Several experimental evidences show that the decision maker has a strong loss aversion [6, 43] leading to the so called endowment effect which means that the loss of utility in giving up an asset is greater than the utility gained in receiving it as a gift. This expression was coined by Thaler [41] to support prospect theory model of riskless choice: he argued that the value seemed to change when a good is embedded in agent's possession. Endowment effect actually refers to two distinct findings: exchange asymmetries and WTA/WTP gaps. The first dates from Knetsch's experiment of mug and candy bar [19] where about 89% of the participants refused to exchange the object received for candy bar or mug given to another group; the second was tested by Kahneman, Knetsch, and Thaler [18] in a series of field experiments conducted in a classroom setting, forming two student groups where a mug was evaluated as a gain by choosers' group and as a loss by sellers' group. Controversial experiments believe that loss aversion is not the only cause of the endowment effect, but there are other factors internal to the individual, or external environment. Knetsch-type experiments were replicated by List [24, 25] at a sports card market. He compared non-dealers and dealers to question the robustness of exchange asymmetries that was strong in the first group, but not in the second: dealers were much more willing to exchange an initial object they were given for another one of similar value. Another controversial experiment stated that subjects perceived the object they were

---

[7] *"Two people are separately asked to name a time and a place in New York City at which to meet each other; each is given the objective of naming exactly the same time and place as the other. Schelling reports that he has given this problem to an "unscientific sample" of respondents; almost all chose 12 noon, and a majority chose Grand Central Station"* - Schelling T. [32] pp. 55-56.

initially given as more valuable, or thinking of the initial object as a gift, one that it would be impolite to exchange [28]. Köszegi and Rabin [20] created a model of reference-dependent preferences, supporting the common view that the endowment effect found in the laboratory is due to loss aversion even if in real world markets there is a difference between buyers and sellers who expected. Loomes, Orr, Sugden [26] found that WTA/WTP disparities occur when individuals are uncertain about their tastes. Repeated experiences of trading a good may reduce these disparities by reducing people's subjective uncertainty about their tastes.

4   New Methods and Future Directions

Although many contributions in literature report experiments on individuals' choices, laboratory experiments can elude some emotions typical of much real life risk taker, as the fear of losing [9]. Methods of behavioral economics that investigate the decision theory are not only field experiments shown up to now, but also computer simulation and brain scan[8]. Certainly, psychological contributions allow to investigate the motivations of decision-making mechanisms in the sphere of the individual and the company[9], as a consequence neuroeconomics opens the door to what many believed before utopian. *"I hesitate to say that men will ever have the means of measuring directly the feelings of the human heart"* Jevons W. [15] p.11.

The difference between economic decisions in case of certain amounts of money and those committed in ambiguous lotteries, is also supported by neuroeconomics: fMRI data show different insula[10] activations, as well as the quick and dirty amygdala response to potential fears. In the wake of the procedural rationality postulated by Simon [36], Rubinstein pioneered the formal modeling of decision making under uncertain as a two step algorithm which considers complexity and mental transaction costs [5]. Moreover Gilboa and Schmedler suggest that decision making under uncertainty is, at least in part, case-based [12]; this approach concerns a choice situation that resembles the previous cases [5]. Economists are influenced by the theory of the decision as to consider the risky choices follow a probability-weighted average of the utility of their possible consequences.

Here I showed a roundup of economic experimental designs in decision process that contrast some of the standard economic assumptions. Note that while the standard theory postulates the utility maximization by the decision maker, data used to test the theory are aggregated over time or across individuals. This approach is usually taken because individual behavior is erratic and heterogeneous, so that the aggregation provides the smoothing necessary in order to detect the pattern. The emotional

---

[8] The main techniques are electro-encephalogram (EEG), positron-emission topography (PET), functional magnetic resonance imaging (fMRI) [8] and eye-tracking analysis.

[9] According to the analogy between brain and company, modern views open the firm's black-box [8]

[10] The insula processes information from the nervous system about bodily states-such as physical pain, hunger, the pain of social exclusion, disgusting odors and choking. This tentative evidence suggests a neural basis for pessimism or "fear of the unknown" influencing choices.

component can make the difference [6, 8, 30], assuming a key role not only for an economic agent but also for a group of them. This is not a limit, in fact brain imaging techniques applied to economic experiments, for example the ultimatum game [31], could help explain some behaviors economic trend, or even have a comparative advantage when other sources of data, such as surveys or self-reports, are unreliable or biased.

References

1. Allais, M.: Le comportement de l'homme rationnel devant le risque, critique des postulats et axiomes de l'école américaine. Econometrica 21, 503–546 (1953)
2. Barberis, N.: Thirty Years of Prospect Theory in Economics: A Review and Assessment. Journal of Economic Perspectives 27(1), 173–196 (2013)
3. Berg, J., Dickhaut, J., McCabe, K.: Trust, reciprocity, and social history. Games and Economic Behavior 10, 122–142 (1995)
4. Buchan, N., Croson, R., Robyn, D.: Swift neighbors and persistent strangers: a cross-cultural investigation of trust and reciprocity in social exchange. American Journal of Sociology 108, 168–206 (2002)
5. Buschena, D., Zilberman, D.: An Empirical Test of Rubinstein's Similarity Definitions for Choice Under Risk (2006)
6. Camerer, C., Babcock, L., Loewenstein, G., Thaler, R.: Labor Supply of New York City Cabdrivers: One Day at a Time. Quarterly Journal of Economics 112(2), 407–441 (1997)
7. Camerer, C., Loewenstein, G., Rabin, M.: Behavioral Economics: Past, Present, Future. Russel Sage Foundation, Princeton University Press, New York (2004)
8. Camerer, C., Loewenstein, G., Prelec, D.: Neuroeconomics: Why Economics Need Brains. The Scandivian Journal of Economics 106(3), 555–579 (2004)
9. Cubitt, R., Sugden, R.: Dynamic Decision-Making Under Uncertainty: An Experimental Investigation of Choices Between Accumulator Gambles. The Journal of Risk and Uncertainty 22(2), 103–128 (2001)
10. Cyert, R.M., Simon, H.A., Trow, D.B.: Observation of a Business Decision. Journal of Business 29, 237–248 (1956)
11. Egidi, M., Rizzello, S.: Cognitive Economics: Foundation and Historical Evolution, Working Paper 04/2003 CESMEP (2003)
12. Gilboa, I., Schmeidler, D.: Case-based decision theory. Quarterly Journal of Economics 110(3), 605–639 (1995)
13. Hsee, C., Leclerc, F.: Will products look more attractive when evaluated jointly or when evaluated separately? Journal of Consumer Research 25, 175–186 (1998)
14. Hsee, C., Loewenstein, G., Blount, S., Bazerman, M.: Preference reversals between joint and separate evaluations of options: a theoretical analysis. Psychological Bulletin 125(5), 576–590 (1999)
15. Jevons, W.: The Theory of Political Economy. Macmillan, London (1871)
16. Kahneman, D., Tversky, A.: Prospect Theory: An analysis of Decision under Risk. Econometrica 47, 263–291 (1979)
17. Kahneman, D., Knetsch, J.L., Thaler, R.: Fairness as a constraint on profit seeking: Entitlements in the market. American Economic Review 76, 728–741 (1986)
18. Kahneman, D., Knetsch, J.L., Thaler, R.H.: Experimental Tests of the Endowment Effect and the Coase Theorem. Journal of Political Economy 98(6), 1325–1348 (1990)
19. Knetsch, J.L.: The Endowment Effect and Evidence of Nonreversible Indifference Curves. American Economic Review 79(5), 1277–1284 (1989)
20. Köszegi, B.: Rabin A model of reference-dependent preferences. The Quarterly Journal of Economics CXXI (4), 1133–1165 (2006)

21. Levinthal, D.A., March, J.G.: The Myopia of Learning. Strategic Management Journal (14), 95–112 (1993)
22. Lichtenstein, S., Slovic, P.: Reversals of preferences between bids and choices in gambling decisions. Journal of Experimental Psychology (89), 46–55 (1971)
23. Lichtenstein, S., Slovic, P.: Response-induced reversals of preference in gambling: An extended replication in Las Vegas. Journal of Experimental Psychology 101, 16–20 (1973)
24. List, J.A.: Does Market Experience Eliminate Market Anomalies? Quarterly Journal of Economics 118(1), 41–71 (2003)
25. List, J.A.: Neoclassical Theory versus Prospect Theory: Evidence from the Marketplace. Econometrica 72(2), 615–625 (2004)
26. Loomes, G., Starmer, C., Sugden, R.: Preference reversals and disparities between willingness to pay and willingness to accept in repeated markets. Journal of Economic Psychology 31, 374–387 (2010)
27. Payne, J.W., Bettman, J.R., Johnson, E.J.: Behavioral decision research: A constructive processing perspective. Annual Review of Psychology 43, 87–131 (1992)
28. Plott, C., Zeiler, K.: Exchange Asymmetries Incorrectly Interpreted as Evidence of Endowment Effect Theory and Prospect Theory? American Economic Review 97(4), 1449–1466 (2007)
29. Rubinstein, A.: Similarity and decision-making under risk (Is there a utility theory resolution to the Allais paradox?). Journal of Economic Theory 46, 145–153 (1988)
30. Rumiati, R.: Decidere Il Mulino (2000)
31. Sanfey, A., Rilling, J., Leigh, E., Nystrom, L., Cohen, J.: The Neural Basis of Economic Decision-Making in the Ultimatum Game. Science New Series 300(5626), 1755–1758 (2003)
32. Schelling, T.: The strategy of conflict. Harvard University Press, Cambridge (1960)
33. Schotter, A.: On the relationship between economic theory and experiments (2009)
34. Simon, H.A.: Models of Man, Social and Rational: Mathematical Essays on Rational Human Behavior in a Social Setting. John Wiley and Sons, New York (1957)
35. Simon, H.A.: Rational Decision-making in business organizations. Nobel Memorial Lecture 8 (1978)
36. Simon, H.A.: The Sciences of the Artificial. The MIT Press (1996)
37. Slovic, P.: The construction of preferences. American Psychologist 50, 364–371 (1995)
38. Starmer, C.: Experiments in Economics: Should We Trust the Dismal Scientists in White Coats? Journal of Economic Methodology 6, 1–30 (1999)
39. Sugden, R., Zamarrón, I.E.: Finding the key: The riddle of focal points. Journal of Economic Psychology 27, 609–621 (2006)
40. Sugden, R.: The Changing Relationship between Theory and Experiment in Economics. Philosophy of Science 75, 621–632 (2008)
41. Thaler, R.: Toward a Positive Theory of Consumer Choice. Journal of Economic Behavior and Organization 1(1), 39–60 (1980)
42. Tversky, A., Kahneman, D.: Judgment under uncertainty: Heuristics and biases. Science 185(4157), 1124–1131 (1974)
43. Tversky, A., Kahneman, D.: Loss Aversion in Riskless Choice: A Reference-Dependent Model. Quarterly Journal of Economics 106(4), 1039–1061 (1991)
44. Viner, J.: The Utility Concept in Value Theory and Its Critics. Journal Political Economy 33, 369–387 (1925)
45. von Neumann, J., Morgenstern, O.: Theory of Games and Economic Behavior. Princeton University Press (S1, 2) (1944)

Shaping an Inductive Macroeconomics
The Heritage of Hyman Minsky

Marcello Silvestri

University of Chieti-Pescara, Department of Philosophical, Pedagogical and
Economic-Quantitative Sciences, Viale Pindaro 42, 65127 Pescara, Italy
`marcello.silvestri@unich.it`

Abstract. In this paper I discuss how to develop an inductive model-
building approach which, in my opinion, could give novel perspectives to
macroeconomic investigation. For this end I explain reasons and ways to
cope with this task. My analysis refers to three different level of thinking under which I try to develop a line of enquiry. Firstly, I focus on
Hyman Minsky's economic thought which offers a novel inductive argument. Secondly, I discuss how this inductive argument fosters a different
use of the experimental method in Economics. Concluding, I argue how
agent based modeling could benefit from a Minskyan vision of Economics
supported by the experimental method.

1 Introduction

Economics likes other sciences has to deal with the conditions under which it is
possible to have scientific knowledge and methods to achieve it. In theoretical
economics there are two forms of logic that helps in dealing with this task:
deduction and induction.

Deduction is widely used in Economics. It involves the process of reasoning
from the universal to the individual. Economists select a problem for enquiry.
They formulate assumptions on the basis of which the problem is to be explored.
Then, they formulate hypotheses (i.e. principles which are assumed to be true)
and verify them on the basis of observed facts. For this purpose economists make
use of statistical analysis and econometrics.

Induction is the process of reasoning from particulars to generals. Like the
deductive method a problem concerning an economic phenomenon is selected.
In the second step there is a collection and analysis of data by using statistical
techniques. Data are used to make observation about particular facts of the
phenomenon under investigation. On the basis of observations, generalisation is
logically derived from particular facts.

Both approaches exhibit merits and demerits. Discussing about the problem
of scientific discovery is left out of the paper. This topic is a highly demanding
task and continuosly debated from the centuries to present days.

I rather discuss the problem of Economics to model the behaviors of agents
which shape the cause and effect relationship between various variables.

A macroeconomic system is usually explained by equations that try to understand how the values of aggregate variables quantitatively change over time and/or as functions of other variables. This method implies that economic phenomena tend towards equilibrium and remain relatively stable over time.

Since Robert Lucas's critique macroeconomy needs microfoundations, a microeconomic analysis of the behavior of agents that underpins such aggregate variables [11].

Economists claimed that the way to achieve this is to look at an individual's objective function (relating to preferences, technology, and resource constraints that are assumed to govern individual behavior) and derive the aggregate behavior of economy; this methodology, based on microeconomic optimizing actions rooted in the axioms of the *homo oeconomicus* (i.e. individual maximizing hypothesis, rational expectations hypothesis, homogeneity hypothesis), is usually referred to as the microeconomic foundations for macroeconomic theory [8].

In contrast with this view, in the next paragraph, I call attention to the work of Hyman Minsky who was particularly averse to derive conclusions from such stylized representation of the agents' economic behavior.

2   Fundamentals of Hyman Minsky's Reasoning

Hyman Minsky develops his reasoning learning on the shoulders of John Maynard Keynes. He gives an original and unprecedent reinterpetation of *The General Theory of Employment, of Interest and Money* [9] which led him to formulate his well known *Financial Instability Hypothesis*, a more detailed endogenous explanation of financial crisis [13,14,15].

In a nutshell, the FIH is a *pre-analytical analysis* which seeks to explain how speculative financial behaviors strongly affect business cycle dynamics and the economy as a whole. The substance of the FIH is that the behaviors of banks and businesses change over time. Firms need to resort to bank financing to start their production processes and to purchase capital assets while banks use short-term deposits (i.e., liabilities) to finance investments in the medium and long term (i.e., assets). This activity is based on the exchange of promises of payment made by customers on a more delayed term through the payment of relative interests. What leads the economy to become more fragile is derived endogenously from inherent speculative behavior of financial nature. During an expansion phase both borrowers and lenders revised upward their expectations of profitability, the first become more eager to borrow and the latter more willing to make loans. The "fuse" is always lit by the pro cyclical increase in the supply of credit that can create the necessary conditions for a strong growth and boom period, which usually takes hold by speculation of the stock market.

Differently from the Walrasian economy where the equilibrium is an attractor, Minsky emphasizes the process of shaping events. The FIH indeed focuses on a capitalist economy that moves through real calendar time in which the past, the present and the future are linked together through financial relationships. This process evolves in an irreversible way and affects socio-economic dynamics.

Although fascinating and at a first glance compelling, the FIH is not supported by sound theoretical and empirical analysis because it is based on concepts and feelings which cannot be handled in rigorous quantitative models.

The issue involves not merely technical mathematical and econometric problems [4], but also ontological [3] and epistemological dilemma [12].

The philosophical conundrum of economics centers around the unfolding of the time. In a deterministic view of the time, most suitable for standard economics, past and future are symmetrical in the sense that they can be treated in a finite way due to our limited knowledge of the laws linking the past to the future. It is assumed that the movement from one equilibrium position to another occurs almost instantaneously by changing the parameters of the models which are underpinned on a set of given quantitative variables. When the parameter is returned to its previous value, the economy returns exactly to its previous position, since that time has not elapsed (stationary processes). Mathematical models are then calibrate through statistical inference. Statistical estimation can be interpreted as predicted correlations of mathematical variables among observed real variables through some counter factual rules. This estimation relies upon the assumption that we can use repeated observations of past events to infer at least some features of the future. The deductive approach and its methodology, although permitting to evaluate the consistency and the generalizability of propositions more easily than verbal representations, get into trouble in understanding the endogenous forces ("monetary instincts") which can affect the transition from one equilibrium position to another.

The vision of Hyman Minsky is rooted precisely in understanding these endogenous forces. To this aim his methodology and epistemology of economics focuses on a dialectical view of time. In a dialectical perspective, the past is fixed and the future is undetermined but human actions in the present affect the process of shaping the future. Time is irreversible [18]. Minsky's ontology is a direct consequence of this perspective and deals with the speculative behavior which is rooted in human nature: *"The fundamental speculative decision of a capitalist economy centers around how much, of the anticipated cash flow from normal operations, a firm, household, or financial institution pledges for the payment of interest and principal on liabilities"*. [15, p.84]

This portfolio choice takes place in a world of fundamental uncertainty in which humans rarely have the knowledge necessary to carry out economic calculations of optimizing nature [19]. To this end, Minsky believes that assumptions about a behavioral model derived from axioms are not consistent with a relevant theory. He prefers to draw his narrower monetary view as a stylized representation of reality derived from reflection on the salient facts that are due to experience. Adopting this perspective, Minsky places at the core of his Financial Instability Hypothesis agents' behaviour over the business cycle.

2.1   The Core of Hyman Minsky's Reasoning

Notwithstanding the peculiarity of Minsky's descriptive representation of a sophisticated monetary economy and its intrinsic methodological limits, a researcher can bypass such complexity moving at the roots of Minsky's argument.

The core of Minsky's analysis centers around two specific features: the process of decision making under uncertainty and the way in which Minsky introduces the balance sheet structure into the apparaturs of Marshallian cost and revenues curves. Minsky argues that uncertainty decisively affects economic behavior, especially in relation to the portfolio decisions made by households, firms, and financial institutions, and in their views regard to the prospective yields of capital assets.

Assets and liabilities involve cash-flows and payment commitments in a delayed future. Economic agents fulfill commitments on liabilities using the cash-flow that will be realized in a delayed term. Each agent must decide the proper mix of assets and liabilities and how to manage their balance sheet structure according to their "enjoyment" of risk. Hence, the peculiarity of money that provides purchasing power in an economy where the payment obligations are certain while future cash flow are uncertain: *"the possession of actual money lulls our disquietude; and the premium which we require to make us part with money is the measure of the degree of our disquietude"*. [10, p.216].

Each agent in accepting a liability structure in order to hold assets (i.e. portfolio choice) is betting that at the future dates the cash payment commitments can be met: it is estimating that such bet is favorable.

The first step in Minsky's analysis consists in resuming the argument of chapter No. 17 of Keyness General Theory with regard to attributes that every assets possess. The crucial element that Keynes does not consider is the liabilities structure. This is what, according to Minsky, brings out the cyclical perspective of Keynes's General Theory. The novelty of Minsky's insights is to place the balance sheet at the background of a Marshallian apparatus of cost and revenue curves. For real physical capital, Minsky shows as its return (q) is the cash flow that the asset will yield from operations. That is the Marshallian quasi-rent. While carrying costs (c) are represented by the interest payments. In this way Minsky introduced his theory of the firm in relation to business cycle theories, that was later to be the core of his analysis of financial fragility, i.e. the notion of the firm as a balance sheet of assets and liabilities.

Minsky argued that the traditional theory of the firm does not explicitly tackle the financial structure issue [16]. Introducing financial considerations, one can argue that beside the usual goal of profit maximization, the firm's management is also committed to avoid bankruptcy. Thus, the liability structure is relevant for the firm survival.

In a capitalist economy every unit is characterized by its portfolio, which consists of assets owned or controlled and liabilities put out to achieve this ownership and control. A portfolio is of necessity speculative, as it reflects yesterdays views and both earn and commit today's and tomorrow's receipts.

An operating firm therefore has to speculate on (q - c). Accordingly to Minsky, the portfolio decision is the fundamental speculative decision of the economy since it depends on payment commitment assumed in the past which will be repaid in a delayed future through uncertain cash flows.

This in turn motivates my intention to bring out the core of Minsky's reasoning by means of the experimental approach. My aim is to build an artificial environment in which subjects may face this kind of speculative decision. However, as explained below, bringing the Minskyan vision into the lab offers premises for a different use of the experimental method in Economics.

3   The Experimental Method

Experimental economists usually start by hypothesizing about an economic problem, which could be related to the economic environment (e.i. the structure of preferences for economic agents) or the institution (e.i. the rules of exchange in a market). In order to evaluate the (null) hypothesis, they design an appropriate "micro-environment" which could be able to isolate the focus variables from the confounding events of the natural world. Performing the experiment in a computerized laboratory, they collect and analyze data in order to accept or reject the null hypothesis which motivated the experiment [17].

One can retrace this way to proceed according to the Galilean method. The Galilean method is essentially a deductive-inductive approach. It starts from observation pure and simple, it is drawn by deduction hypothesis and is checked by induction through a further phase of observation/experimentation.

However, to perform this kind of investigation in economics one should have a theoretical hypothesis of subjects behavior. In this way experimental economists put experimental subjects in a theoretical context to verify whether their microeconomic behavior conforms to those predicted by the theoretical model.

The peculiarity of Minsky's reasoning is that he never defined theoretical hypothesis to model the behavior of individuals. Essentially Minsky has an inductive mind [2]. This motivates my intention to bring the core of Minsky's reasoning into the lab without imposing neither theoretical outcome of subject behavior nor testing any kind of behavioral hypotheses.

I suggest to build experiments as a "discovery-driven research" [1] and perform data mining in searching for experimental patterns rather than to assess the validity of deductive theoretical results.

Recently, some scholars have encouraged the use of experiments as tool for investigating empirical regularities hoping an inductive turn in experimental economics from highly abstract and formal theorizing towards empirical investigations [1].

---

[1] See recent advances of exploratory methodology in biology [5].

3.1 Bringing the Core of Hyman Minsky's Reasoning into the Lab

In this section I show the "interactive model" that echoes the core of Minsky's argument (for a detailed and technical presenatation see [6]).The Graphical User Interface is shown in figure 1.

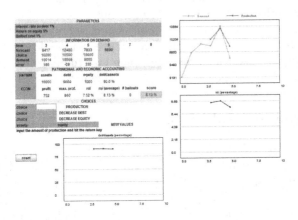

Fig. 1. The Graphical User Interface

Essentially entrepreneurs in each time period have to manage a virual firm making a portfolio decision under risk and uncertainty. The user has a set of forecasting service and three last past realization of demand. In each time period s/he has to insert in the production choice text field the value of production anticipating the level of realization of demand. The value of production is the assets of the firm. Once a production choice has been inputted, the level of demand is revealed and an economic result is calculated. The economic result is a profit equation as follows

$$\pi_{j,t} = y_{j,t} - c_{j,t}^p - c_{j,t}^{ad} - c_{j,t}^f \tag{1}$$

where $y_{j,t}$ is production of entrepreneur j at time t and $c_{j,t}^p$, $c_{j,t}^{ad}$ and $c_{j,t}^f$ denotes production, adjustment and financing costs respectively.

Adjustment costs are modeled by a quadratic function which is built upon the difference between the production setted by the user and the level of demand. Then, the economic result is a decreasing function of entrepreneur error in anticipating the demand (both shortgage and excess) and it is affecting by the different costs of financial liabilities (i.e. the cost of debt is cheaper than the cost of equity).

In a nutshell, entrepreneurs finance their production (assets) with two source of funds (debt and equity) in attempt to satisfy the future realization of demand. If they correctly anticipate the level of demand they reach the maximum profit otherwise the profit is a decreasing function of their errors. The idea is that at the beginning of the period, entrepreneurs set up the production scale. Production can be seen as temporary full capacity. If, at the end of the period, demand is lower/higher than it, entrepreneurs are forced to a fast adaptation which is costly. In our setting demand adapts to production. If entrepreneurs get a loss but the equity is enough to cover it their firm survive. If the loss is higher than equity a bailout procedure is activated. In any case the amount of the loss is withdrawn from assets and equity.

If entrepreneurs get a profit they have to decide the amount of profit to reduce debt. The residual amount does not affect balance sheet items, it can be seen as a virtual managerial reward. On the other hand they have to decide to decrease the level of equity if they want. Due to balance sheet identity changes in debt imply changes in equity and vice versa.

I assign to entrepreneurs the goal of maximizing a score given by the difference between the average roi (profit/assets) and the number of bailouts (i.e. each bailout gives a penalty to the users, it reduces their score of 1%). This goal takes care of two aspects: i) maximize the entrepreneurs reward and ii) minimize the number of bailouts. This gives a key role to the equity ratio. In fact, a high level of the ratio favours the achievement of ii) while it makes harder the achievement of i) because the cost of equity is higher than the cost of debt.

These financial decisions underpin the entrepreneurs' speculative financial behaviors. Debt is a cheaper source of finance than equity, so if they finance their production with more debt they have more probability to increase their score. On the other hand they have to deal with uncertain future level of demand[2] upon which it depends the survival of their firm. Equity is more expensive than debt but it is a margin of safety against losses.

4   Conclusions

In this paper I provide a different methodological contribution to the Hyman Minsky's Economics. Minsky himself was aware of the difficulties in modeling, or better yet, formalizing his FIH by the deductive approach. He enthusiastically embraced the new possibilities opened by simulation studies as alternative investigation methodologies to induce the financial instability.

For this end, I highlight how Minsky's Financial Instability Hypothesis can be used as a conceptual framework to build virtual economies with which any agents (entrepreneurs, households, banks) interacts by taking speculative decisions in a dynamic and uncertain environments. By building such virtual economies, in my opinion, researchers can also use the methodology of experimental economics to search for inductive microfoundations rather than to test deductive behavioral model. Such inductive microfoundations can be particularly useful in modeling an agent-based model wherein the behavioral rules of artificial agents are induced

---

[2] The forecasting service as well as demand trends are built upon a smooth business cycle benchmark pattern. Starting from the benchmark it is possible to generate different patterns as the realization of a gaussian distribution. Tuning the standard deviation of such gaussian distribution is possible to modify the smoothness or volatility of the forecasts and demand time series.

from experimental data (seminal results on this issue can be found in [7]. This in turn helps to build more lifelike and interactive artificial macroeoconomic model.

References

1. Bardsley, N., Cubitt, R., Loomes, G., Moffatt, P., Starmer, C., Sugden, R.: Experimental Economics. Rethinking the Rules. Princeton University Press, Princeton (2010)
2. Dimsky, G., Pollin, R.: New Perspective in Monetary Macroeconomics: Exploration in the Tradition of Hyman P. Minsky. University of Michigan Press (1994)
3. Dow, S.: Cognition, Market Sentiment, and Financial Instability. Cambridge Journal of Economics 35(2), 233–249 (2011)
4. Foley, D.K.: Hyman Minsky and the Dilemmas of Contemporary Economic Method. In: Papadimitriou, D., Wray, R.W. (eds.) The Elgar Companion to Hyman Minsky, ch. 9, pp. 169–181. Edward Elgar, Cheltenham and Northampton ((2010)
5. Franklin, L.R.: Exploratory Experiments. Philosophy of Science 72(1), 888–899 (2005)
6. Giulioni, G., Bucciarelli, E., Silvestri, M.: Modeling firms financial behavior in an evolving economy by using artificial intelligence. In: Pérez, J.B., et al. (eds.) Highlights on PAAMS. AISC, vol. 156, pp. 333–340. Springer, Heidelberg (2012)
7. Giulioni, G., Bucciarelli, E., Silvestri, M., D'Orazio, P.: Towards a Multi-avatar Macroeconomic System. In: Demazeau, Y., Ishida, T., Corchado, J.M., Bajo, J. (eds.) PAAMS 2013. LNCS, vol. 7879, pp. 97–109. Springer, Heidelberg (2013)
8. Janssen, M.C.: Microfoundations. In: Durlauf, S.N., Blume, L.E. (eds.) The New Palgrave Dictionary of Economics. Palgrave Macmillan, Basingstoke (2008)
9. Keynes, J.M.: The General Theory of Employment, Interest, and Money. Macmillan, London (1936)
10. Keynes, J.M.: The General Theory of Employment. The Quarterly Journal of Economics 51(2), 209–223 (1937)
11. Lucas, R.J.: Econometric policy evaluation: A critique. Carnegie-Rochester Conference Series on Public Policy 1(1), 19–46 (1976)
12. Mehrling, P.: The Vision of Hyman P. Minsky. Journal of Economic Behavior & Organization 39(2), 129–158 (1999)
13. Minsky, H.P.: John Maynard Keynes. Columbia University Press, New York (1975)
14. Minsky, H.P.: Can 'It' Happen Again? Essays on Instability and Finance. M.E. Sharpe, New York (1982)
15. Minsky, H.P.: Stabilizing an Unstable Economy. Yale University Press, New Haven (1986)
16. Minsky, H.P.: Induced Investment and Business Cycle. Edward Elgar Publishing, Northampton (2004); Publication of Minsky's 1954 PhD thesis. Book edited by Dimitri B. Papadimitriou
17. Plott, C., Smith, V.L.: Handbook Experimental Economics Results. Elsevier (2005)
18. Prigogine, I.: Time, Structure, and Flluctuations. Science 201(4358), 777–785 (1978)
19. Simon, H.A.: Reason in Human Affairs. Stanford University Press, Stanford (1983)

Author Index

Printed in the United States
By Bookmasters